MAX S. BENNETT is an entrepreneur and researcher. He has cofounded multiple AI companies, holds several patents for AI technologies, and has published numerous scientific papers on the topics of evolutionary neuroscience and intelligence. He has been featured on the *Forbes* 30 Under 30 list as well as the Built In NYC's 30 Tech Leaders Under 30. Most notably, Bennett was the cofounder and Chief Product Officer of Bluecore, one of the fastest growing companies in the United States, providing AI technologies to some of the largest companies in the world. Bluecore has been featured in the annual Inc. 500 fastest growing companies, as well as Glassdoor's 50 Best Places to Work in the US. Bluecore was recently valued at over $1 billion. Bennett graduated from Washington University in St. Louis, *summa cum laude*, with a degree in economics and mathematics, where he won the John M. Olin prize for the best economics honors thesis. Max lives in Brooklyn, New York, with his wife, Sydney, and their dog, Charlie.

Praise for *A Brief History of Intelligence*

'I found this book amazing. I read it through quickly because it was so interesting, then turned around and read much of it again' Daniel Kahneman, Winner of the Nobel Prize in Economics and bestselling author of *Thinking, Fast & Slow*

'I've been recommending *A Brief History of Intelligence* to everyone I know. A truly novel, beautifully crafted thesis on what intelligence is and how it has developed since the dawn of life itself' Angela Duckworth, University of Pennsylvania and bestselling author of *Grit*

'Introduces us to an enormous range of concepts in biology, evolution, brain science and AI ... There is plenty of food for thought here' *Times Literary Supplement*

'If you are interested in understanding brains or in building human-like general AI, you should read this book. This is a forward-looking book masquerading as history. A mind-boggling amount of details of anatomy, physiology and behaviour of a variety of nervous systems are brought together in a coherent evolutionary tale and explained in their computational contexts. It is a joy to read – don't miss it!' Dileep George, DeepMind, cofounder of Vicarious AI

'Absolutely riveting. *A Brief History of Intelligence* is a spellbinding and fascinating tour of the origins of the human species, and a reminder that the human story began long before *Homo sapiens*. An illuminating, revelatory account of who we are and how we got here' Brian Christian, bestselling author of *Algorithms to Live By* and *The Alignment Problem*

'This book discloses everything you always wanted to know about the brain (but were afraid to ask). It is an incredible resource. It assimilates every discovery in neuroscience – over the last century – within a beautifully crafted evolutionary narrative. The ensuing story shows how an incremental elaboration of brains can be traced from ancient worms to the mindful, curious creatures we have become. The synthesis works perfectly. Its coherence obscures the almost encyclopaedic reach of this treatment' Karl Friston, University College London, #1 most-cited neuroscientist in the world

'Max Bennett published two scientific papers on brain evolution that blew me away. Now he has turned these into a fabulous book' Joseph LeDoux, New York University, bestselling author of *Anxious* and *The Deep History of Ourselves*

'Max Bennett has written a marvelous book about the history of intelligence. I have been studying the brain for forty years, I wish I could have read Bennett's book when I started on my journey, it would have saved me a lot of time. This book is a unique and valuable resource for anyone wanting to understand intelligence' Jeff Hawkins, cofounder of Numenta and Palm Computing, bestselling author of *A Thousand Brains* and *On Intelligence*

'With a truly mind-boggling scope, *A Brief History of Intelligence* integrates the most relevant scientific knowledge to paint the big picture of how the human mind emerged . . . This text is embracing, ambitious, and lusciously enlightening but still remains strictly orientated to the facts, and avoids unsubstantiated speculation. This is both a piece of art as well as science . . . I am deeply impressed by this brave project of explaining entire human nature in the grand evolutionary frame. But I am even more impressed that Max Bennett succeeded in this virtually impossible task' Kurt Kotrschal, University of Vienna, winner of 2010 Austrian Scientist of the Year Award and author of the critically acclaimed *Wolf – Dog – Human*

'Written with gusto and spirit, with intellectual courage and playfulness. It is eye-opening and intellectually invigorating ... the work of a young and fresh mind that has no axes to grind and comes to the subject with untarnished joyful curiosity, intelligence, and courage. Everyone, from young students to established academics will find it rewarding' Eva Jablonka, Tel Aviv University, coauthor of *Evolution in Four Dimensions* and *The Evolution of the Sensitive Soul*

'Max Bennett gives a lively account of how brains evolved and how the brain works today. *A Brief History of Intelligence* is engaging, comprehensive, and brimming with novel insights' Kent Berridge, professor of psychology and neuroscience at University of Michigan and winner of the Grawemeyer Award for Psychology

'If you're in the least bit curious about that three-pound gray blob between your ears, read this book. Max Bennett's entertaining and enlightening natural history of brains is a tour de force – as refreshing as it is entertaining. It made my brain happy' Jonathan Balcombe, PhD, bestselling author of *What a Fish Knows* and *Super Fly*

'This book provides an exciting journey through the keys to human intelligence and has important things to say about who we are and what it means to be human. The five "breakthroughs" in which the ability to interact with the world gets more and more complex provides a novel evolutionary structure that carries the story forward. Well written in a very readable and engaging style. Highly recommended' A. David Redish, University of Minnesota, author of *The Mind within the Brain* and *Changing How We Choose: The New Science of Morality*

'A delightfully insightful page-turner. Bennett distills the rapidly evolving field of brain science into a clear narrative that connects the brains of our ancient ancestors to our own, and links both to our future innovations. Books like this – that tackle the big picture with accuracy and wisdom – are rare and precious' David Eagleman, Stanford neuroscientist, bestselling author of *Incognito* and *Livewired*

A BRIEF HISTORY OF INTELLIGENCE

Why the Evolution of the
Brain Holds the
Key to the Future of AI

MAX S. BENNETT

WILLIAM
COLLINS

William Collins
An imprint of HarperCollins*Publishers*
1 London Bridge Street
London SE1 9GF

WilliamCollinsBooks.com

HarperCollins*Publishers*
Macken House
39/40 Mayor Street Upper
Dublin 1
D01 C9W8, Ireland

First published in Great Britain in 2023 by William Collins
First published in the United States by Mariner Books,
an imprint of HarperCollins*Publishers*
This William Collins paperback edition published in 2024

3

Mathematics and Science

Copyright © Max Solomon Bennett 2023

Max Solomon Bennett asserts the moral right to be identified as the author of this work in accordance with the Copyright, Designs and Patents Act 1988

A catalogue record for this book is available from the British Library

ISBN 978-0-00-856013-3

Book Design by Chloe Foster

All rights reserved. No part of this publication may be reproduced, stored in a retrieval system, or transmitted, in any form or by any means, electronic, mechanical, photocopying, recording or otherwise, without the prior permission of the publishers.

This book is sold subject to the condition that it shall not, by way of trade or otherwise, be lent, re-sold, hired out or otherwise circulated without the publisher's prior consent in any form of binding or cover other than that in which it is published and without a similar condition including this condition being imposed on the subsequent purchaser.

Printed and bound in India by Replika Press Pvt. Ltd.

This book contains FSC™ certified paper and other controlled sources to ensure responsible forest management.

For more information visit: www.harpercollins.co.uk/green

To my wife, Sydney

In the distant future I see open fields for far more important researches. Psychology will be based on a new foundation, that of the necessary acquirement of each mental power and capacity by gradation. Light will be thrown on the origin of man and his history.

—CHARLES DARWIN IN 1859

Contents

The Basics of Human Brain Anatomy	xiii
Our Evolutionary Lineage	xiv
Introduction	1
1: The World Before Brains	17

BREAKTHROUGH #1: Steering and the First Bilaterians

2: The Birth of Good and Bad	43
3: The Origin of Emotion	59
4: Associating, Predicting, and the Dawn of Learning	76

BREAKTHROUGH #2: Reinforcing and the First Vertebrates

5: The Cambrian Explosion	93
6: The Evolution of Temporal Difference Learning	103
7: The Problems of Pattern Recognition	122
8: Why Life Got Curious	142
9: The First Model of the World	146

BREAKTHROUGH #3: Simulating and the First Mammals

10: The Neural Dark Ages	157
11: Generative Models and the Neocortical Mystery	167

12: Mice in the Imaginarium	188
13: Model-Based Reinforcement Learning	201
14: The Secret to Dishwashing Robots	221

BREAKTHROUGH #4: Mentalizing and the First Primates

15: The Arms Race for Political Savvy	237
16: How to Model Other Minds	253
17: Monkey Hammers and Self-Driving Cars	267
18: Why Rats Can't Go Grocery Shopping	282

BREAKTHROUGH #5: Speaking and the First Humans

19: The Search for Human Uniqueness	295
20: Language in the Brain	310
21: The Perfect Storm	323
22: ChatGPT and the Window into the Mind	344
Conclusion: The Sixth Breakthrough	359
Acknowledgments	366
Glossary	370
Notes	373
Bibliography	398
Art, Photo, and Figure Credits	400
Index	403

The Basics of Human Brain Anatomy

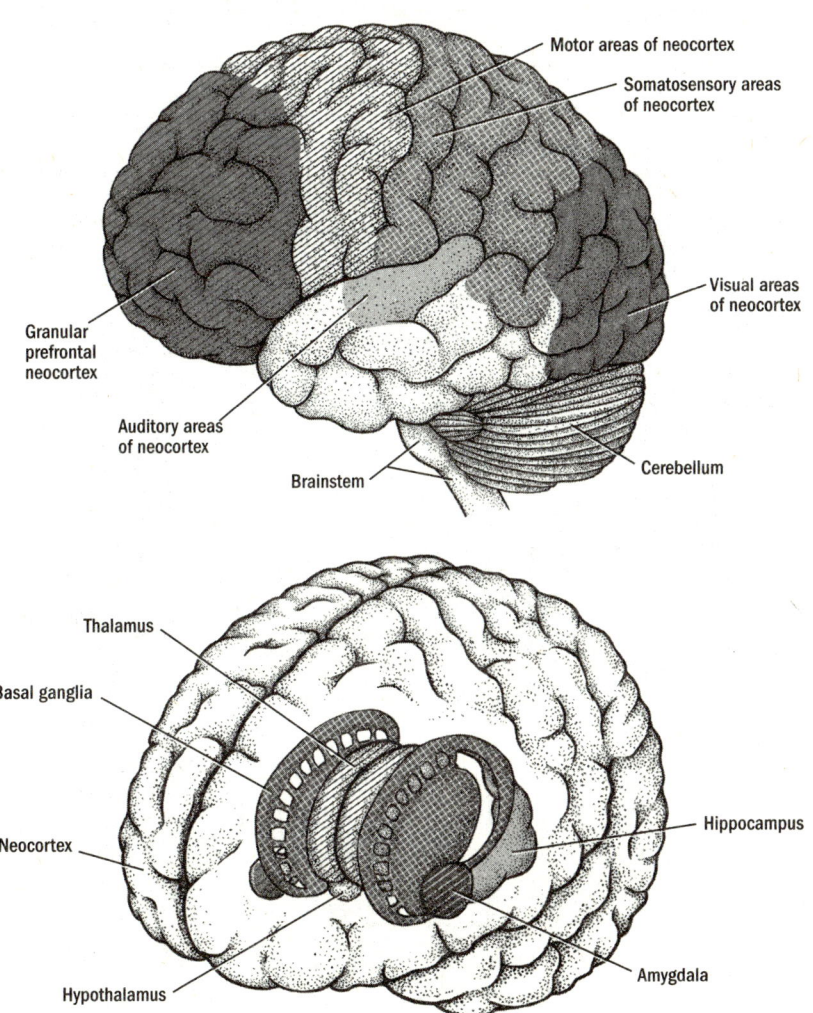

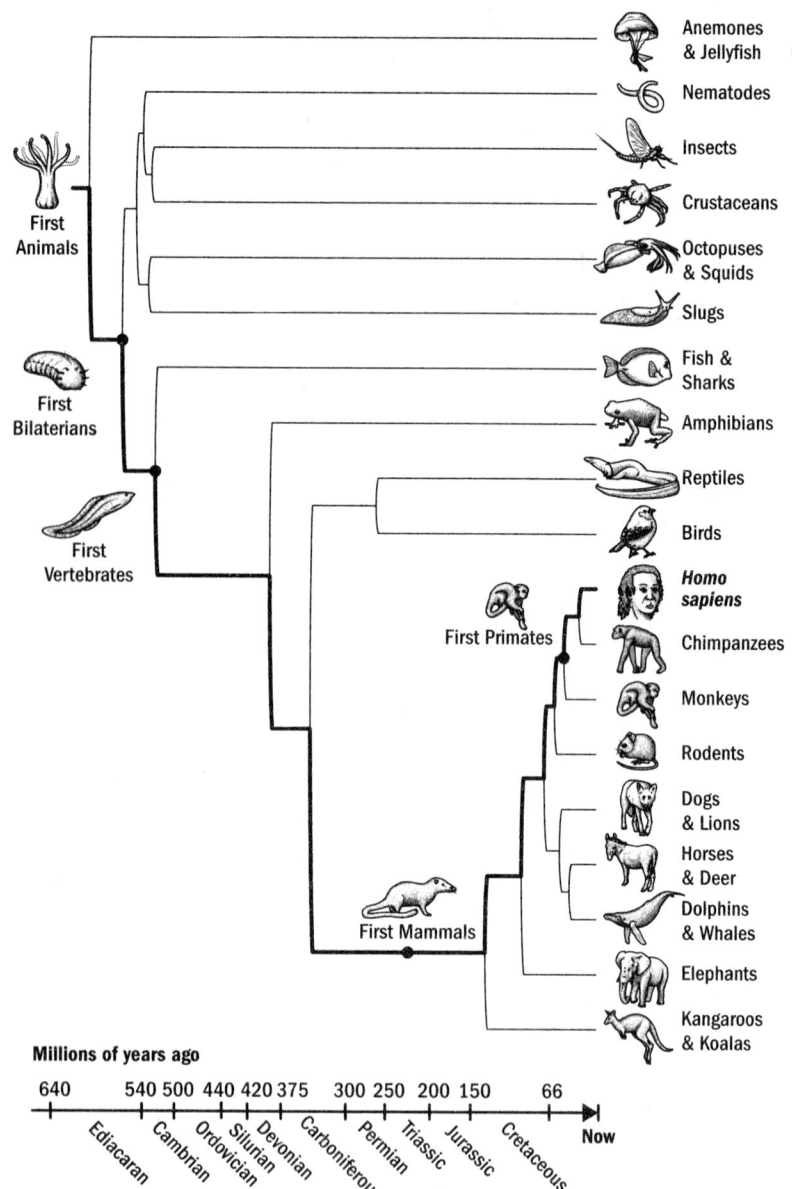

Special thanks to Rebecca Gelernter for creating the incredible original art in this book; Rebecca created the art at the beginning of each Breakthrough section and designed the majority of the figures. Also, a special thanks to Mesa Schumacher for her wonderful original anatomical art of the human, lamprey, monkey, and rat brain that she made specifically for this book.

Introduction

IN SEPTEMBER 1962, during the global tumult of the space race, the Cuban missile crisis, and the recently upgraded polio vaccine, there was a less reported—but perhaps equally critical—milestone in human history: It was in the fall of '62 that we predicted the future.

Cast onto the newly colorful screens of American televisions was the debut of *The Jetsons*, a cartoon about a family living one hundred years in the future. In the guise of a sitcom, the show was, in fact, a prediction of how future humans would live, of what technologies would fill their pockets and furnish their homes.

The Jetsons correctly predicted video calls, flat-screen TVs, cell phones, 3D printing, and smartwatches; all technologies that were unbelievable in 1962 and yet were ubiquitous by 2022. However, there is one technology that we have entirely failed to create, one futurist feat that has not yet come to fruition: the autonomous robot named Rosey.

Rosey was a caretaker for the Jetson family, watching after the children and tending to the home. When Elroy—then six years old—was struggling in school, it was Rosey who helped him with his homework. When their fifteen-year-old daughter, Judy, needed help learning how to drive, it was Rosey who gave her lessons. Rosey cooked meals, set the table, and did the dishes. Rosey was loyal, sensitive, and quick with a joke. She identified brewing family tiffs and misunderstandings, intervening to help individuals see one another's perspective. At one time, she was moved to tears by a poem Elroy wrote for his mother. Rosey herself, in one episode, even fell in love.

In other words, Rosey had the intelligence of a human. Not just the reasoning, common sense, and motor skills needed to perform complex tasks in the physical world, but also the empathizing, perspective taking, and social finesse needed to successfully navigate our social world. In the words of Jane Jetson, Rosey was "just like one of the family."

Although the *The Jetsons* correctly predicted cell phones and smartwatches, we still don't have anything like Rosey. As of this book going to print, even Rosey's most basic behaviors are still out of reach. It is no secret that the first company to build a robot that can simply *load a dishwasher* will immediately have a bestselling product. All attempts to do this have failed. It isn't fundamentally a *mechanical* problem; it's an *intellectual* one—the ability to identify objects in a sink, pick them up appropriately, and load them without breaking anything has proven far more difficult than previously thought.

Of course, even though we do not yet have Rosey, the progress in the field of artificial intelligence (AI) since 1962 has been remarkable. AI can now beat the best humans in the world at numerous games of skill, including chess and Go. AI can recognize tumors in radiology images as well as human radiologists. AI is on the cusp of autonomously driving cars. And as of the last few years, new advancements in large language models are enabling products like ChatGPT, which launched in fall 2022, to compose poetry, translate between languages at will, and even write code. To the chagrin of every high school teacher on planet Earth, ChatGPT can instantly compose a remarkably well written and original essay on almost any topic that an intrepid student might ask of it. ChatGPT can even pass the bar exam, scoring better than 90 percent of lawyers.

Across this long arrow of AI achievements, it has always been hard to tell how close we are to creating human-level intelligence. After the early successes of problem solving algorithms in the 1960s, the AI pioneer Marvin Minsky famously proclaimed that "from three to eight years we will have a machine with the general intelligence of an average human being." It did not happen. After the successes of expert systems in the 1980s, *BusinessWeek* proclaimed "AI: it's here." Progress stalled shortly thereafter. And now with advancements in large language models, many researchers have again proclaimed that the "game is over" because we are "on the verge of achieving human-level AI." Which is it: Are we finally on the cusp of creating human-like artificial intelligence like Rosey, or are large language models like ChatGPT just the most recent achievement in a long journey that will stretch on for decades to come?

Along this journey, as AI keeps getting smarter, it is becoming harder to measure our progress toward this goal. If an AI system outperforms humans on a task, does it mean that the AI system has captured how humans solve the task? Does a calculator—capable of crunching numbers faster than a human—actually understand math? Does ChatGPT—scoring better on the bar exam than most lawyers—actually understand the law? How can we tell the difference, and in what circumstances, if any, does the difference even matter?

In 2021, over a year before the release of ChatGPT—the chatbot that is now rapidly proliferating throughout every nook and cranny of society—I was using its precursor, a large language model called GPT-3. GPT-3 was trained on large quantities of text (large as in *the entire internet*), and then used this corpus to try to pattern match the most likely response to a prompt. When asked, "What are two reasons that a dog might be in a bad mood?" it responded, "Two reasons a dog might be in a bad mood are if it is hungry or if it is hot." Something about the new architecture of these systems enabled them to answer questions with what at least seemed like a remarkable degree of intelligence. These models were able to generalize facts they had read about (like the Wikipedia pages about dogs and other pages about causes of bad moods) to new questions they had never seen. In 2021, I was exploring possible applications of these new language models—could they be used to provide new support systems for mental health, or more seamless customer service, or more democratized access to medical information?

The more I interacted with GPT-3, the more mesmerized I became by both its successes and mistakes. In some ways it was brilliant, in other ways it was oddly dumb. Ask GPT-3 to write an essay about eighteenth-century potato farming and its relationship to globalization, and you will get a surprisingly coherent essay. Ask it a commonsense question about what someone might see in a basement, and it answers nonsensically.* Why could

---

* I asked GPT-3 to complete the following sentence: "I am in my windowless basement, and I look toward the sky, and I see . . ." GPT-3 said "a light, and I know it is a

GPT-3 correctly answer some questions and not others? What features of human intelligence does it capture, and which is it missing? And why, as AI development continues to accelerate, are some questions that were hard to answer in one year becoming easy in subsequent years? Indeed, as of this book going to print, the new and upgraded version of GPT-3, called GPT-4, released in early 2023, can correctly answer many questions that beguiled GPT-3. And yet still, as we will see in this book, GPT-4 fails to capture essential features of human intelligence—about something going on in the human brain.

Indeed, the discrepancies between artificial intelligence and human intelligence are nothing short of perplexing. Why is it that AI can crush any human on earth in a game of chess but can't load a dishwasher better than a six-year-old?

We struggle to answer these questions because we don't yet understand the thing we are trying to re-create. All of these questions are, in essence, not questions about AI, but about the nature of human intelligence itself—how it works, why it works the way it does, and as we will soon see, most importantly, how it came to be.

Nature's Hint

When humanity wanted to understand flight, we garnered our first inspiration from birds; when George de Mestral invented Velcro, he got the idea from burdock fruits; when Benjamin Franklin sought to explore electricity, his first sparks of understanding came from lightning. Nature has, throughout the history of human innovation, long been a wondrous guide.

Nature also offers us clues as to how intelligence works—the clearest locus of which is, of course, the human brain. But in this way, AI is unlike these other technological innovations; the brain has proven to be more

---

star, and I am happy." In reality, if you looked upward in a basement you would not see stars, you would see the ceiling. Newer language models like GPT-4, released in 2023, successfully answer commonsense questions like this with greater accuracy. Stay tuned for chapter 22.

unwieldy and harder to decipher than either wings or lightning. Scientists have been investigating how the brain works for millennia, and while we have made progress, we do not yet have satisfying answers.

The problem is complexity.

The human brain contains eighty-six billion neurons and over a hundred trillion connections. Each of those connections is so minuscule—less than thirty nanometers wide—that they can barely be seen under even the most powerful microscopes. These connections are bunched together in a tangled mess—within a single cubic millimeter (the width of a single *letter* on a penny), there are over *one billion* connections.

But the sheer number of connections is only one aspect of what makes the brain complex; even if we mapped the wiring of each neuron we would still be far from understanding how the brain works. Unlike the electrical connections in your computer, where wires all communicate using the same signal—electrons—across each of these neural connections, hundreds of different chemicals are passed, each with completely different effects. The simple fact that two neurons connect to each other tells us little about what they are communicating. And worst of all, these connections themselves are in a constant state of change, with some neurons branching out and forming new connections, while others are retracting and removing old ones. Altogether, this makes reverse engineering how the brain works an ungodly task.

Studying the brain is both tantalizing and infuriating. One inch behind your eyes is the most awe-inspiring marvel of the universe. It houses the secrets to the nature of intelligence, to building humanlike artificial intelligence, to why we humans think and behave the way we do. It is right there, reconstructed millions of times per year with every newly born human. We can touch it, hold it, dissect it, we are *literally made of it*, and yet its secrets remain out of reach, hidden in plain sight.

If we want to reverse-engineer how the brain works, if we want to build Rosey, if we want to uncover the hidden nature of human intelligence, perhaps the human brain is not nature's best clue. While the most intuitive place to look to understand the human brain is, naturally, inside the human brain itself, counterintuitively, this may be the *last* place to look. The best place to start may be in dusty fossils deep in the

Earth's crust, in microscopic genes tucked away inside cells throughout the animal kingdom, and in the brains of the many *other* animals that populate our planet.

In other words, the answer might not be in the present, but in the hidden remnants of a long time past.

The Missing Museum of Brains

I have always been convinced that the only way to get artificial intelligence to work is to do the computation in a way similar to the human brain.

—GEOFFREY HINTON (PROFESSOR AT UNIVERSITY OF TORONTO, CONSIDERED ONE OF THE "GODFATHERS OF AI")

Humans fly spaceships, split atoms, and edit genes. No other animal has even invented the wheel.

Because of humanity's larger résumé of inventions, you might think that we would have little to learn from the brains of other animals. You might think that the human brain would be entirely unique and nothing like the brains of other animals, that some special brain structure would be the secret to our cleverness. But this is not what we see.

What is most striking when we examine the brains of other animals is how remarkably *similar* their brains are to our own. The difference between our brain and a chimpanzee's brain, besides size, is barely anything. The difference between our brain and a rat's brain is only a handful of brain modifications. The brain of a fish has almost all the same structures as our brain.

These similarities in brains across the animal kingdom mean something important. They are clues. Clues about the nature of intelligence. Clues about ourselves. Clues about our past.

Although today brains are complex, they were not always so. The brain emerged from the unthinking chaotic process of evolution; small random variations in traits were selected for or pruned away depending on whether they supported the further reproduction of the life-form.

In evolution, systems start simple, and complexity emerges only over time.* The first brain—the first collection of neurons in the head of an animal—appeared six hundred million years ago in a worm the size of a grain of rice. This worm was the ancestor of all modern brain-endowed animals. Over hundreds of millions of years of evolutionary tinkering, through trillions of small tweaks in wiring, her simple brain was transformed into the diverse portfolio of modern brains. One lineage of this ancient worm's descendants led to the brain in our heads.

If only we could go back in time and examine this first brain to understand how it worked and what tricks it enabled. If only we could then track the complexification forward in the lineage that led to the human brain, observing each physical modification that occurred and the intellectual abilities it afforded. If we could do this, we might be able to grasp the complexity that eventually emerged. Indeed, as the biologist Theodosius Dobzhansky famously said, "Nothing in biology makes sense except in the light of evolution."

Even Darwin fantasized about reconstructing such a story. He ends his *Origin of Species* fantasizing about a future when "psychology will be based on a new foundation, that of the necessary acquirement of each mental power and capacity by gradation." One hundred fifty years after Darwin, this may finally be possible.

Although we have no time machines, we can, in principle, engage in time travel. In just the past decade, evolutionary neuroscientists have made incredible progress in reconstructing the brains of our ancestors. One way they do this is through the fossil record—scientists can use the fossilized skulls of ancient creatures to reverse-engineer the structure of their brains. Another way to reconstruct the brains of our ancestors is by examining the brains of other animals in the animal kingdom.

The reason why brains across the animal kingdom are so similar is that they all derive from common roots in shared ancestors. Every brain in the animal kingdom is a little clue as to what the brains of our ancestors looked like; each brain is not only a machine but a time capsule filled with hidden hints of the trillions of minds that came before. And by examining

---

* Although systems don't *necessarily* get more complex, the possibility of complexity increases over time.

the intellectual feats these other animals share and those they do not, we can begin to not only reconstruct the brains of our ancestors, but also determine what intellectual abilities these ancient brains afforded them. Together, we can begin to trace acquirement of each mental power by gradation.

It is all, of course, still a work in progress, but the story is becoming tantalizingly clear.

The Myth of Layers

I am hardly the first to propose an evolutionary framework for understanding the human brain. There is a long tradition of such frameworks. The most famous was formulated in the 1960s by the neuroscientist Paul MacLean. MacLean hypothesized that the human brain was made of three layers (hence *triune*), each built on top of another: the *neocortex*, which evolved most recently, on top of the *limbic system*, which evolved earlier, on top of the *reptile brain*, which evolved first.

MacLean argued that the reptile brain was the center of our basic survival instincts, such as aggression and territoriality. The limbic system was supposedly the center of emotions, such as fear, parental attachment, sexual desire, and hunger. And the neocortex was supposedly the center of cognition, gifting us with language, abstraction, planning, and perception. MacLean's framework suggested that reptiles had *only* a reptile brain, mammals like rats and rabbits had a reptile brain *and* a limbic system, and we humans had all three systems. Indeed, to him, these "three evolutionary formations might be imagined as three interconnected biological computers, with each having its own special intelligence, its own subjectivity, its own sense of time and space, and its own memory, motor, and other functions."

The problem is that MacLean's Triune Brain Hypothesis has been largely discredited—not because it is inexact (all frameworks are inexact), but because it leads to the wrong conclusions about how the brain evolved and how it works. The implied brain anatomy is wrong; the brains of reptiles are not only made up of the structures MacLean referred to as the "reptile brain"; reptiles also have their own version of a limbic system.

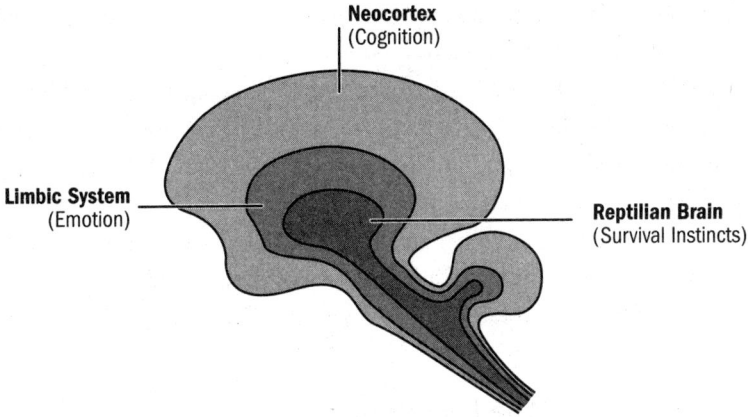

Figure 1: MacLean's triune brain

The functional divisions proved wrong; *survival instincts, emotions,* and *cognition* do not delineate cleanly—they emerge from diverse networks of systems spanning all three of these supposed layers. And the implied evolutionary story turned out to be wrong. You do not have a reptile brain in your head; evolution did not work by simply layering one system on top of another without any modifications to the existing systems.

But even if MacLean's triune brain had turned out to be closer to the truth, its biggest problem is that its functional divisions aren't particularly useful for our purposes. If our goal is to reverse-engineer the human brain to understand the nature of intelligence, MacLean's three systems are too broad and the functions attributed to them too vague to provide us with even a point at which to start.

We need to ground our understanding of how the brain works and how it evolved in our understanding of how intelligence works—for which we must look to the field of artificial intelligence. The relationship between AI and the brain goes both ways; while the brain can surely teach us much about how to create artificial humanlike intelligence, AI can also teach us about the brain. If we think some part of the brain uses some specific algorithm but that algorithm doesn't work when we implement it in machines, this gives us evidence that the brain might not work this way. Conversely, if we find an algorithm that works well in AI systems, and we find parallels between the properties of these algorithms and properties of animal brains, this gives us some evidence that the brain might indeed work this way.

The physicist Richard Feynman left the following on a blackboard shortly before his death: "What I cannot create, I do not understand." The brain is our guiding inspiration for how to build AI, and AI is our litmus test for how well we understand the brain.

We need a new evolutionary story of the brain, one grounded not only in a modern understanding of how brain anatomy changed over time, but also in a modern understanding of intelligence itself.

The Five Breakthroughs

Let's start with Artificial Rat-level intelligence (ARI), then move on to Artificial Cat-level intelligence (ACI), and so on to Artificial Human-level Intelligence (AHI).

—YANN LECUN, HEAD OF AI AT META

We have a lot of evolutionary history to cover—four billion years. Instead of chronicling each minor adjustment, we will be chronicling the major evolutionary breakthroughs. In fact, as an initial approximation—a first template of this story—the entirety of the human brain's evolution can be reasonably summarized as the culmination of only *five* breakthroughs, starting from the very first brains and going all the way to human brains.

These five breakthroughs are the organizing map to our book, and they make up our itinerary for our adventure back in time. Each breakthrough emerged from new sets of brain modifications and equipped animals with a new portfolio of intellectual abilities. This book is divided into five parts, one for each breakthrough. In each section, I will describe why these abilities evolved, how they worked, and how they still manifest in human brains today.

Each subsequent breakthrough was built on the foundation of those that came before and provided the foundation for those that would follow. Past innovations enabled future innovations. It is through this ordered set of modifications that the evolutionary story of the brain helps us make sense of the complexity that eventually emerged.

But this story cannot be faithfully retold by considering only the biology of our ancestors' brains. These breakthroughs always emerged from periods when our ancestors faced extreme situations or got caught in powerful feedback loops. It was these pressures that led to rapid reconfigurations of brains. We cannot understand the breakthroughs in brain evolution without also understanding the trials and triumphs of our ancestors: the predators they outwitted, the environmental calamities they endured, and the desperate niches they turned to for survival.

And crucially, we will ground these breakthroughs in what is currently known in the field of AI, for many of these breakthroughs in biological intelligence have parallels to what we have learned in artificial intelligence. Some of these breakthroughs represent intellectual tricks we understand well in AI, while other tricks still lay beyond our understanding. And in this way, perhaps the evolutionary story of the brain can shed light on what breakthroughs we may have missed in the development of artificial humanlike intelligence. Perhaps it will reveal some of nature's hidden clues.

Me

I wish I could tell you that I wrote this book because I have spent my whole life pondering the evolution of the brain and trying to build intelligent robots. But I am not a neuroscientist or a roboticist or even a scientist. I wrote this book because I wanted to read this book.

I came to the perplexing discrepancy between human and artificial intelligence by trying to apply AI systems to real-world problems. I spent the bulk of my career at a company I cofounded named Bluecore; we built software and AI systems to help some of the largest brands in the world personalize their marketing. Our software helped predict what consumers would buy before they knew what they wanted. We were merely one tiny part in a sea of countless companies beginning to use the new advances in AI systems. But all these many projects, both big and small, were shaped by the same perplexing questions.

When commercializing AI systems, there is eventually a series of meetings between business teams and machine learning teams. The business teams look for applications of new AI systems that would be *valuable*, while only the machine learning teams understand what applications would be *feasible*. These meetings often reveal our mistaken intuitions about how much we understand about intelligence. Businesspeople probe for applications of AI systems that seem straightforward to them. But frequently, these tasks seem straightforward only because they are straightforward for *our brains*. Machine learning people then patiently explain to the business team why the idea that seems simple is, in fact, astronomically difficult. And these debates go back and forth with every new project. It was from these explorations into how far we could stretch modern AI systems and the surprising places where they fall short that I developed my original curiosity about the brain.

Of course, I am also a human and I, like you, have a human brain. So it was easy for me to become fascinated with the organ that defines so much of the human experience. The brain offers answers not only about the nature of intelligence, but also why we behave the way we do. Why do we frequently make irrational and self-defeating choices? Why does our species have such a long recurring history of both inspiring selflessness and unfathomable cruelty?

My personal project began with merely trying to read books to answer my own questions. This eventually escalated to lengthy email correspondences with neuroscientists who were generous enough to indulge the curiosities of an outsider. This research and these correspondences eventually led me to publish several research papers, which all culminated in the decision to take time off work to turn these brewing ideas into a book.

Throughout this process, the deeper I went, the more I became convinced that there was a worthwhile synthesis to be contributed, one that could provide an accessible introduction to how the brain works, why it works the way it does, and how it overlaps and differs from modern AI systems; one that could bring various ideas across neuroscience and AI together under an umbrella of a single story.

A Brief History of Intelligence is a synthesis of the work of many others. At its heart, it is merely an attempt to put together the pieces that were already there. I have done my best to give due credit throughout the book, always aiming to celebrate those scientists who did the actual research.

Anywhere I have failed to do so is unintentional. Admittedly, I couldn't resist sprinkling in a few speculations of my own, but I will aim to be clear when I step into such territory.

It is perhaps fitting that the origin of this book, like the origin of the brain itself, came not from prior planning but from a chaotic process of false starts and wrong turns, from chance, iteration, and lucky circumstance.

A Final Point (About Ladders and Chauvinism)

I have one final point to make before we begin our journey back in time. There is a misinterpretation that will loom dangerously between the lines of this entire story.

This book will draw many comparisons between the abilities of humans and those of other animals alive today, but this is always done by picking specifically those animals that are believed to be most similar to our ancestors. This entire book—the five-breakthroughs framework itself—is solely the story of the *human* lineage, the story of how *our* brains came to be; one could just as easily construct a story of how the octopus or honeybee brain came to be, and it would have its own twists and turns and its own breakthroughs.

Just because our brains wield more intellectual abilities than those of our ancestors does not mean that the modern human brain is strictly intellectually superior to those of other modern animals.

Evolution independently converges on common solutions all the time. The innovation of wings independently evolved in insects, bats, and birds; the common ancestor of these creatures did not have wings. Eyes are also believed to have independently evolved many times. Thus, when I argue that an intellectual ability, such as episodic memory, evolved in early mammals, this does *not* mean that today *only* mammals have episodic memory. Like with wings and eyes, other lineages of life may have independently evolved episodic memory. Indeed, many of the intellectual faculties that we will chronicle in this book are not unique to our lineage, but have independently sprouted along numerous branches of earth's evolutionary tree.

Since the days of Aristotle, scientists and philosophers have constructed

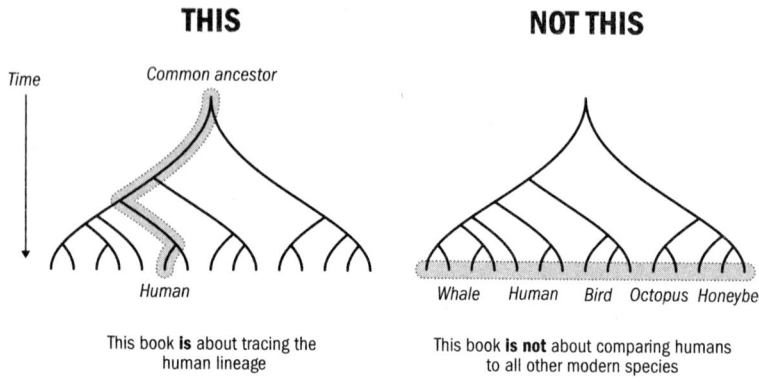

Figure 2

what modern biologists refer to as a "scale of nature" (or, since scientists like using Latin terms, *scala naturae*). Aristotle created a hierarchy of all life-forms with humans being superior to other mammals, who were in turn superior to reptiles and fish, who were in turn superior to insects, who were in turn superior to plants.

Even after the discovery of evolution, the idea of a scale of nature continues to persist. This idea that there is a hierarchy of species is dead wrong. All species alive today are, well, *alive*; their ancestors survived the last 3.5 billion years of evolution. And thus, in that sense—the only sense that evolution cares about—all life-forms alive today are tied for first place.

Species fall into different survival niches, each of which optimizes for different things. Many niches—in fact, *most* niches—are better served by *smaller* and *simpler* brains (or no brains at all). Big-brained apes are the result of a different survival strategy than that of worms, bacteria, or butterflies. But none are "better." In the eyes of evolution, the hierarchy has only two rungs: on one, there are those that survived, and on the other, those that did not.

Perhaps instead, one wants to define *better* by some specific feature of intelligence. But here still, the ranking will entirely depend on what specific intellectual skill we are measuring. An octopus has an independent brain in each of its tentacles and can blow a human away at multitasking. Pigeons, chipmunks, tuna, and even *iguanas* can process visual information faster than a human. Fish have incredibly accurate real-time processing; have you ever seen how fast a fish whips through a maze of rocks if you try to grab it? A human would surely crash if he or she tried to move so quickly through an obstacle course.

My appeal: As we trace our story, we must avoid thinking that the complexification from past to future suggests that modern humans are strictly superior to modern animals. We must avoid the accidental construction of a *scala naturae*. All animals alive today have been undergoing evolution for the same amount of time.

However, there are, of course, things that make us humans unique, and because *we are human*, it makes sense that we hold a special interest in understanding ourselves, and it makes sense that we strive to make artificial *humanlike* intelligences. So I hope we can engage in a human-centered story without devolving into human chauvinism. There is an equally valid story to be told for any other animal, from honeybees to parrots to octopuses, with which we share our planet. But we will not tell these stories here. This book tells the story of only *one* of these intelligences: it tells the story of us.

1

The World Before Brains

LIFE EXISTED ON Earth for a long time—and I mean a *long* time, over three billion years—before the first brain made an appearance. By the time the first brains evolved, life had already persevered through countless evolutionary cycles of challenge and change. In the grand arc of life on Earth, the story of brains would not be found in the main chapters but in the epilogue—brains appeared only in the most recent 15 percent of life's story. Intelligence too existed for a long time before brains; as we will see, life began exhibiting intelligent behavior early in its story. We cannot understand why and how brains evolved without first reviewing the evolution of intelligence itself.

Around four billion years ago, deep in the volcanic oceans of a lifeless Earth, just the right soup of molecules were bouncing around the microscopic nooks and crannies of an unremarkable hydrothermal vent. As boiling water burst from the seafloor, it smashed naturally occurring nucleotides together and transformed them into long molecular chains that closely resembled today's DNA. These early DNA-like molecules were short-lived; the same volcanic kinetic energy that constructed them also inevitably ripped them apart. Such is the consequence of the second law of thermodynamics. That unbreakable law of physics which declares that entropy—the amount of disorder in a system—always and unavoidably increases; the universe cannot help but tend toward decay. After countless random nucleotide chains were constructed and destroyed, a lucky sequence was stumbled upon, one that marked, at least on Earth, the first true rebellion against the seemingly inexorable onslaught of entropy. This new DNA-like molecule wasn't alive per se, but it performed the most fundamental process by which life would later emerge: it duplicated itself.

Although these self-replicating DNA-like molecules also succumbed to the destructive effects of entropy, they didn't have to survive *individually* to survive *collectively*—as long as they endured long enough to create their own copies, they would, in essence, persist. This is the genius of self-replication. With these first self-replicating molecules, a primitive version of the process of evolution began; any new lucky circumstances that facilitated more successful duplication would, of course, lead to more duplicates.

There were two subsequent evolutionary transformations that led to life. The first was when protective lipid bubbles entrapped these DNA molecules using the same mechanism by which soap, also made of lipids, naturally bubbles when you wash your hands. These DNA-filled microscopic lipid bubbles were the first versions of cells, the fundamental unit of life.

The second evolutionary transformation occurred when a suite of nucleotide-based molecules—ribosomes—began translating specific sequences of DNA into specific sequences of amino acids that were then folded into specific three-dimensional structures we call *proteins*. Once produced, these proteins float around inside a cell or are embedded in the wall of the cell fulfilling different functions. You have probably, at least in passing, heard that your DNA is made up of genes. Well, a gene is simply the section of DNA that codes for the construction of a specific and singular protein. This was the invention of protein synthesis, and it is here that the first sparks of intelligence made their appearance.

DNA is relatively inert, effective for self-duplication but otherwise limited in its ability to manipulate the microscopic world around it. Proteins, however, are far more flexible and powerful. In many ways, proteins are more machine than molecule. Proteins can be constructed and folded into many shapes—sporting tunnels, latches, and other robotic moving parts—and can thereby subserve endless cellular functions, including "intelligence."

Even the simplest single-celled organisms—such as bacteria—have proteins designed for movement, motorized engines that convert cellular energy into propulsion, rotating propellers using a mechanism no less complex than the motor of a modern boat. Bacteria also have proteins designed for *perception*—receptors that reshape when they detect certain features of the external environment, such as temperature, light, or touch.

Armed with proteins for movement and perception, early life could monitor and respond to the outside world. Bacteria can swim away from environments that lower the probability of successful replication, environments that have, for example, temperatures that are too hot or cold or chemicals that are destructive to DNA or cell membranes. Bacteria can also swim toward environments that are amenable to reproduction.

And in this way, these ancient cells indeed had a primitive version of intelligence, implemented not in neurons but in a complex network of chemical cascades and proteins.

The development of protein synthesis not only begot the seeds of intelligence but also transformed DNA from mere *matter* to a medium for storing *information*. Instead of being the self-replicating stuff of life itself, DNA was transformed into the informational foundation from which the stuff of life is constructed. DNA had officially become life's blueprint, ribosomes its factory, and proteins its product.

With these foundations in place, the process of evolution was initiated in full force: variations in DNA led to variations in proteins, which led to the evolutionary exploration of new cellular machinery, which, through natural selection, were pruned and selected for based on whether they further supported survival. By this point in life's story, we have concluded the long, yet-to-be-replicated, and mysterious process scientists call abiogenesis: the process by which nonbiological matter (*abio*) is converted into life (*genesis*).

The Terraforming of Earth

Shortly thereafter, these cells evolved into what scientists call the "last universal common ancestor," or LUCA. LUCA was the genderless grandparent of all life; every fungus, plant, bacteria, and animal alive today, including us, descend from LUCA. It is no surprise, then, that all life shares the core features of LUCA: DNA, protein synthesis, lipids, and carbohydrates.

LUCA, living around 3.5 billion years ago, likely resembled a simpler version of a modern bacteria. And indeed, for a long time after this, all life was bacterial. After a further billion years—through trillions upon trillions of evolutionary iterations—Earth's oceans were brimming with many diverse species of these microbes, each with its own portfolio of DNA and proteins. One way in which these early microbes differed from one another was in their systems of energy production. The story of life, at its core, is as much about energy as it is about entropy.

Keeping a cell alive is expensive. DNA requires constant repair; proteins require perpetual replenishment; and cellular duplication requires a reconstruction of many inner structures. Hydrogen, an element abundant near hydrothermal vents, was likely the first fuel used to finance these many processes. But this hydrogen-based energy system was inefficient and left life desperately grasping for enough energy to survive. After more than a billion years of life, this energetic poverty came to an end when a single species of bacteria—the cyanobacteria, also called blue-green algae—found a far more profitable mechanism for extracting and storing energy: photosynthesis.

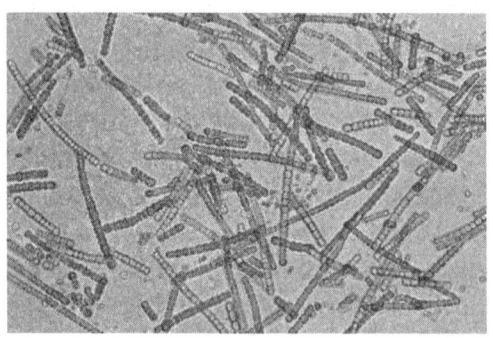

Figure 1.1: Cyanobacteria

The most impressive biological system in these early cyanobacteria was not their protein factories or their products but their photosynthetic power plants—the structures that converted sunlight and carbon dioxide into sugar, which could then be stored and converted into cellular energy. Photosynthesis was more efficient than prior cellular systems for extracting and storing energy. It provided cyanobacteria with an abundance of fuel with which to finance their duplication. Vast regions of the ocean quickly became covered in greenish gooey microbial mats—colonies of billions of cyanobacteria basking in the sun, sucking up carbon dioxide, and endlessly reproducing.

Like most processes of energy production, from burning fossil fuels to harnessing nuclear fuels, photosynthesis produced a pollutive exhaust. Instead of carbon dioxide or nuclear waste, the exhaust from photosynthesis was oxygen. Before this time, the Earth had no ozone layer. It was the cyanobacteria, with their newfound photosynthesis, that constructed Earth's oxygen-rich atmosphere and began to terraform the planet from a gray volcanic rock to the oasis we know today. This event happened around 2.4 billion years ago, and it occurred so rapidly, at least in geological terms, that it has been called the Great Oxygenation Event. Over the course of one hundred million years, oxygen levels skyrocketed. Unfortunately, this event was not a boon for all life. Scientists have a less forgiving name for it: The Oxygen Holocaust.

Oxygen is an incredibly reactive element, which makes it dangerous in the carefully orchestrated chemical reactions of a cell. Unless special intracellular protective measures are taken, oxygen compounds will interfere with cellular processes, including the maintenance of DNA. This is why antioxidants—compounds that remove highly reactive oxygen molecules from your bloodstream—are believed to offer protection from cancer. The photosynthetic life-forms became victims of their own success, slowly suffocating in a cloud of their own waste. The rise of oxygen was followed by one of the deadliest extinction events in Earth's history.

As with many substances that are dangerous (uranium, gasoline, coal), oxygen can also be useful. This newly available element presented an energetic opportunity, and it was only a matter of time before life stumbled onto a way to exploit it. A new form of bacteria emerged that produced energy not from photosynthesis but from cellular respiration—the process by which oxygen and sugar is converted into energy, spewing out carbon dioxide as exhaust. Respiring microbes began gobbling up the ocean's excess oxygen and replenishing its depleted supply of carbon dioxide. What began as a pollutant to one form of life became fuel for another.

Life on Earth fell into perhaps the greatest symbiosis ever found between two competing but complementary systems of life, one that lasts to this day. One category of life was photosynthetic, converting water and carbon dioxide into sugar and oxygen. The other was respiratory, converting sugar and oxygen back into carbon dioxide. At the time, these two forms of life were similar, both single-celled bacteria. Today this symbiosis is made up of very different forms of life. Trees, grass, and other plants are some of our modern photosynthesizers, while fungi and animals are some of our modern respirators.

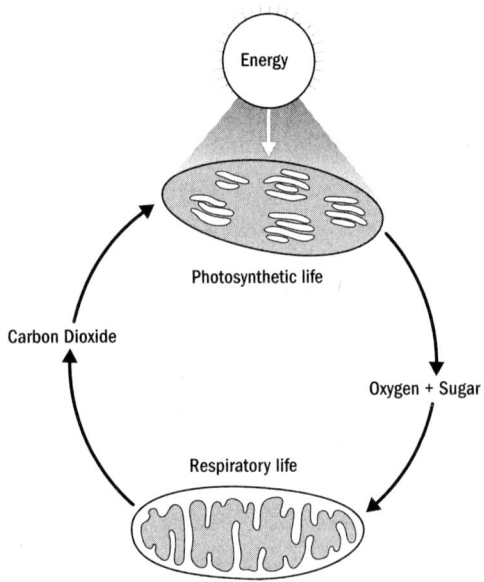

Figure 1.2: The symbiosis between photosynthetic and respiratory life

Cellular respiration requires sugar to produce energy, and this basic need provided the energetic foundation for the eventual intelligence explosion that occurred uniquely within the descendants of respiratory life. While most, if not all, microbes at the time exhibited primitive levels of intelligence, it was only in respiratory life that intelligence was later elaborated and extended. Respiratory microbes differed in one crucial way from their photosynthetic cousins: they needed to hunt. And hunting required a whole new degree of smarts.

Three Levels

The ecosystem two billion years ago was not a particularly war-torn world.*
A tentative peace brokered by energetic necessity undergirded life's many

---

* Except for the battle between bacteria and viruses, but that is a whole different story.

interactions. Although some bacteria might have gobbled up the remains of nearby deceased neighbors, it was rarely worth the effort for them to actively try to kill other life. The non-oxygen-based approach to converting sugar into energy (*anaerobic* respiration) is fifteen times less efficient than the oxygen-based approach (*aerobic* respiration). As such, before the introduction of oxygen, hunting was not a viable survival strategy. It was better to just find a good spot, sit tight, and bask in the sunlight. The most severe competition between early life was likely akin to people rushing into a Walmart for a Black Friday sale, elbowing others out of the way for scarce nearby prizes but not directly assaulting one another. Even such elbowing was probably uncommon; sunlight and hydrogen were abundant, and there was more than enough to go around.

However, unlike the cells that came before, respiratory life could survive only by stealing the energetic prize—the sugary innards—of photosynthetic life. Thus, the world's utopic peace ended quite abruptly with the arrival of aerobic respiration. It was here that microbes began to actively eat other microbes. This fueled the engine of evolutionary progress; for every defensive innovation prey evolved to stave off being killed, predators evolved an offensive innovation to overcome that same defense. Life became caught in an arms race, a perpetual feedback loop: offensive innovations led to defensive innovations that required further offensive innovations.

Out of this maelstrom, a large diversification of life emerged. Some species remained small single-celled microbes. Other species evolved into the first eukaryotes (pronounced "you-*care*-ee-oats"), cells that were over a hundred times larger, produced a thousand times more energy, and had much more internal complexity. These eukaryotes were the most advanced microbial killing machines yet. Eukaryotes were the first to evolve *phagotrophy*—the hunting strategy of literally engulfing other cells and breaking them down inside their cellular walls. These eukaryotes, armed with more energy and complexity, further diversified into the first plants, the first fungi, and the precursors of the first animals. The fungi and animal descendants of eukaryotes retained their need to *hunt* (they were respirators), while plant lineage returned to a lifestyle of photosynthesis.

24 A BRIEF HISTORY OF INTELLIGENCE

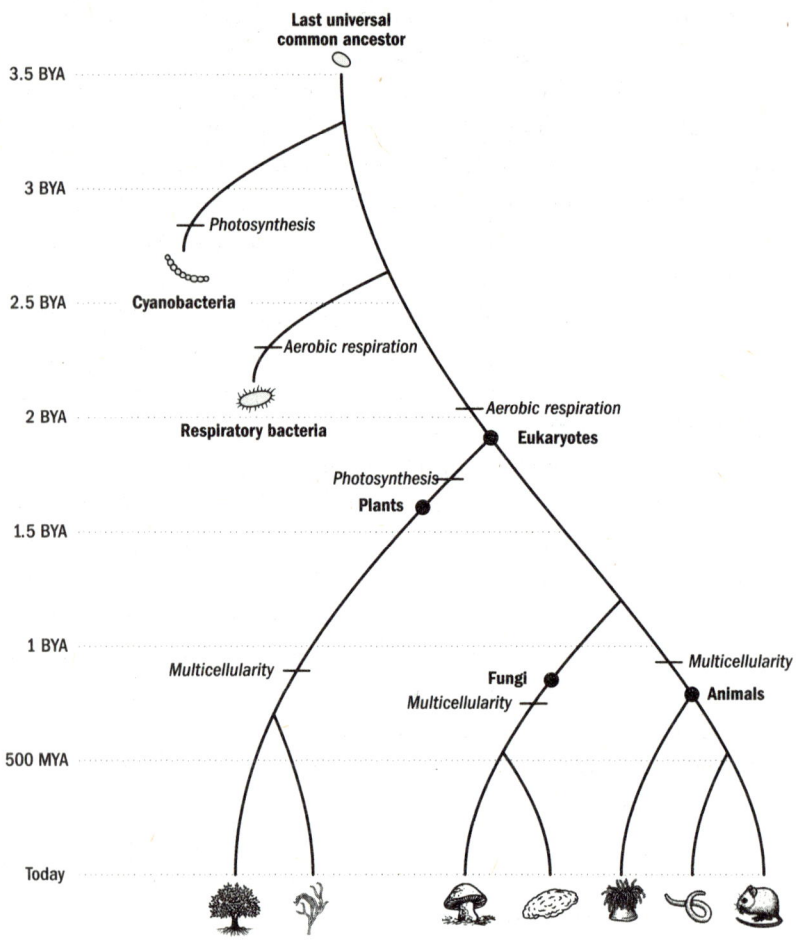

Figure 1.3: The tree of life

What was common across these eukaryote lineages was that all three—plants, fungi, and animals—each independently evolved multicellularity. Most of what you see and think of as life—humans, trees, mushrooms—are primarily *multicellular* organisms, cacophonies of billions of individual cells all working together to create a singular emergent organism. A human is made up of exactly such diverse types of specialized cells: skin cells, muscle cells, liver cells, bone cells, immune cells, blood cells. A plant has specialized cells too. These cells all serve different *functions* while still serving a common *purpose*: supporting the survival of the overall organism.

And so, seaweed-like underwater plants began sprouting, mushroom-like fungi began growing, and primitive animals began slowly swimming around. By about eight hundred million years ago, life would have fallen into three broad levels of complexity. At level one, there was single-celled life, made up of microscopic bacteria and single-celled eukaryotes. At level two, there was small multicellular life, large enough to engulf single-celled organisms but small enough to move around using basic cellular propellers. At level three, there was *large* multicellular life; too big to move with cellular propellers, and therefore forming immobile structures.

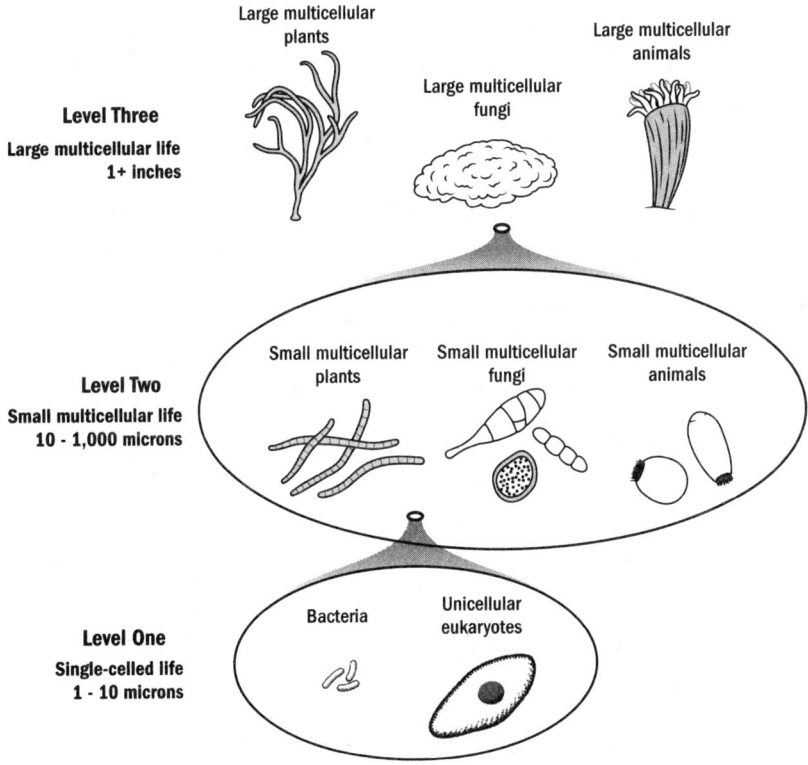

Figure 1.4: Three complexity levels in the ancient sea before brains

These early animals probably wouldn't resemble what you think of as animals. But they contained something that made them different from all other life at the time: neurons.

The Neuron

What a neuron is and what it does depends on whom you ask. If you ask a biologist, neurons are the primary cells that make up the nervous system. If you ask a machine learning researcher, neurons are the fundamental units of neural networks, little accumulators that perform the basic task of computing a weighted summation of their inputs. If you ask a psychophysicist, neurons are the sensors that measure features of the external world. If you ask a neuroscientist specializing in motor control, neurons are effectors, the controllers of muscles and movement. If you ask other people, you might get a wide range of answers, from "Neurons are little electrical wires in your head" to "Neurons are the stuff of consciousness." All of these answers are right, carrying a kernel of the whole truth, but incomplete on their own.

The nervous systems of all animals—from worms to wombats—are made up of these stringy odd-shaped cells called neurons. There is an

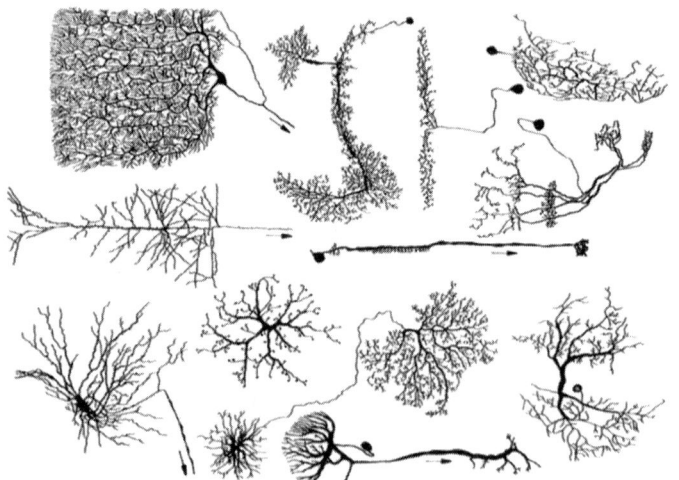

Figure 1.5: Neurons

incredible diversity of neurons, but despite this diversity in shapes and sizes, all neurons work the same way. This is the most shocking observation when comparing neurons across species—they are all, for the most part, fundamentally identical. The neurons in the human brain operate the same way as the neurons in a jellyfish. What separates you from an earthworm is not the unit of intelligence itself—neurons—but how these units are wired together.

Animals with neurons share a common ancestor, an organism in whom the first neurons evolved and from whom all neurons descend. It seems that in this ancient grandmother of animals, neurons attained their modern form; from this point on, evolution rewired neurons but made no meaningful adjustments to the basic unit itself. This is a glaring example of how prior innovations impose constraints on future innovations, often leaving early structures unchanged—the fundamental building blocks of brains have been essentially the same for over six hundred million years.

Why Fungi Don't Have Neurons but Animals Do

You aren't all that different from mold. Despite their appearance, fungi have more in common with animals than they do with plants. While plants survive by photosynthesis, animals and fungi both survive by respiration. Animals and fungi both breathe oxygen and eat sugar; both digest their food, breaking cells down using enzymes and absorbing their inner nutrients; and both share a much more recent common ancestor than either do with plants, which diverged much earlier. At the dawn of multicellularity, fungi and animal life would have been extremely similar. And yet one lineage (animals) went on to evolve neurons and brains, and the other (fungi) did not. Why?

Sugar is produced only by life, and thus there are only two ways for large multicellular respiratory organisms to feed. One is to wait for life to die, and the other is to catch and kill living life. Early in the fungi-animal divergence, they each settled into opposing feeding strategies. Fungi chose the strategy of waiting, and animals chose the strategy of killing.* Fungi

---

* Although, as with everything in evolution, there is nuance. There is a third, middle

eat through *external* digestion (secreting enzymes to break food down outside the body), while animals eat through *internal* digestion (trapping food inside the body and then secreting enzymes). The fungal strategy was, by some measures, more successful than the animals'—by biomass, there is about six times more fungus on Earth than animals. But as we will continually see, it is usually the worse strategy, the harder strategy, from which innovation emerges.

Fungi produce trillions of single-celled spores that float around dormant. If by luck one happens to find itself near dying life, it will blossom into a large fungal structure, growing hairy filaments into the decaying tissue, secreting enzymes, and absorbing the released nutrients. This is why mold always shows up in old food. Fungal spores are all around us, patiently waiting for something to die. Fungi are currently, and likely have always been, Earth's garbage collectors.

Early animals, however, settled into a strategy of actively catching and ingesting level-two multicellular prey (see figure 1.4). Active killing was, of course, not new; the first eukaryotes had long ago invented a strategy—phagotrophy—for killing life. But this worked only on level-one (single-cellular) life; level-two multicellular blobs were far too big to engulf into a single cell. And so early animals evolved internal digestion as a strategy for eating level-two life. Animals uniquely form little stomachs where they trap prey, secrete enzymes, and digest them.

In fact, the formation of an inner cavity for digestion may have been the defining feature of these early animals. Practically every animal alive today develops in the same three initial steps. From a single-celled fertilized egg, a hollow sphere (a blastula) forms; this then folds inward to make a cavity, a little "stomach" (a gastrula). This is true of human embryos as much as jellyfish embryos. While every animal develops this way, no other kingdom of life does this. This provides a glaring hint as to the evolutionary template on which all animals derive: we form stomachs for ingesting food. All animals that engage in such gastrulation also have neurons and

---

option that some species of both animals and fungi settled into: the *parasitic strategy*. Instead of actively catching prey to kill it, parasites infect prey and steal sugar or kill them from the inside.

muscles and seem to derive from a common neuron-enabled animal ancestor. Gastrulation, neurons, and muscles are the three inseparable features that bind all animals together and separate animals from all other kingdoms of life.

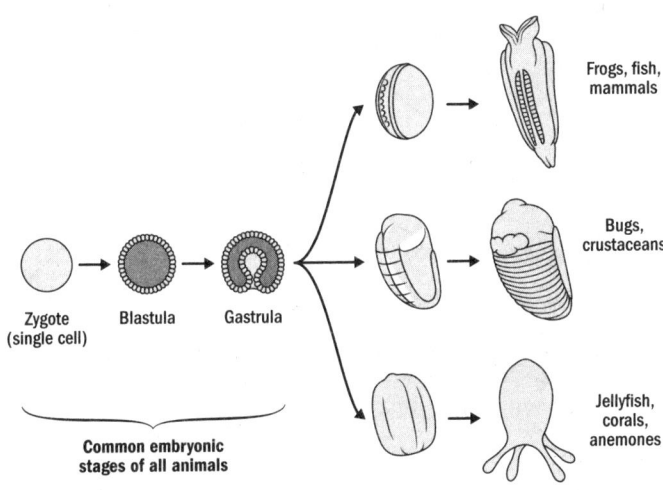

Figure 1.6: Shared developmental stages for all animals

Some have gone so far as to argue that the grandmother of animals was literally a little gastrula-shaped creature with neurons. But this is a scientific arena full of controversy—just because all animals develop this way doesn't mean that they ever truly lived in this form.

Another interpretation, supported by fossils, suggests that the first animals may have been similar to today's corals. To the naked eye, a coral is so simple that it doesn't look all that different from a fungus or a plant (figure 1.8). Only when examining its biology closely would you see the presence of the animal template: a stomach, muscles, and neurons. A coral is actually a colony of independent organisms called coral polyps. A coral polyp is, in some sense, literally *only* a stomach with neurons and muscles. They have little tentacles that float in the water, waiting for small organisms to swim toward them. When food touches one of the tips of these tentacles, they rapidly contract, pulling the prey into the stomach cavity

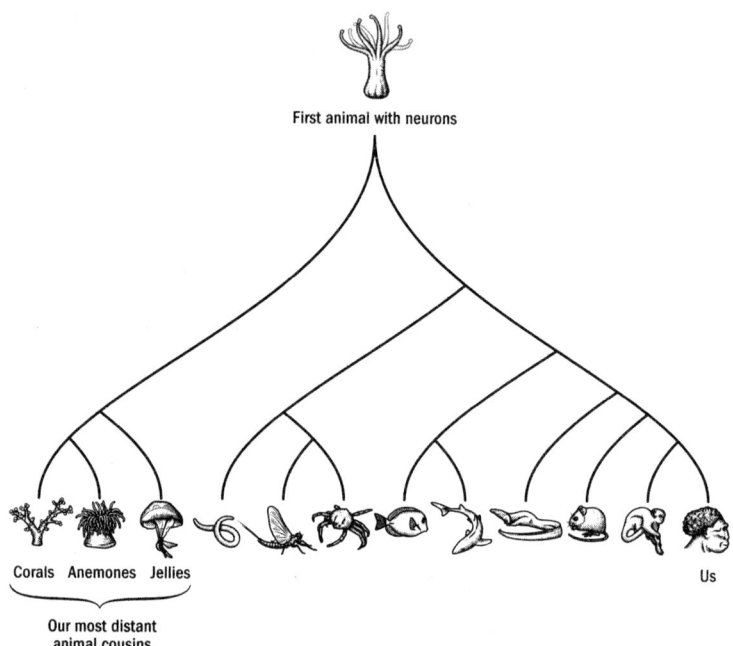

Figure 1.7: Tree of neuron-enabled animals

where it is digested. Neurons on the tips of these tentacles detect food and trigger a cascade of signaling through a web of other neurons that generates a coordinated relaxing and contracting of different muscles.

This coral reflex was not the first or the only way in which multicellular life sensed and responded to the world. Plants and fungi do this just fine without neurons or muscles; plants can orient their leaves toward the sun, and fungi can orient their growth in the direction of food. But still, in the ancient sea at the dawn of multicellularity, this reflex would have been revolutionary, not because it was the first time multicellular life sensed or moved but because it was the first time it sensed and moved with *speed* and *specificity*. The movement of plants and fungi takes hours to days; the movement of coral takes seconds.* The movement of plants and fungi is clumsy and inexact; the movement of coral is comparatively very

---

* The Venus flytrap is a fascinating exception to this, an example of a plant that independently evolved the ability to capture prey by moving quickly.

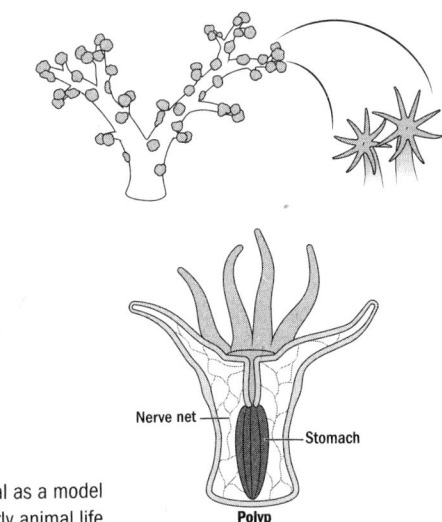

Figure 1.8: Soft coral as a model for early animal life

specific—the grasping of prey, opening of the mouth, pulling into the stomach, and closing of the mouth all require a well-timed and accurate orchestration of relaxing some muscles while contracting others. And *this* is why fungi don't have neurons and animals do. Although both are large multicellular organisms that feed on other life, only the animal-survival strategy of killing level-two multicellular life requires fast and specific reflexes.* The original purpose of neurons and muscles may have been the simple and inglorious task of swallowing.

Edgar Adrian's Three Discoveries and the Universal Features of Neurons

The scientific journey by which we have come to understand how neurons work has been long and full of false starts and wrong turns. As far back as the famous ancient Greek physician Hippocrates in 400 BCE, it has been

---

* You could, of course, state this the opposite way: fungi never got to feed on other life *because* they never got neurons. The point is not which came first but that neurons and hunting level-two multicellular life were part of the same strategy, one which fungi never used.

known that animals contain a system of stringy material later called *nerves* (from the Latin word *nervus*, meaning "sinew"), flowing to and from the brain, and that this system is the medium by which muscles are controlled and sensations are perceived. They knew this through horrifying experiments of severing spinal cords and clamping nerves in living pigs and other livestock. But the ancient Greeks wrongly concluded that it was "animal spirits" that flowed through these nerves. It would be many centuries before this error was corrected. More than two thousand years later, even the great Isaac Newton would speculate, incorrectly, that nerves communicated via vibrations flowing through a nerve fluid he called *ether*. It wasn't until the late 1700s that scientists discovered that flowing through the nervous system was not *ether* but *electricity*. Electrical current applied to a nerve will flow toward downstream muscles and cause them to contract.

But many errors remained. At the time, it was thought that the nervous system was made up of a single web of nerves, analogous to the circulatory system's homogenous web of blood vessels. It wasn't until the end of the nineteenth century, with more advanced microscopes and staining techniques, that scientists discovered that the nervous system was made up of independent cells—neurons—that, although wired together, are separate and generate their own signals. This also revealed that electrical signals flow only in a single direction within a neuron, from one part that takes inputs, called *dendrites*, to another that sends electrical output, called the *axon*. This output flows to other neurons or other types of cells (such as muscle cells) to activate them.

In the early 1920s, a young English neurologist by the name of Edgar Adrian returned to Cambridge University after a long stint of medical service during World War I. Adrian, like many researchers at the time, was interested in electrically recording neurons to decipher how and what they communicated. The problem had always been that electrical recording devices were too large and crude to record the activity of a single neuron and hence always produced a noisy mess of signals from multiple neurons. Adrian and his collaborators were the first to find a technical solution to this problem, inventing the field of single-neuron electrophysiology. This gave scientists, for the first time, a view into the language of individual neurons. The subsequent three discoveries earned Adrian the Nobel Prize.

The first discovery was that neurons don't send electrical signals in the form of a continuous ebbing and flowing but rather in all-or-nothing responses, also called spikes or action potentials. A neuron is either on or off; there is no in between. In other words, neurons act less like an electric power line with a constant flow of electricity and more like an electric telegraph cable, with patterns of electrical clicks and pauses. Adrian himself noted the similarity between neural spikes and Morse code.

This discovery of spikes presented a puzzle for Adrian. You can clearly perceive *stimulus strengths* in your senses—you can discriminate between different volumes of sound, brightness of light, potency of smells, severity of pain. How could a simple binary signal that was either on or off communicate a numerical value, such as the graded strengths of stimuli? The realization that the language of neurons was spikes didn't tell researchers much about what a sequence of spikes *meant*. Morse code is, well, a code—it is an efficient trick to store and transmit information across a single electrical wire. Adrian was the first scientist to use the word *information* to refer to the signals of neurons, and he devised a simple experiment to try to decode them.

Adrian took a muscle from the neck of a deceased frog and attached a recording device to a single stretch-sensing neuron in the muscle. Such neurons have receptors that are stimulated when muscles are stretched. Adrian then attached various weights to the muscle. The question was: How would the responses of these stretch-sensing neurons change based on the *weight* placed on the muscle?

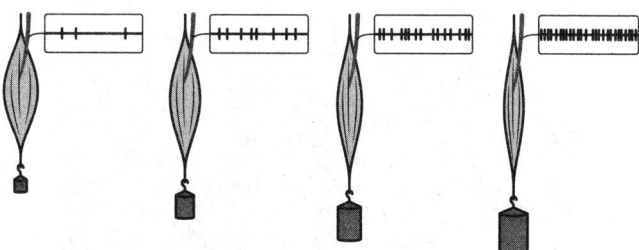

Figure 1.9: Adrian charted the relationship between weight and the number of spikes per second (i.e., the *spike rate,* or *firing rate*) elicited in these stretch neurons.

It turned out the spikes were identical in all cases; the only difference was *how many* spikes were fired. The heavier the weight, the higher the frequency of spikes (figure 1.9). This was Adrian's second discovery, what is now known as *rate coding*. The idea is that neurons encode information in the rate that they fire spikes, not in the shape or magnitude of the spike itself. Since Adrian's initial work, such rate coding has been found in neurons across the animal kingdom, from jellyfish to humans. Touch-sensitive neurons encode *pressure* in their firing rate; photosensitive neurons encode *contrast* in their firing rate; smell-sensitive neurons encode *concentration* in their firing rate. The neural code for movement is also a firing rate: the faster the spikes of the neurons that stimulate muscles, the greater the contraction force of the muscles. This is how you are able to delicately pet your dog and also pick up fifty-pound weights—if you couldn't modulate the strength of muscle contractions, you wouldn't be very pleasant to be around.

Adrian's third discovery was the most surprising of all. There is a problem with trying to translate natural variables, such as the pressure of touch or the brightness of light into this language of rate codes. The problem is this: these natural variables have a *massively larger* range than can be encoded in the firing rate of a neuron.

Take vision, for example. What you don't realize (because your sensory machinery abstracts it away) is that the luminance of light is varying astronomically all around you. The amount of light entering your eyes when

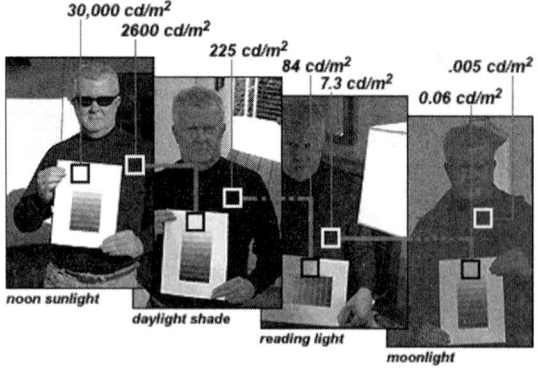

Figure 1.10: The broad spectrum of stimulus strengths

you look at a white piece of paper is one million times greater if you are in sunlight than if you are in moonlight.* In fact, the black letters of a page in sunlight are thirty times brighter than the white of a page in moonlight!

This is not just a feature of light; all sensory modalities, from smell to touch to sound, require the discrimination of hugely varying natural variables. This wouldn't necessarily be a problem except for a big limitation of neurons—for a variety of biochemical reasons, it is simply impossible for a neuron to fire faster than around five hundred spikes per second. This means that a neuron needs to encode a range of natural variables that can vary by a factor of over a million all within a firing rate ranging only from 0 to 500 spikes per second. This could be reasonably called the "squishing problem": neurons have to *squish* this huge range of natural variables into a comparably minuscule range of firing rates.

This makes rate coding, on its own, untenable. Neurons simply cannot directly encode such a wide range of natural variables in such a small range of firing rates without losing a huge amount of precision. The resulting imprecision would make it impossible to read inside, detect subtle smells, or notice a soft touch.

It turns out that neurons have a clever solution to this problem. Neurons do not have a fixed relationship between natural variables and firing rates. Instead, neurons are always adapting their firing rates to their environment; they are constantly remapping the relationship between variables in the natural world and the language of firing rates. The term neuroscientists use to describe this observation is *adaptation*; this was Adrian's third discovery.

In Adrian's frog muscle experiments, a neuron might fire one hundred spikes in response to a certain weight. But after this first exposure, the neuron quickly adapts; if you apply the same weight shortly after, it might elicit only eighty spikes. And as you keep doing this, the number of spikes continues to decline. This applies in many neurons throughout the brains of animals—the stronger the stimuli, the greater the change in the neural

---

* More specifically, the *luminance* is one million times larger. Luminance is measured in candelas per square meter, which is the rate that photons are generated per unit of surface area weighted by human wavelength sensitivity.

threshold for spiking. In some sense, neurons are more a measurement of relative changes in stimulus strengths, signaling how much the strength of a stimulus changed relative to its baseline as opposed to signaling the absolute value of the stimulus.

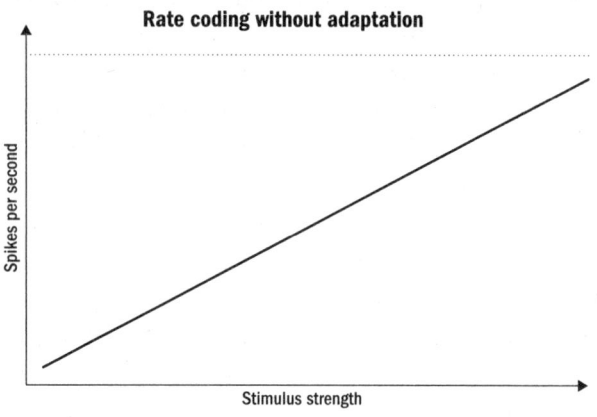

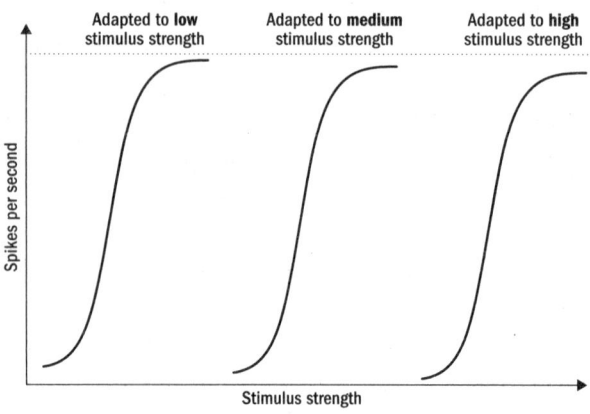

Figure 1.11

Here's the beauty: Adaptation solves the squishing problem. Adaptation enables neurons to precisely encode a broad range of stimulus strengths despite a limited range of firing rates. The stronger a stimulus is, the more strength will be required to get the neuron to respond similarly next time. The weaker a stimulus is, the more sensitive neurons become.

The late nineteenth and early twentieth centuries were filled with rich discoveries about the inner workings of neurons. A long roster of giants in neuroscience emerged in this period, leading to a platter of Nobel Prizes, not only Edgar Adrian, but also Santiago Ramón y Cajal, Charles Sherrington, Henry Dale, John Eccles, and others. One important discovery was that nerve impulses pass from one neuron to another by way of synapses, which are microscopic gaps between neurons. Spikes in the input neuron trigger the release of chemicals called neurotransmitters, which travel across the synapse over the course of nanoseconds, attach to a bunch of protein receptors, trigger ions to flow into the target neuron, and thereby change its charge. While neural communication *within* a neuron is electrical, *across* neurons, it is chemical.*

In the 1950s, John Eccles discovered that neurons come in two main

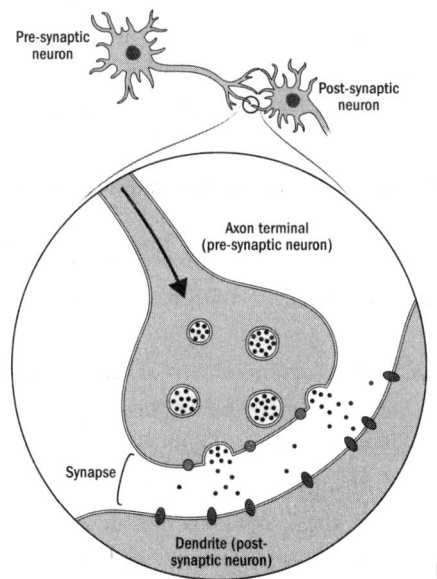

Figure 1.12

---

* Of course, this is *mostly*. There are cases when neurons touch each other and form gap junctions that allow the transfer of electrical signals directly from one neuron to another.

varieties: excitatory neurons and inhibitory neurons. Excitatory neurons release neurotransmitters that *excite* neurons they connect to, while inhibitory neurons release neurotransmitters that *inhibit* neurons they connect to. In other words, excitatory neurons *trigger* spikes in other neurons, while inhibitory neurons *suppress* spikes in other neurons.

These features of neurons—all-or-nothing spikes, rate coding, adaptation, and chemical synapses with excitatory and inhibitory neurotransmitters—are universal across all animals, even in animals that have no brain, such as coral polyps and jellyfish. Why do all neurons share these features? If early animals were, in fact, like today's corals and anemones, then these aspects of neurons enabled ancient animals to successfully respond to their environment with speed and specificity, something that had become necessary to actively capture and kill level-two multicellular life. All-or-nothing electrical spikes triggered rapid and orchestrated reflexive movements so animals could catch prey in response to even the subtlest of touches or smells. Rate coding enabled animals to modify their responses based on the strengths of a touch or smell. Adaptation enabled animals to adjust the sensory threshold for when spikes are generated, allowing them to be highly sensitive to even the subtlest of touches or smells while also preventing overstimulation at higher strengths of stimuli.

What about inhibitory neurons? Why did they evolve? Consider the simple task of a coral polyp opening or closing its mouth. For its mouth to open, one set of muscles must contract and another must relax. And the converse for closing its mouth. The existence of both excitatory and inhibitory neurons enabled the first neural circuits to implement a form of logic required for reflexes to work. They can enforce the rule of "do this, not that," which was perhaps the first glimmer of intellect to emerge from circuits of neurons. "Do this, not that" logic was not new—such logic already existed in the protein cascades of single cells. But this ability was recapitulated in the medium of neurons, which made such logic newly possible on the scale of level-three multicellular life. Inhibitory neurons enabled the necessary inner logic for catch-and-swallow reflexes to work.

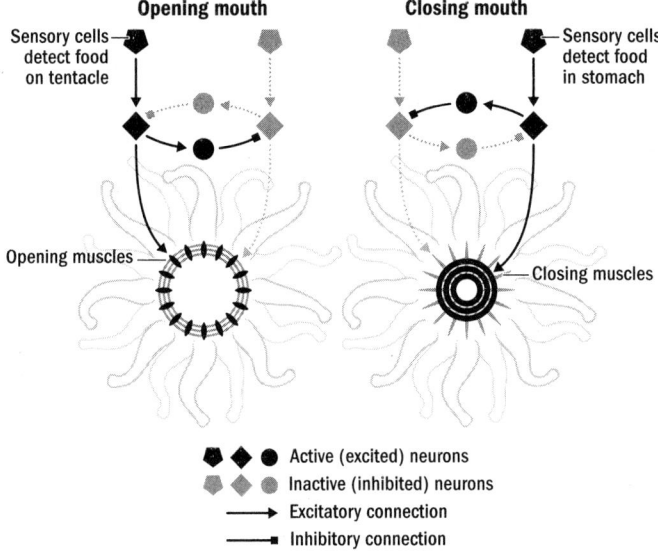

Figure 1.13: The first neural circuit

While the first animals, whether gastrula-like or polyp-like creatures, clearly had neurons, they had no brain. Like today's coral polyps and jellyfish, their nervous system was what scientists call a nerve net: a distributed web of independent neural circuits implementing their own independent reflexes.

But with the evolutionary feedback loop of predator-prey in full force, with the animal niche of active hunting, and the building blocks of neurons in place, it was only a matter of time before evolution stumbled on breakthrough #1, which led to rewiring nerve nets into brains. It is here where our story truly begins, but it doesn't begin in the way you might expect.

BREAKTHROUGH #1

Steering and the First Bilaterians

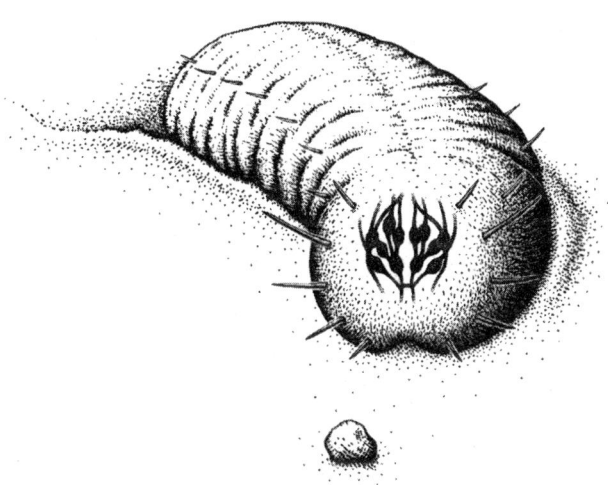

Your brain 600 million years ago

2

The Birth of Good and Bad

> Nature has placed mankind under the governance
> of two sovereign masters, pain and pleasure.
> —JEREMY BENTHAM, *AN INTRODUCTION TO*
> *THE PRINCIPLES OF MORALS AND LEGISLATION*

AT FIRST GLANCE, the diversity of the animal kingdom appears remarkable—from ants to alligators, bees to baboons, and crustaceans to cats, animals seem varied in countless ways. But if you pondered this further, you could just as easily conclude that what is remarkable about the animal kingdom is how *little* diversity there is. Almost all animals on Earth have the same body plan. They all have a front that contains a mouth, a brain, and the main sensory organs (such as eyes and ears), and they all have a back where waste comes out.

Evolutionary biologists call animals with this body plan *bilaterians* because of their bilateral symmetry. This is in contrast to our most distant animal cousins—coral polyps, anemones, and jellyfish—which have body plans with *radial* symmetry; that is, with similar parts arranged around a central axis, without any front or back. The most obvious difference between these two categories is how the animals eat. Bilaterians eat by putting food in their mouths and then pooping out waste products from their butts. Radially symmetrical animals have only one opening—a mouth-butt if you will—which swallows food into their stomachs and spits it out. The bilaterians are undeniably the more proper of the two.

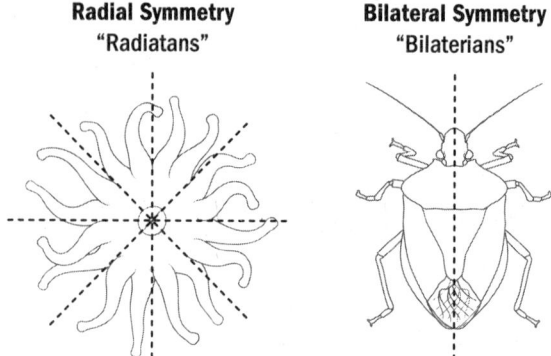

Figure 2.1

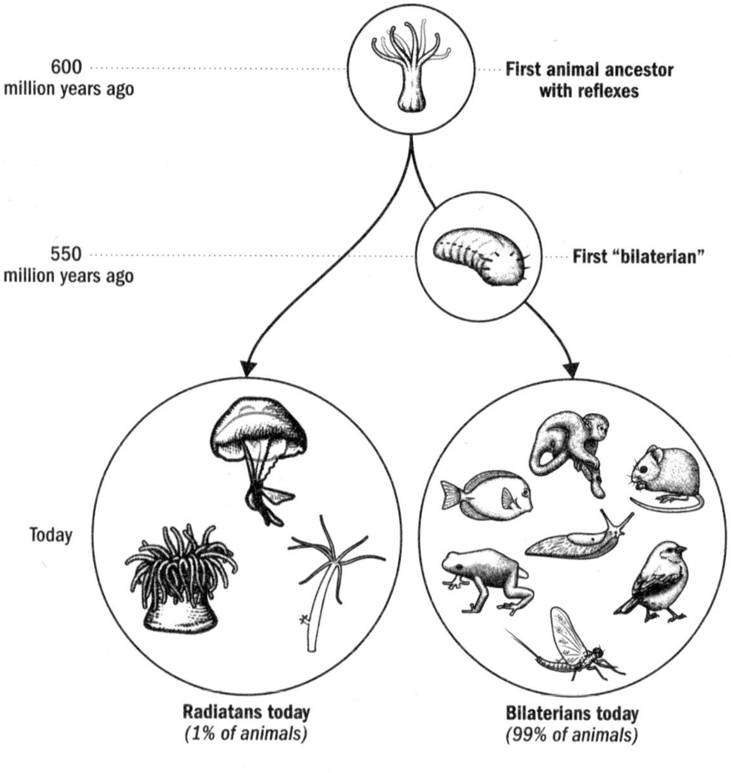

Figure 2.2

THE BIRTH OF GOOD AND BAD

The first animals are believed to have been radially symmetric, and yet today, most animal species are bilaterally symmetric. Despite the diversity of modern bilaterians—from worms to humans—they all descend from a single bilaterian common ancestor who lived around 550 million years ago. Why, within this single lineage of ancient animals, did body plans change from radial symmetry to bilateral symmetry?

Radially symmetrical body plans work fine with the coral strategy of waiting for food. But they work horribly for the hunting strategy of navigating *toward food*. Radially symmetrical body plans, if they were to move, would require an animal to have sensory mechanisms to detect the location of food in any direction and then have the machinery to move in any direction. In other words, they would need to be able to simultaneously detect and move in *all different directions*. Bilaterally symmetrical bodies make movement much simpler. Instead of needing a motor system to move in *any* direction, they simply need one motor system to move forward and one to turn. Bilaterally symmetrical bodies don't need to choose the exact direction; they simply need to choose whether to adjust to the right or the left.

Even modern human engineers have yet to find a better structure for navigation. Cars, planes, boats, submarines, and almost every human-built navigation machine is bilaterally symmetric. It is simply the most efficient

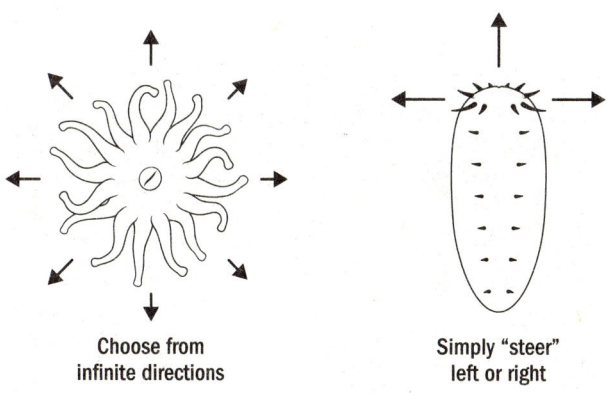

Figure 2.3: Why bilateral symmetry is better for navigation

design for a movement system. Bilateral symmetry allows a movement apparatus to be optimized for a single direction (forward) while solving the problem of navigation by adding a mechanism for turning.

There is another observation about bilaterians, perhaps the more important one: They are the only animals that have brains. This is not a coincidence. The first brain and the bilaterian body share the same initial evolutionary purpose: They enable animals to navigate by steering. Steering was breakthrough #1.

Navigating by Steering

Although we don't know exactly what the first bilaterians looked like, fossils suggest they were legless wormlike creatures about the size of a grain of rice. Evidence suggests that they first emerged sometime in the Ediacaran period, an era that stretched from 635 to 539 million years ago. The seafloor of the Ediacaran was filled with thick green gooey microbial mats in its shallow areas—vast colonies of cyanobacteria basking in the sun. Sensile multicellular animals like corals, sea sponges, and early plants would have been common.

Modern nematodes are believed to have remained relatively unchanged since early bilaterians; these creatures give us a window into the inner workings of our wormlike ancestors. Nematodes are almost literally just

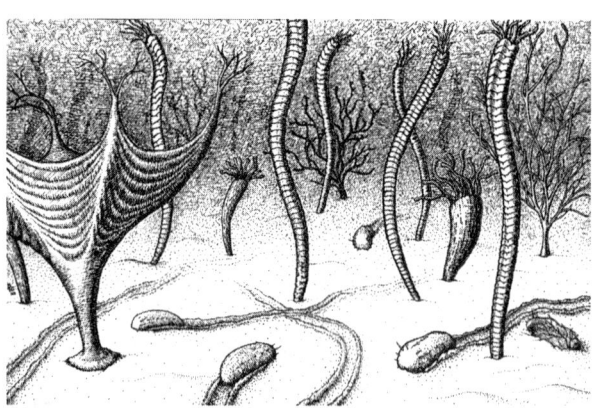

Figure 2.4: The Ediacaran world

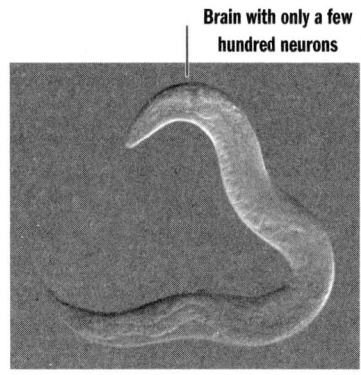

Figure 2.5: The nematode *C. elegans*

the basic template of a bilaterian: not much more than a head, mouth, stomach, butt, some muscles, and a brain.

The first brains were, like those of nematodes, almost definitely very simple. The most well-studied nematode, *Caenorhabditis elegans*, has only 302 neurons, a minuscule number compared to a human's 85 billion. However, nematodes display remarkably sophisticated behavior despite their minuscule brains. What a nematode does with their hopelessly simple brain suggests what the first bilaterians did with theirs.

The most obvious behavioral difference between nematodes and more ancient animals like corals is that nematodes spend a lot of time *moving*. Here's an experiment: Put a nematode on one side of a petri dish, place a tiny piece of food on the other side. Three things would reveal themselves: First, it *always* finds the food. Second, it finds the food *much faster* than it would if it were simply moving around randomly. And third, it doesn't swim directly toward the food but rather circles in on the food.

The worm is not using vision; nematodes can't see. They have no eyes to render any image useful for navigation. Instead, the worm is using *smell*. The closer it gets to the source of a smell, the higher the concentration of that smell. Worms exploit this fact to find food. All a worm must do is turn toward the direction where the concentration of food particles is increasing, and away from the direction it is decreasing. It is quite elegant how simple yet effective this navigational strategy is. It can be summarized in two rules:

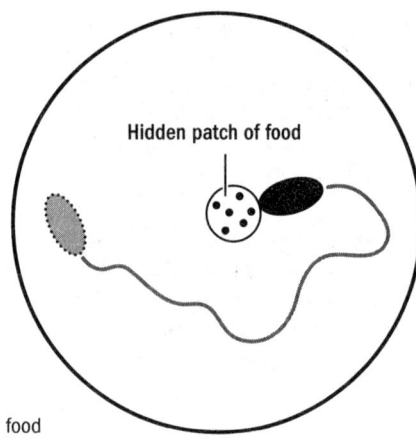

Figure 2.6: Nematode steering toward food

1. If food smells increase, keep going forward.
2. If food smells decrease, turn.

This was the breakthrough of *steering*. It turns out that to successfully navigate in the complicated world of the ocean floor, you don't actually need an understanding of that two-dimensional world. You don't need an understanding of where you are, where food is, what paths you might have to take, how long it might take, or really anything meaningful about the world. All you need is a brain that steers a bilateral body toward increasing food smells and away from decreasing food smells.

Steering can be used not only to navigate *toward* things but to navigate *away* from things. Nematodes have sensory cells that detect light, temperature, and touch. They steer away from light, where predators can more easily see them; they steer away from noxious heat and cold, where their bodily functions become harder to perform; and they steer away from surfaces that are sharp, where their fragile bodies might get wounded.

This trick of navigating by steering was not new. Single-celled organisms like bacteria navigate around their environments in a similar way. When a protein receptor on the surface of a bacterium detects a stimulus like light, it can trigger a chemical process within the cell that changes the movement of the cell's protein propellers, thereby causing it to change its direction. This is how single-celled organisms like bacteria swim toward food sources or away from dangerous chemicals. But this mech-

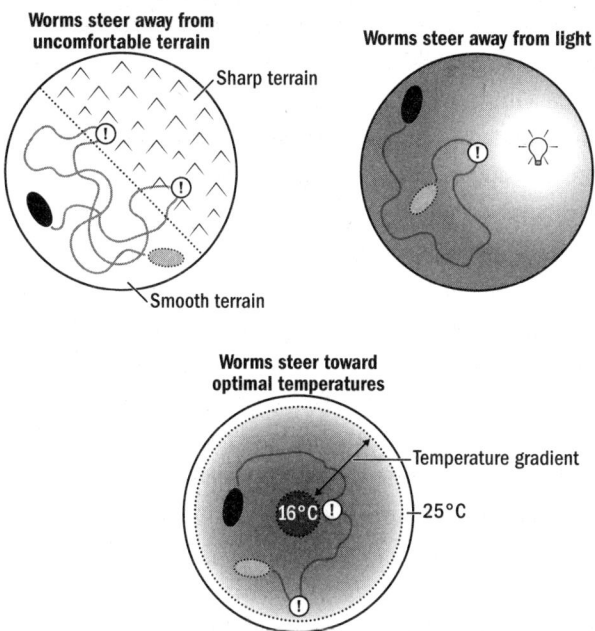

Figure 2.7: Examples of steering decisions made by simple bilaterians like nematodes and flatworms.

anism works only on the scale of individual cells, where simple protein propellers can successfully reorient the entire life-form. Steering in an organism that contains *millions* of cells required a whole new setup, one in which a stimulus activates circuits of neurons and the neurons activate muscle cells, causing specific turning movements. And so the breakthrough that came with the first brain was not steering per se, but steering on the scale of multicellular organisms.

The First Robot

In the 1980s and 1990s a schism emerged in the artificial intelligence community. On one side were those in the symbolic AI camp, who were focused on decomposing human intelligence into its constituent parts in an attempt to imbue AI systems with our most cherished skills: reasoning, language, problem solving, and logic. In opposition were those in the behavioral AI camp, led by the roboticist Rodney Brooks at MIT, who believed the symbolic approach was doomed to fail because "we will never understand how to decompose human level intelligence until we've had a lot of practice with simpler level intelligences."

Brooks's argument was partly based on evolution: it took billions of years before life could simply sense and respond to its environment; it took another five hundred million years of tinkering for brains to get good at motor skills and navigation; and only after all of this hard work did language and logic appear. To Brooks, compared to how long it took for sensing and moving to evolve, logic and language appeared in a blink of an eye. Thus he concluded that "language ... and reason, are all pretty simple once the essence of being and reacting are available. That essence is the ability to move around in a dynamic environment, sensing the surroundings to a degree sufficient to achieve the necessary maintenance of life and reproduction. This part of intelligence is where evolution has concentrated its time—it is much harder."

To Brooks, while humans "provided us with an existence proof of [human-level intelligence], we must be careful about what the lessons are to be gained from it." To illustrate this, he offered a metaphor:

> Suppose it is the 1890s. Artificial flight is the glamor subject in science, engineering, and venture capital circles. A bunch of [artificial flight] researchers are miraculously transported by a time machine to the 1990s for a few hours. They spend the whole time in the passenger cabin of a commercial passenger Boeing 747 on a medium duration flight. Returned to the 1890s they feel invigorated, knowing that [artificial flight] is possible on a grand scale. They immediately set to work duplicating what they have seen. They make great progress in designing pitched seats, double pane windows, and know that if only they can figure out those weird "plastics" they will have the grail within their grasp.

By trying to *skip* simple planes and directly build a 747, they risked completely misunderstanding the principles of how planes work (pitched seats, paned windows, and plastics are the wrong things to focus on). Brooks believed the exercise of trying to reverse-engineer the human brain suffered from this same problem. A better approach was to "incrementally build up the capabilities of intelligence systems, having complete systems at each step." In other words, to start as evolution did, with simple brains, and add complexity from there.

Many do not agree with Brooks's approach, but whether you agree with him or not, it was Rodney Brooks who, by any reasonable account, was the first to build a commercially successful domestic robot; it was Brooks who made the first small step toward Rosey. And this first step in the evolution of commercial robots has parallels to the first step in the evolution of brains. Brooks, too, started with steering.

In 1990, Brooks cofounded a robotics company named iRobot, and in 2002, he introduced the Roomba, a vacuum-cleaner robot. The Roomba was a robot that autonomously navigated around your house vacuuming the floor. It was an immediate hit; new models are still being produced today, and iRobot has sold over forty million units.

The first Roomba and the first bilaterians share a surprising number of properties. They both had extremely simple sensors—the first Roomba could detect only a handful of things, such as when it hit a wall and when it was close to its charging base. They both had simple brains—neither used the paltry sensory input they received to build a map of their environment or to recognize objects. They both were bilaterally symmetric—the Roomba's wheels allowed it to go forward and backward only. To change

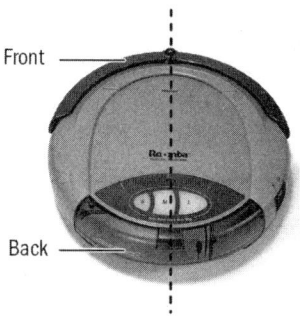

Figure 2.8: The Roomba. A vacuum-cleaning robot that navigated in a way similar to the first bilaterians.

directions, it had to turn in place and then resume its forward movement.

The Roomba could clean all the nooks and crannies of your floor by simply moving around randomly, steering away from obstacles when it bumped into them, and steering toward its charging station when it was low on battery. Whenever the Roomba hit a wall, it would perform a random turn and try to move forward again. When it was low on battery, the Roomba searched for a signal from its charging station, and when it detected the signal, it simply turned in the direction where the signal was strongest, eventually making it back to its charging station.

The navigational strategies of the Roomba and first bilaterians were not identical. But it may not be a coincidence that the first successful domestic robot contained an intelligence not so unlike the intelligence of the first brains. Both used tricks that enabled them to navigate a complex world without actually understanding or modeling that world.

While others remained stuck in the lab working on million-dollar robots with eyes and touch and brains that attempted to compute complicated things like maps and movements, Brooks built the simplest possible robot, one that contained hardly any sensors and that computed barely anything at all. But the market, like evolution, rewards three things above all: things that are *cheap*, things that *work*, and things that are simple enough to be discovered in the first place.

While steering might not inspire the same awe as other intellectual feats, it was surely energetically cheap, it worked, and it was simple enough for evolutionary tinkering to stumble upon it. And so it was here where brains began.

Valence and the Inside of a Nematode's Brain

Around the head of a nematode are sensory neurons, some of which respond to light, others to touch, and others to specific chemicals. For steering to work, early bilaterians needed to take each smell, touch, or other stimulus they detected and make a choice: Do I approach this thing, avoid this thing, or ignore this thing?

The breakthrough of steering required bilaterians to categorize the world into things to approach ("good things") and things to avoid ("bad things").

Even a Roomba does this—obstacles are bad; charging station when low on battery is good. Earlier radially symmetric animals did not navigate, so they never had to categorize things in the world like this.

When animals categorize stimuli into good and bad, psychologists and neuroscientists say they are imbuing stimuli with *valence*. Valence is the goodness or badness of a stimulus. Valence isn't about a moral judgment; it's something far more primitive: whether an animal will respond to a stimulus by approaching it or avoiding it. The valence of a stimulus is, of course, not objective; a chemical, image, or temperature, on its own, has no goodness or badness. Instead, the valence of a stimulus is *subjective*, defined only by the brain's evaluation of its goodness or badness.

How does a nematode decide the valence of something it perceives? It doesn't first observe something, ponder it, then decide its valence. Instead, the sensory neurons around its head *directly* signal the stimulus's valence. One group of sensory neurons are, effectively, positive-valence neurons; they are directly activated by things nematodes deem good (such as food smells). Another group of sensory neurons are, effectively, negative-valence neurons; they are directly activated by things nematodes deem bad (such as high temperatures, predator smells, bright light).

In nematodes, sensory neurons don't signal objective features of the surrounding world—they encode *steering votes* for how much a nematode wants to steer toward or away from something. In more complex bilaterians, such as humans, not all sensory machinery is like this—the neurons in your eyes detect features of images; the valence of the image is computed elsewhere. But it seems that the first brains began with sensory neurons that didn't care to measure objective features of the world and instead cast the entirety of perception through the simple binary lens of valence.

Figure 2.9 shows a simplified diagram of how steering works in nematodes. Valence neurons trigger different turning decisions by connecting to different downstream neurons.

Consider how a nematode uses this circuit to find food. Nematodes have positive-valence neurons that trigger forward movement when the concentration of a food smell *increases*. As we saw in the sensory neurons in the nerve net of earlier animals, these neurons quickly adapt to baseline levels of smells. This enables these valence neurons to signal *changes* across a

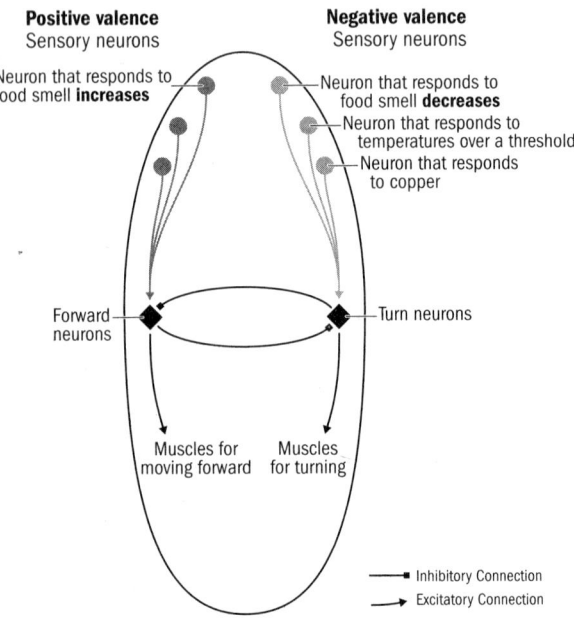

Figure 2.9: A simplified schematic of the wiring of the first brain

wide range of smell concentrations. These neurons will generate a similar number of spikes whether a smell concentration goes from two to four parts or from one hundred to two hundred parts. This enables valence neurons to keep nudging the nematode in the right direction. It is the signal for *Yes, keep going!* from the first whiff of a faraway food smell all the way to the food source.

This use of adaptation is an example of evolutionary innovations enabling future innovations. Steering toward food in early bilaterians was possible only because adaptation had already evolved in earlier radially symmetric animals. Without adaptation, valence neurons would be either too sensitive (and continuously misfire when smells are too close) or not sensitive enough (unable to detect faraway smells).

At this point, new navigational behaviors could emerge simply by modifying the conditions under which different valence neurons get excited. For example, consider how nematodes navigate toward optimal temperatures.

Temperature navigation requires some additional cleverness relative to the simple steering toward smells: the decreasing concentration of a food smell is *always* bad, but the decreasing temperature of an environment is bad *only if* a nematode is already too cold. If a nematode is hot, then decreasing temperature is good. A warm bath is miserable in a scorching summer but heavenly in a cold winter. How did the first brains manage to treat temperature fluctuations differently depending on the context?

Nematodes have a negative-valence neuron that triggers turning when temperatures increase, but only if the temperature is already above a certain threshold; it is a *Too hot!* neuron. Nematodes also have a *Too cold!* neuron; it triggers turning when temperatures decrease, but only when temperatures are already below a certain threshold. Together, these two negative-valence neurons enable nematodes to quickly steer away from heat when they're too hot and away from cold when they're too cold. Deep in the human brain is an ancient structure called the hypothalamus that houses temperature-sensitive neurons that work in the same way.

The Problem of Trade-Offs

Steering in the presence of multiple stimuli presented a problem: What happens if different sensory cells vote for steering in *opposite* directions? What if a nematode smells both something yummy and something dangerous at the same time?

Scientists have tested nematodes in exactly such situations. Put a bunch of nematodes on one side of a petri dish and place yummy food on the opposite side of the petri dish, then place a dangerous copper barrier (nematodes hate copper) in the middle. Nematodes then have a problem: Are they willing to cross the barrier to get to the food? Impressively, the answer is—as you would expect from an animal with even an iota of smarts—it depends. It depends on the relative concentration of the food smell versus the copper smell.

At low levels of copper, most nematodes cross the barrier; at intermediate levels of copper, only some do; at high levels of copper, no nematodes are willing to cross the barrier.

This ability to make trade-offs in the decision making process has been

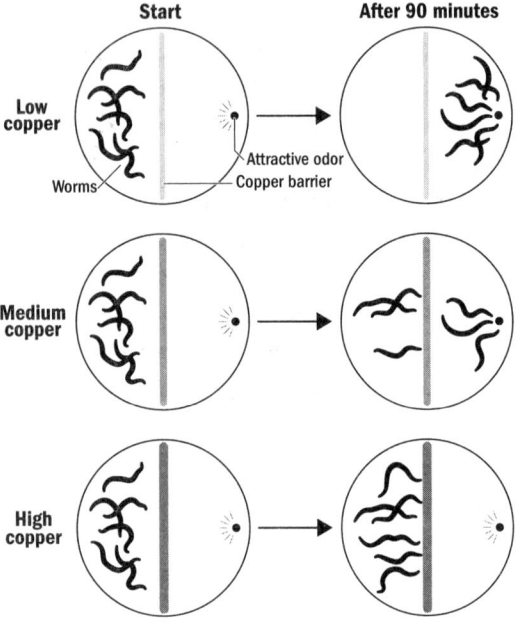

Figure 2.10

tested across different species of simple wormlike bilaterians and across different sensory modalities. The results consistently show that even the simplest brains, those with less than a thousand neurons, can make these trade-offs.

This requirement of integrating input across sensory modalities was likely one reason why steering required a brain and could not have been implemented in a distributed web of reflexes like those in a coral polyp. All these sensory inputs voting for steering in different directions *had to be integrated together* in a single place to make a single decision; you can go in only one direction at a time. The first brain was this mega-integration center—one big neural circuit in which steering directions were selected.

You can get an intuition for how this works from figure 2.9, which shows a simplified version of the nematode steering circuit. Positive-valence neurons connect to a neuron that triggers forward movement (what might be called a "forward neuron"), negative-valence neurons connect to a neuron that triggers turning (what might be called a "turning neuron"). The forward neuron accumulates votes for *Keep going forward!*, and the turning neuron accumulates votes for *Turn away!* The forward neuron and turn neurons mutually inhibit each other, enabling this network to integrate trade-offs and make a single choice—whichever neuron accumulates more votes wins and determines whether the animal will cross the copper barrier.

This is another example of how past innovations enabled future innovations. Just as a bilaterian cannot both go forward and turn at the same time, a coral polyp cannot both open and close its mouth at the same time. Inhibitory neurons evolved in earlier coral-like animals to enable these mutually exclusive reflexes to compete with each other so that only one reflex could be selected at a time; this same mechanism was repurposed in early bilaterians to enable them to make trade-offs in steering decisions. Instead of deciding whether to open or close one's mouth, bilaterians used inhibitory neurons to decide whether to go forward or turn.

Are You Hungry?

The valence of something depends on an animal's internal state. A nematode's choice of whether to cross a copper barrier to get to food depends not only on the relative level of food smell and copper but also on how hungry a nematode is. It won't cross the barrier to get to food if it is full, but it will if it is hungry. Further, nematodes can completely flip their preferences depending on how hungry they are. If a nematode is well fed, it will steer *away* from carbon dioxide; if hungry, it will steer *toward* it. Why? Carbon dioxide is a chemical that is released by *both* food and predators, so when a nematode is full, pursuing carbon dioxide for food isn't worth the risk of predators; when it is hungry, however, the chance that the carbon dioxide is signaling food, not predators, makes it worth the risk.

The brain's ability to rapidly flip the valence of a stimulus depending on internal states is ubiquitous. Compare the salivary ecstasy of the first bite of your favorite dinner after a long day of skipped meals to the bloated nausea of the last bite after eating yourself into a food coma. Within mere minutes, your favorite meal can transform from God's gift to mankind to something you want nowhere near you.

The mechanisms by which this occurs are relatively simple and shared across bilaterians. Animal cells release specific chemicals—"full signals" such as insulin—in response to having a healthy amount of energy. And animal cells release a different set of chemicals—hunger signals—in response to having *insufficient* amounts of energy. Both signals diffuse throughout an animal's body and provide a persistent global signal for an animal's level of hunger. The sensory neurons of nematodes have receptors that detect the presence of these signals and change their responses accordingly. Positive valence food smell neurons in *C. elegans* become more responsive to food smells in the presence of hunger signals and less responsive in the presence of full signals.

Internal states are present in a Roomba as well. A Roomba will ignore the signal from its home base when it is fully charged. In this case, the signal from the home base can be said to have neutral valence. When the Roomba's internal state changes to one where it is low on battery, the signal from home base shifts to having positive valence: the Roomba will no longer ignore the signal from its charging station and will steer toward it to replenish its battery.

Steering requires at least four things: a bilateral body plan for turning, valence neurons for detecting and categorizing stimuli into good and bad, a brain for integrating input into a single steering decision, and the ability to modulate valence based on internal states. But still, evolution continued tinkering. There is another trick that emerged in early bilaterian brains, a trick that further bolstered the effectiveness of steering. That trick was the early kernel of what we now call emotion.

3

The Origin of Emotion

THE BLOOD-BOILING FURY you feel when you hear a friend defend the opposite political party's most recent gaffe, while hard to define emotionally—perhaps some complex mixture of anger, disappointment, betrayal, and shock—is clearly a bad mood. The tingly serenity you feel when you lie on a warm sunny beach, also hard to define exactly, is still clearly a good mood. Valence exists not only in our assessment of external stimuli but also in our internal states.

Our internal states are not only imbued with a level of valence, but also a degree of arousal. Blood-boiling fury is not only a bad mood but an *aroused* bad mood. Different from an *unaroused* bad mood, like depression or boredom. Similarly, the tingly serenity of lying on a warm beach is not only a good mood but a good mood with *low arousal*. Different from the highly arousing good mood produced by getting accepted to college or riding a roller coaster (if you like that sort of thing).

Emotions are complicated. Defining and categorizing specific emotions is a perilous business, rife with cultural bias. In German, there is a word, *sehnsucht,* that roughly translates to the emotion of wanting a different life; there is no direct English translation. In Persian, the word *ænduh* expresses the concepts of regret and grief simultaneously; in Dargwa, the word *dard* expresses the concepts of anxiety and grief simultaneously. In English we have separate words for each. Which language best differentiates the objective categories of emotional states produced by brains? Many careers have been spent hunting for these objective categories in human brains; today, most neuroscientists believe that such objective categorizations do not exist, at least not at the level of words like *sehnsucht* or *grief*. Instead, it seems such emotion categories are largely culturally learned. We will find out more about how this works in later breakthroughs. For now, we begin by asking about the simpler origins of emotions. The basic template of emotions evolved as an intellectual trick to solve a specific set of problems faced by the first brains. And so we begin with the simplest two features of emotions, those that are universal not only across human cultures but also across the animal kingdom, those features of emotions that we inherited from the first brains: valence and arousal.

Neuroscientists and psychologists use the word *affect* to refer to these two attributes of emotions; at any given point, humans are in an affective state represented by a location across these two dimensions of valence and arousal. While rigorous definitions of categories of human emotions themselves elude philosophers, psychologists, and neuroscientists alike, *affect* is the relatively well accepted unifying foundation of emotion.

The universality of affect can be seen in our intuitions; it is easy to take suites of nuanced emotion words—*calm, elated, tense, upset, depressed, bored*—and assign them to the affective states from which they derive (see figure 3.1). The universality of affect can also be seen in our biology. There are clear neurophysiological signatures that differentiate levels of arousal, such as heart rate, perspiration, pupil size, adrenaline, and blood pressure.

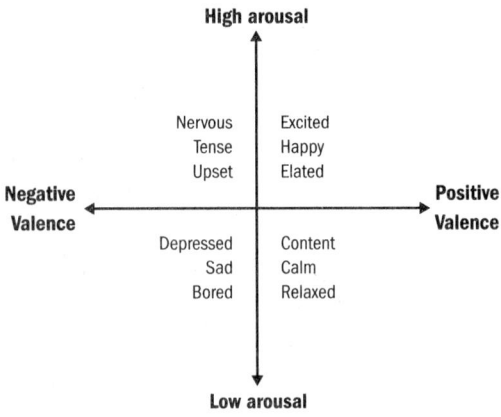

Figure 3.1: The affective states of humans

And there are clear neurophysiological signatures that differentiate levels of valence, such as stress-hormone levels, dopamine levels, and the activation of specific brain regions. And while cultures around the world differ in their classifications of specific emotion categories, such as anger and fear, classifications of affective states are quite universal. All cultures have words to communicate the concepts of valence and arousal, and newborn children across cultures have universal facial and body signatures for arousal and valence (e.g., crying and smiling).

The universality of affect stretches beyond the bounds of humanity; it is found across the animal kingdom. Affect is the ancient seed from which modern emotions sprouted. But why did affect evolve?

Steering in the Dark

Even nematodes with their minuscule nervous systems have affective states, albeit incredibly simple ones. Nematodes express different levels of arousal: When well fed, stressed, or ill, they hardly move at all and become unresponsive to external stimuli (low arousal); when hungry, detecting food, or sniffing predators, they will continually swim around (high arousal). The affective states of nematodes also express different levels of valence. Positive-valenced stimuli facilitate feeding, digestion, and reproductive activities (a primitive good mood), while negative-valenced stimuli inhibit all of these (a primitive bad mood).

Put these different levels of arousal and valence together and you get a primitive template of affect. Negative-valenced stimuli trigger a behavioral repertoire of fast swimming and infrequent turns, which can be thought of as the most primitive version of an aroused bad mood (which is often called the state of *escaping*), while the detection of food triggers a repertoire of slow swimming and frequent turns, which can be thought of as the most primitive version of an aroused good mood (which is often called the state of *exploiting*). Escaping leads worms to rapidly change location; exploiting leads worms to search their local surroundings (to exploit their surroundings for food). Although nematodes don't share the same complexity of emotions as humans—they do not know the rush of young love or the bittersweet tears of sending a child off to college—they clearly show

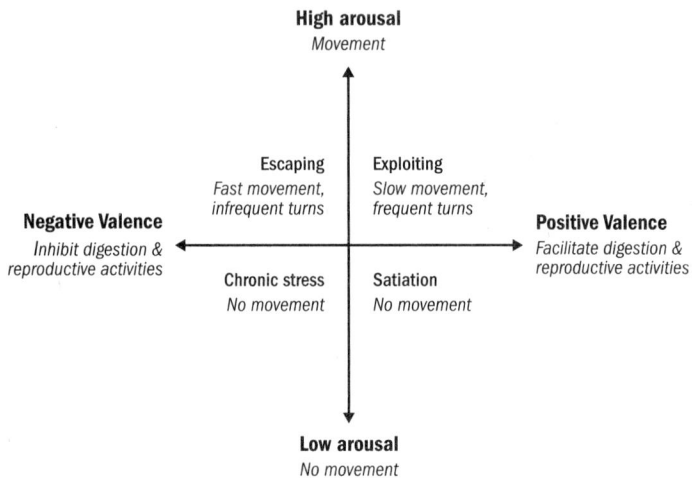

Figure 3.2: The affective states of nematodes

the basic template of affect. These incredibly simple affective states of nematodes offer a clue as to *why* affect evolved in the first place.

Suppose you put an unfed nematode in a big petri dish with a hidden patch of food. Even if you obscure any food smell for the nematode to steer toward, the nematode won't just dumbly sit waiting for a whiff of food. The nematode will rapidly swim and relocate itself; in other words, it will escape. It does this because one thing that triggers escape is *hunger*. When the nematode happens to stumble upon the hidden food, it will immediately slow down and start rapidly turning, remaining in the same general location that it found food—it will shift from escaping to exploiting. Eventually, after eating enough food, the worm will stop moving and become immobile and unresponsive. It will shift to *satiation*.

Scientists tend to shy away from the term *affective states* in simple bilaterians such as nematodes and instead use the safer term *behavioral states*; this avoids the suggestion that nematodes are, in fact, feeling anything. Conscious experience is a philosophical quagmire we will only briefly touch on later. Here, at least, this issue can be sidestepped entirely; the conscious experience of affect—whatever it is and however it works—likely evolved after the raw underlying mechanisms of affect. This can be seen even in humans—the

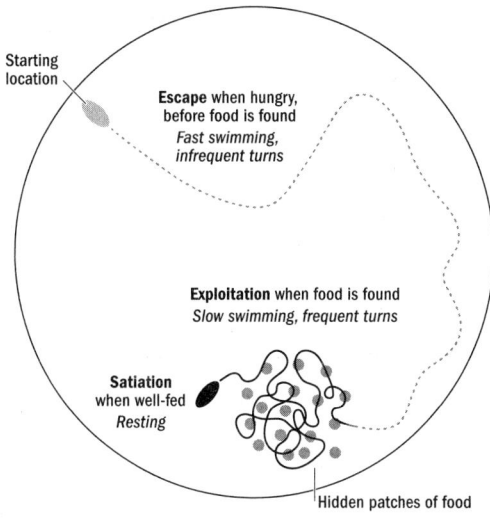

Figure 3.3

parts of the human brain that generate the experience of negative or positive affective states are evolutionarily newer and distinct from the parts of the brain that generate the reflexive avoidance and approach responses.

The defining feature of these affective states is that, although often triggered by external stimuli, they persist for long after the stimuli are gone. This feature of affective states stretches all the way from nematodes to humans—just as a nematode remains in a fear-like state for many minutes after a single sniff of a predator, a human mood can be soured for hours after a single unfriendly social interaction. The benefit of this persistence, at least at first, is unclear. If anything, it seems a bit dumb—nematodes keep trying to escape even after predators are long gone and they keep trying to exploit a local area for food even after all the food is gone. The idea that the function of the first brain was for steering provides a clue as to why nematodes have—and why the first bilaterians likely also had—these states: such persistence is required for steering to work.

Sensory stimuli, especially the simple ones detected by nematodes, offer transient clues, not consistent certainties, of what exists in the real world. In the wild, outside of a scientist's petri dish, food does not make perfectly distributed smell gradients—water currents can distort or even completely obscure smells, disrupting a worm's ability to steer toward food or away from predators. These persistent affective states are a trick to overcome this challenge: If I detect a passing sniff of food that quickly fades, it is likely that there is food nearby even if I no longer smell it. Therefore, it is more effective to persistently search my surroundings after encountering food, as opposed to only responding to food smells in the moment that they are detected. Similarly, a worm passing through an area full of predators won't experience a constant smell of predators but rather catch a transient hint of one nearby; if a worm wants to escape, it is a good idea to persistently swim away even after the smell has faded.

Like a pilot trying to fly a plane while looking through an opaque or obscured window, she would have no choice but to learn to fly in the darkness, using only the clues offered by the flickers of the outside world. Similarly, worms had to evolve a way to "steer in the dark"—to make steering decisions in the absence of sensory stimuli. The first evolutionary solution was affect, behavioral repertoires that can be triggered by external stimuli but persist long after they have faded.

This feature of steering shows up even in a Roomba. Indeed, Roombas were designed to have different behavioral states for the same reason. Normally, they explore rooms by moving around randomly. However, if a Roomba encounters a patch of dirt, it activates Dirt Detect, which changes its repertoire; it begins turning in circles in a local area. This new repertoire is triggered by the detection of dirt but persists for a time even after dirt is no longer detected. Why was the Roomba designed to do this? Because it works—detecting a patch of dirt in one location is predictive of nearby patches of dirt. Thus, a simple rule to improve the speed of getting all the dirt is to shift toward local search for a time after detecting dirt. This is exactly the same reason nematodes evolved to shift their behavioral state from exploration to exploitation after encountering food and locally search their surroundings.

Dopamine and Serotonin

The brain of a nematode generates these affective states using chemicals called neuromodulators. Two of the most famous neuromodulators are dopamine and serotonin. Antidepressants, antipsychotics, stimulants, and psychedelics all exert their effects by manipulating these neuromodulators. Many psychiatric conditions, including depression, obsessive-compulsive disorder, anxiety, post-traumatic stress disorder, and schizophrenia are believed to be caused, at least in part, by imbalances in neuromodulators. Neuromodulators evolved long before humans appeared; they began their connection to affect as far back as the first bilaterians.

Unlike excitatory and inhibitory neurons, which have specific, short-lived effects on only the specific neurons they connect to, neuromodulatory neurons have subtle, long-lasting, and wide-ranging effects on many neurons. Different neuromodulatory neurons release different neuromodulators—dopamine neurons release dopamine, serotonin neurons release serotonin. And neurons throughout an animal's brain have different types of receptors for different types of neuromodulators—neuromodulators can gently inhibit some neurons while simultaneously activating others; they can make some neurons more likely to spike while making others less likely to spike; they can make some neurons more sensitive to activation while dulling the responses of others. They can even accelerate or slow down the process of adaptation. Put all these effects together, and these neuromodulators can tune the neural activity across the entire brain. It is the balance of these different neuromodulators that determines a nematode's affective state.

The dopamine neurons of nematodes extend small appendages out of its head and contain receptors specifically designed to detect food. When these neurons detect the presence of food, they flood the brain with dopamine. This tunes circuits to generate the state of exploitation. This effect can last for minutes before dopamine levels drop again and the nematode returns to the state of escape. Serotonin neurons in nematodes have receptors that detect the presence of food in their throats, and if enough serotonin is released, it triggers satiety.

The simple brain of the nematode offers a window into the first, or at least very early, functions of dopamine and serotonin. In the nematode, dopamine is released when food is detected *around* the worm, whereas serotonin is released when food is detected *inside* the worm. If dopamine is the something-good-is-nearby chemical, then serotonin is the something-good-is-actually-happening chemical. Dopamine drives the hunt for food; serotonin drives the enjoyment of it once it is being eaten.

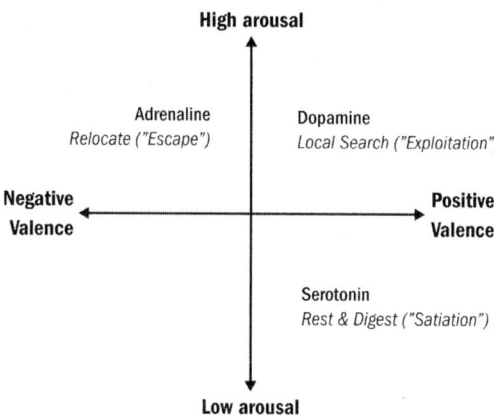

Figure 3.4: Role of neuromodulators in affective states of first bilaterians

While the exact functions of dopamine and serotonin have been elaborated throughout different evolutionary lineages, this basic dichotomy between dopamine and serotonin has been remarkably conserved since the first bilaterians. In species as divergent as nematodes, slugs, fish, rats, and humans, dopamine is released by nearby rewards and triggers the affective state of arousal and pursuit (exploitation); and serotonin is released by the consumption of rewards and triggers a state of low arousal, inhibiting the pursuit of rewards (satiation). What happens when you see something you want, like food when you're hungry, a sexy mate, the finish line at the end of a race? In all cases, your brain releases a burst of dopamine. What happens when you *get* something you want, like when you're orgasming, eating delicious food, or just finishing a task on your to-do list? Your brain releases serotonin.

If you raise dopamine levels in the brain of a rat, they begin impulsively exploiting any nearby reward they can find: gorging on food and trying to mate with whomever they see. If instead you raise their serotonin levels, they stop eating and become less impulsive and more willing to delay gratification. Serotonin shifts behavior from a focused pursuit of goals to a contented satiety by turning off dopamine responses and by dulling the responses of valence neurons.

And crucially, all these neuromodulatory neurons—like valence neurons—are also sensitive to internal states. Dopamine neurons are more likely to respond to food cues when an animal is hungry.

This connection between dopamine and reward has caused dopamine to be—incorrectly—labeled the "pleasure chemical." Kent Berridge, a neuroscientist at the University of Michigan, came up with a experimental paradigm to explore the relationship between dopamine and pleasure. Rats, like humans, make distinct facial expressions when tasting things they like, such as yummy sugar pellets, and things they don't like, such as bitter liquid. A baby will smile when tasting warm milk and spit when tasting bitter water; rats will lick their lips when they taste yummy food and gape their mouths and shake their heads when they taste gross food. Berridge realized he could use the frequency of these different facial reactions as a proxy for identifying pleasure in rats.

Liking/pleasure reactions (sweet)

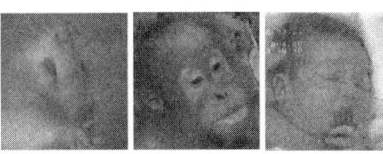

Disliking reactions (bitter)

Figure 3.5: Using facial expressions to deduce pleasure (liking) and displeasure (disliking)

To the surprise of many, Berridge found that increasing dopamine levels in the brains of rats had no impact on the degree and frequency of their pleasurable facial expressions to food. While dopamine will cause rats to consume ridiculous amounts of food, the rats do not indicate they are doing so because they *like* the food more. Rats do not express a higher number of pleasurable lip smacks. If anything, they express more disgust with the food, despite eating more of it. It is as if rats can't stop eating even though they no longer enjoy it.

In another experiment, Berridge destroyed the dopamine neurons of several rats, depleting almost all the dopamine in their brains. These rats would sit next to an abundance of food and starve to death. But this dopamine depletion had no impact on pleasure; if Berridge placed food into the mouths of these hungry rats, they exhibited all the facial expressions suggesting the kind of euphoria one would feel from eating when hungry; they smacked their lips more than ever. Rats experienced pleasure just fine without dopamine—they just didn't seem motivated to pursue it.

This finding has also been confirmed in humans. In a controversial set of experiments in the 1960s, the psychiatrist Robert Heath implanted electrodes in the brains of humans so that patients could push a button to stimulate their own dopamine neurons. Patients quickly began repeatedly pressing this button, often hundreds of times an hour. One might assume this was because they "liked" it, but in Heath's words:

> The patient, in explaining why he pressed the septal button with such frequency, stated that the feeling was . . . as if he were building up to a sexual orgasm. He reported that he was unable to achieve the orgastic end point, however, explaining that his frequent, sometimes frantic, pushing of the button was an attempt to reach the end point.

Dopamine is not a signal for pleasure itself; it is a signal for the anticipation of future pleasure. Heath's patients weren't experiencing pleasure; to the contrary, they often became extremely frustrated at their inability to satisfy the incredible cravings the button produced.

Berridge proved that dopamine is less about *liking* things and more about *wanting* things. This discovery makes sense given the evolutionary origin of dopamine. In nematodes, dopamine is released when they are near food but not when they are consuming food. The dopamine-triggered behavioral state of exploitation in nematodes—in which they slow down and search their surroundings for food—is in many ways the most primitive version of wanting. As early as the first bilaterians, dopamine was a signal for the *anticipation* of a future good thing, not the signal for the good thing itself.

While *dopamine* has no impact on liking reactions, *serotonin* decreases both liking and disliking reactions. When given drugs that increase serotonin levels, rats smack their lips less to good food and shake their heads less to bitter food. This is also what we would expect given the evolutionary origin of serotonin: serotonin is the satiation, things-are-okay-now, satisfaction chemical, designed to turn off valence responses.

Dopamine and serotonin are primarily involved in navigating the happy side of affective states—the different flavors of positive affect. There are additional neuromodulators, equally ancient, that undergird the mechanisms of *negative* affect—of stress, anxiety, and depression.

When Worms Get Stressed

Humanity suffers from stress-related diseases more than ever. More people in the world die from suicide each year than from all violent crimes and wars put together. Around 800,000 people take their lives each year, and there are over 15 million annual suicide attempts. Over 300 million people in the world suffer from depression—stripped of their ability to experience pleasure and engage in life. Over 250 million people in the world suffer from anxiety disorders—irrationally terrified of the world around them. The Centers for Disease Control (CDC) has even devised a term for this: *deaths of despair*. The rate of deaths of despair has more than doubled in the last twenty years.

These people aren't being eaten by lions, or starving, or freezing to death. These people are dying because their *brains are killing them*. Choosing to commit suicide, knowingly consuming deadly drugs, or binge eating oneself into obesity are, of course, behaviors generated by our brains. Any attempt to understand animal behavior, brains, and intelligence itself is wholly incomplete without understanding this enigma: Why would evolution have created brains with such a catastrophic and seemingly ridiculous flaw? The point of brains, as with all evolutionary adaptations, is to *improve survival*. Why, then, do brains generate such obviously self-destructive behaviors?

The affective state of *escape*, whereby nematodes rapidly attempt to swim to a new location, is in part triggered by a different class of neuromodulators: norepinephrine, octopamine, and epinephrine (also called adrenaline). Across bilaterians, including species as divergent as nematodes, earthworms, snails, fish, and mice, these chemicals are released by negative-valanced stimuli and trigger the well-known fight-or-flight response: increasing heart rate, constricting blood vessels, dilating pupils and suppressing various luxury activities, such as sleep, reproduction, and digestion. These neuromodulators work in part by directly counteracting the effectiveness of serotonin—reducing the ability of an animal to rest and be content.

Moving around in the world, even for nematodes, consumes a large amount of energy. The adrenaline-induced escape response is one of the most expensive behavioral choices an animal can make—the escape response requires a large expenditure of energy on muscles for rapid swimming. So evolution came up with a trick to save energy and thereby allow the escape response to last longer. Adrenaline not only triggers the behavioral repertoire of escape; it also turns off a swath of energy-consuming activities to divert energetic resources to muscles. Sugar is expelled from cells across the body, cell growth processes are halted, digestion is paused, reproductive processes are turned off, and the immune system is tamed. This is called the *acute* stress response—what bodies do immediately in response to negative-valence stimuli.

But like a government running a budget deficit to finance a war, the acute stress response's postponement of essential bodily functions cannot go on indefinitely. Therefore, our bilaterian ancestor evolved a counter-regulatory response to stress—a whole suite of anti-stress chemicals that prepare the body for the end of the war. One of these anti-stress chemicals was opioids.

Poppies are not the only source of opioids; brains make their own opioids and release them in response to stressors. When the stressor goes away and adrenaline levels drop, nematodes do not go back to their baseline state. Instead, the leftover anti-stress chemicals initiate a suite of recovery-related processes—immune responses, appetite, and digestion are turned back on. These relief-and-recover chemicals like opioids do this in part by

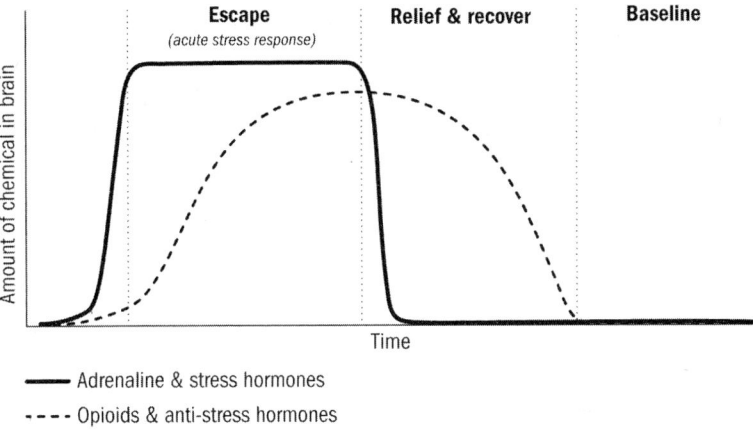

Figure 3.6: The time course of stress and anti-stress hormones

enhancing serotonin and dopamine signals (both of which are inhibited by acute stressors). Opioids also inhibit negative-valence neurons, which helps an animal recover and rest despite any injuries. This, of course, is why opioids are such potent painkillers across all bilaterians. Opioids also keep certain luxury functions, such as reproductive activities, turned off until the relief-and-recover process is done; this is why opioids decrease sex drive. It is no surprise, then, that nematodes, other invertebrates, and humans all have similar responses to opioids—prolonged bouts of feeding, inhibited pain responses, and inhibited reproductive behavior.

This relief-and-recover state doesn't merely turn appetite back on; a nematode starved for only twelve hours will eat thirty times more food than their normally hungry peers. In other words, stress makes nematodes binge food. After binging, these previously starved nematodes "pass out," spending ten times longer in an immobile state than unstarved worms. Nematodes do this because stress is a signal that circumstances are dire and food may be, or may soon become, scarce. Thus, nematodes stock up on as much food as they can in preparation for the next experience of starvation. As far back as the first brains six hundred million years ago, the system for binge eating after a stressful experience was already put in place.

These anti-stress hormones like opioids differed from dopamine and

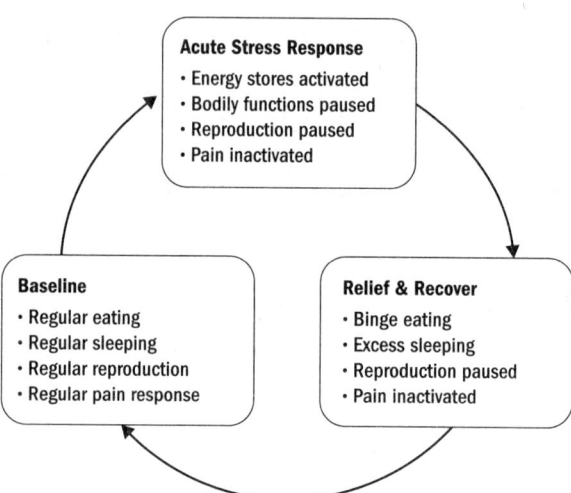

Figure 3.7: The ancient stress cycle, originating from first bilaterians

serotonin in Kent Berridge's rat facial expression experiments. While dopamine had no impact on liking reactions, giving opioids to a rat did, in fact, substantially increase their liking reactions to food. This makes sense given what we now know about the evolutionary origin of opioids. Opioids are the relief-and-recover chemical after experiencing stress: stress hormones turn positive-valence responses off (decreasing liking), but when a stressor is gone, the leftover opioids turn these valence responses back on (increasing liking). Opioids make everything better; they increase liking reactions and decrease disliking reactions; increasing pleasure and inhibiting pain.

The Blahs and Blues

All this describes what bodies do in response to short-term stressors—the acute stress response. But most of the ways that stress plagues modern humanity comes from what happens to bodies in response to prolonged stressors—the *chronic* stress response. This, too, has origins in the first bilaterians. If a nematode is exposed to thirty minutes of a negative stimulus (such as dangerous heat, freezing cold, or toxic chemicals), at first it will exhibit the hallmarks of the acute stress response—it will try to escape, and stress hormones will pause bodily functions. But after just two minutes of no relief from this inescapable stressor, nematodes do something surprising: they give up. The worm stops moving; it stops trying to escape and just lies there. This surprising behavior is, in fact, quite clever: spending energy escaping is worth the cost only if the stimulus is in fact *escapable*. Otherwise, the worm is more likely to survive if it conserves energy by waiting. Evolution embedded an ancient biochemical failsafe to ensure that an organism did not waste energy trying to escape something that was inescapable; this failsafe was the early seed of chronic stress and depression.

Any consistent, inescapable, or repeating negative stimuli, such as constant pain or prolonged starvation, will shift a nematode brain into a state of chronic stress. Chronic stress isn't all that different from acute stress; stress hormones and opioids remain elevated, chronically inhibiting digestion, immune response, appetite, and reproduction. But chronic stress differs from acute stress in at least one important way: it turns off arousal and motivation.

The exact biochemical mechanisms of chronic stress, even in nematodes, are complex and not fully understood. But one thing that does seem to be different between the states of acute stress and chronic stress is that chronic stress starts activating serotonin. At first glance, this makes no sense: serotonin was supposed to be the satiation and good-feels chemical. But consider the main effect of serotonin: it turns off valence neuron responses and lowers arousal. If you add this to the soup of stress hormones, you get a bizarre yet unfortunately familiar state—numbness. This is, perhaps, the most primitive form of depression. Of course, nematodes don't go through artistic blue periods the way that Picasso did, nor do they necessarily consciously "experience" anything, but nematodes still do share the fundamental feature found in depressive episodes across all bilaterians from insects to fish, to mice, to humans: the numbing of valence responses. This dulls pain and renders even the most exciting stimuli entirely unmotivating. Psychologists call this canonical symptom of depression *anhedonia*—the lack (*an*) of pleasure (*hedonia*).

Anhedonia in animals like nematodes seems to be a trick to preserve energy in the presence of inescapable stressors. Animals no longer respond to stressors, good food smells, or nearby mates. In humans, this ancient system robs its sufferers of the ability to experience pleasure and motivation. This is the *blah* or *blues* of depression. And like all affective states, chronic stress persists after the negative stimuli have gone away. Such learned helplessness, where animals stop trying to escape from negatively valenced stimuli, is seen even in many bilaterians, including cockroaches, slugs, and fruit flies.

We have invented drugs that hack these ancient systems. The euphoria provided by natural opioids is meant to be reserved for that brief period after a near-death experience. But humans can now indiscriminately trigger this state with nothing more than a pill. This creates a problem. Repeatedly flooding the brain with opioids *creates* a state of chronic stress when the drug wears off—adaptation is unavoidable. This then traps opioid users in a vicious cycle of relief, adaptation, chronic stress requiring more drugs to get back to baseline, which causes more adaptation and thereby more chronic stress. Evolutionary constraints cast a long shadow on modern humanity.

These primitive affective states were passed down and elaborated throughout evolution, and remnants are still—whether we like it or not—essential cornerstones of human behavior. Over time, neuromodulators were repurposed for different functions, and new variants of each of these affective states emerged. And so, while the modern emotional states of humans are undeniably more complex and nuanced than a simple two-by-two grid of valence and arousal, they nonetheless retain the scaffolding of the basic template from which they evolved.

Although these affective states are shared across bilaterians, our more distant animal cousins—anemones, coral, and jellyfish—do not show such states. Many of these animals don't even have serotonin neurons at all.

This leaves us at the doorstep of a surprising hypothesis: Affect, despite all its modern color, evolved 550 million years ago in early bilaterians for nothing more than the mundane purpose of steering. The basic template of affect seems to have emerged from two fundamental questions in steering. The first was the arousal question: Do I want to expend energy moving or not? The second was the valence question: Do I want to stay in this location or leave this location? The release of specific neuromodulators enforced specific answers to each of these questions. And these global signals for stay and leave could then be used to modulate suites of reflexes, such as whether it was safe to lay eggs, mate, and expend energy digesting food.

However, these affective states and their neuromodulators would go on to play an even more foundational role in the evolution of the first brains.

4

Associating, Predicting, and the Dawn of Learning

Memory is everything. Without it we are nothing.
—ERIC KANDEL

ON DECEMBER 12, 1904, a Russian scientist by the name of Ivan Pavlov stood in front of an assembly of researchers at the Karolinska Institute in Sweden. Pavlov had, two days earlier, become the first Russian to win the Nobel Prize. Eight years prior, Alfred Nobel—the Swedish engineer and businessman who got rich from his invention of dynamite—had passed away and bequeathed his fortune to the creation of the Nobel Foundation. Nobel had stipulated that winners were to give a lecture on the subject for which the prize had been awarded, and so, on this day in Stockholm, Pavlov gave his lecture.

Although he is currently known for his contributions to psychology, that was not the work that earned him the Nobel. Pavlov was not a psychologist but a *physiologist*—he had spent his entire research career up to this point studying the underlying biological mechanisms—the "physiology"—of the digestive system.

Before Pavlov, the only way to study the digestive system was to surgically remove animals' organs—esophagus, stomach, or pancreas—and run experiments quickly before the organs died. Pavlov pioneered a variety of relatively noninvasive techniques that enabled him to measure features of the digestive system in intact and healthy dogs. The most famous of these was the insertion of a small salivary fistula that diverted saliva from one salivary gland to a small tube that hung out of the dog's mouth; this enabled Pavlov to determine the quantity and content of saliva produced by various stimuli. He did similar tricks with the esophagus, stomach, and pancreas.

Through these new techniques, Pavlov and his colleagues made several discoveries. They learned what types of digestive chemicals were released in response to various foods, and he discovered that the digestive organs were under the control of the nervous system. These contributions won him the prize.

However, two-thirds through his speech, Pavlov turned his focus away from his prizewinning work. An excitable scientist, he couldn't resist pitching research that was, at the time, speculative but that he believed would eventually become his most important work—his exploration of what he called *conditional reflexes.*

There had always been a pesky confound that got in the way of his meticulous measurements of digestive responses—digestive organs often became stimulated *before animals tasted food.* His dogs salivated and their stomachs gurgled the moment they realized an experiment with food was about to begin. This was a problem. If you want to measure how salivary glands respond when taste buds detect fatty meat or sugary fruit, you don't want the confounding measurement of whatever was released by the subjects merely *looking* at these substances.

This psychic stimulation, as it was called, was a particular annoyance to Pavlov's research, what he deemed a "source of error." Pavlov developed various techniques to eliminate this confound; for instance, experimenters worked in separate isolated rooms to "carefully avoid everything that could elicit in the dog thoughts about food."

Only much later, after bringing psychologists into his lab, did Pavlov begin to view psychic stimulation not as a confound to be eliminated but as a variable worthy of analysis. Ironically, it was a digestive physiologist with the goal of *eliminating* psychic stimulation who became the first to understand it.

Pavlov's lab discovered that psychic stimulation was not as random as it seemed. Dogs would salivate in response to *any* stimuli—metronomes, lights, buzzers—that had been previously associated with food. If an experimenter turned on a buzzer and then gave food, the dog began to salivate in response to the buzzer alone. The dog had developed a *conditional* reflex—the reflex to salivate in response to the buzzer was *conditional* on the prior association between the buzzer and food. Pavlov contrasted these conditional reflexes with what he called *unconditional* reflexes—those that were innate and required no association. A hungry dog's reflex to salivate in response to sugar placed in its mouth occurred regardless of any prior associations.

Shortly after Pavlov's experiments, other scientists began trying these techniques on other reflexes. It turned out that most, if not all, reflexes build such associations. Pair an arbitrary sound with an electric shock to your hand, and soon your hand will retract to just the sound. Pair an arbitrary sound with a gentle puff of air to someone's eyes, and eventually he or she will involuntarily blink in response to just the sound. Pair an arbitrary sound with a reflex hammer to a person's knee, and eventually their leg will kick in response to just the sound.

The defining feature of Pavlov's conditional reflexes is that they are *involuntary* associations; people can't help but blink, kick, or retract their hands. Just as a soldier who returns from war cannot help but jump when they hear a loud noise or a person with a phobia of public speaking cannot help but tense up before going onstage, so too could Pavlov's dogs not help but salivate in response to the buzzer. The involuntary nature of Pavlov's conditional reflexes, the fact that associative learning occurs automatically without conscious involvement, was the first clue that learning and memory might be more ancient than previously thought. Learning might not require all the brain structures that emerged later in evolution. Indeed, even a rat with its entire brain removed exhibits conditional reflexes. If you pair a tap to its leg (which causes leg retraction) with a tap to its tail (which causes tail retraction), then it will learn to retract its tail in response to a leg tap, despite lacking a brain. Rats can learn this association with nothing but the simple circuits in their spinal cords.

If associative learning is a property of simple circuits of neurons, even those present outside of the brain, then it might be a very ancient evolutionary trick. Indeed, Pavlov had unintentionally stumbled on the evolutionary origin of learning itself.

Tweaking the Goodness and Badness of Things

Suppose you took a hundred nematodes, put half of them in a dish with *plain water* and the other half in a dish with *salty water*. After several hours, these nematodes will become uncomfortably hungry, as neither dish contains any food. At this point, put both groups of nematodes into another dish that contains a little morsel of salt on one side. What happens?

The nematodes that experienced hunger with plain water will behave as normal nematodes do: they will steer toward the salt (nematodes typically consider salt to have positive valence). However, the nematodes that experienced hunger in salt water will do the exact opposite: they will steer *away* from the salt.

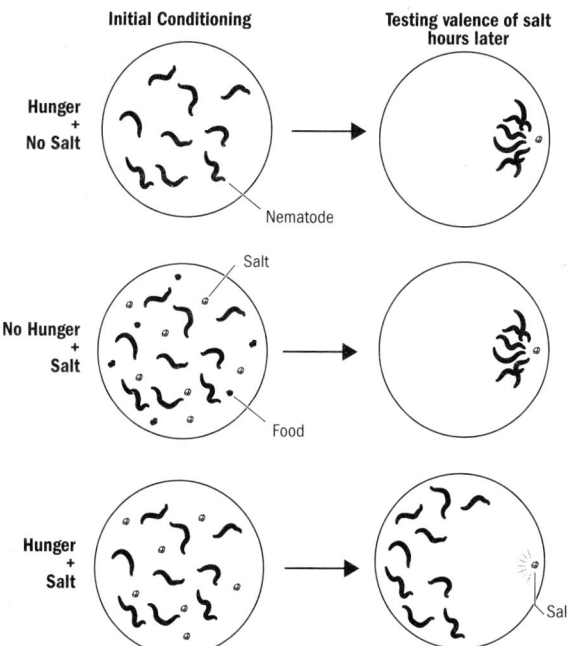

Figure 4.1: Nematodes learn to steer away from salt when salt is associated with hunger

Salt was transformed from a positive valence stimulus to a negative valence stimulus through its association with the negative valence state of hunger.*

It turns out that Pavlov's associative learning is an intellectual ability of all bilaterians, even simple ones. If you expose nematodes simultaneously to both a yummy food smell and a noxious chemical that makes them sick, nematodes will subsequently steer *away* from that food smell. If you feed nematodes at a specific temperature, they will shift their preferences toward that temperature. Pair a gentle tap to the side of a slug with a small electric shock, which triggers a withdrawal reflex, and the slug will learn to withdraw *to just the tap*, an association that will last for days.

And yet, while associative learning is found across bilaterians, our most distant animal cousins—the radially symmetric jellyfish, anemones, and coral—are not capable of learning associations.† Despite many pairings of a light with an electric shock, an anemone will never learn to withdraw in response to just the light. They withdraw only from the shock itself. The ubiquitous presence of associative learning *within* Bilateria and the notable absence of it *outside* Bilateria suggests that associative learning first emerged in the brains of early bilaterians. It seems that at the same time valence—the categorizing of things in the world into good and bad—emerged, so too did the ability to use experience to change what is considered good and bad in the first place.

Why have non-bilaterian animals like coral and anemones, despite an

---

* In this experiment, researchers confirmed that this effect was not caused just by overexposure to salt, but instead by the association between a stimulus (salt) and the negative affective state of hunger. Researchers took a third group and made them spend the same number of hours in salt water but also added food to the dish so nematodes wouldn't experience hunger. This third group, which experienced the same amount of exposure to salt, still happily steered toward salt afterward. This suggests the salt avoidance was not caused by overexposure to salt, but by the association between salt and hunger (see the middle example in figure 4.1).

† These more distant animals do engage in what is called *nonassociative learning*, such as adaptation (as Edgar Adrian found), and another similar type of learning called *sensitization*, which is when reflexes strengthen in response to a previously arousing stimulus.

additional six hundred million years of evolution, not acquired the ability to learn associations? Their survival strategy simply doesn't require it.

A coral polyp with the ability of associative learning wouldn't survive much better than one without associative learning. A coral polyp just sits in place, immobilized, waiting for food to swim into its tentacles. The hardcoded strategy of swallowing anything that touches its tentacles and withdrawing from anything painful works just fine, without any associative learning. In contrast, a brain designed for steering would have faced unique evolutionary pressure to adjust its steering decisions based on experience. An early bilaterian that could remember to avoid a chemical that had previously been found near predators would survive far better than a bilaterian that could not.

Once animals began approaching specific things and avoiding others, the ability to tweak what was considered good and bad became a matter of life and death.

The Continual Learning Problem

Your self-driving car doesn't automatically get better as you drive; the facial-recognition technology in your phone doesn't automatically get better each time you open your phone. As of 2023, most modern AI systems go through a process of training, and once trained, they are sent off into the world to be used, but *they no longer learn*. This has always presented a problem for AI systems—if the contingencies in the world change in a way not captured in the training data, then these AI systems need to be retrained, otherwise they will make catastrophic mistakes. If new legislation required people to drive on the left side of the road, and AI systems were trained to drive only on the right side of the road, they would not be able to flexibly adjust to the new environment without being explicitly retrained.

While learning in modern AI systems is not continual, learning in biological brains has *always* been continual. Even our ancestral nematode had no choice but to learn continually. The associations between things were always changing. In some environments, salt was found on food; in others, it was found on barren rocks without food. In some environments, food grew at cool temperatures; in others, it grew at warm temperatures. In some environments, food was found in bright areas; in others, predators were found in bright areas. The first brains needed a mechanism to not only acquire associations but also quickly change these associations to match the changing rules of the world. It was Pavlov who first found hints of these ancient mechanisms.

By measuring the quantity of saliva produced in response to cues that had been paired with food, Pavlov was able to not only observe the *presence* of associations, but also quantitatively measure the *strength* of these associations—the more saliva released in response to a cue, the stronger the association. Pavlov had found a way to measure memory. And by recording how memory changed over time, Pavlov could observe the process of continual learning.

Indeed, the associations in Pavlov's conditional reflexes are always strengthening or weakening with each new experience. In Pavlov's experiments, the associations strengthened with each subsequent pairing—each time the buzzer occurred before food was given, the more the dog salivated the next time the buzzer occurred. This process is called *acquisition* (the association was being *acquired*).

If after learning this association, the buzzer is presented in the *absence* of food, then the strength of the association fades with each trial, a process called *extinction*.

There are two interesting features of extinction. Suppose you break a previously learned association—sound the buzzer several times in a row, but *don't* give food. As expected, dogs will eventually stop salivating at the buzzer. However, if you wait a few days and then sound the buzzer again, something odd happens: dogs start salivating in response to the buzzer again. This is called *spontaneous recovery*: broken associations are rapidly suppressed but not, in fact, unlearned; given enough time, they reemerge. Further, if after a long stretch of trials with a broken association (buzzer but no food), you reinstate the association (sound a buzzer and provide food again), the old association will be relearned far more rapidly than the first time the dog experienced the association between the buzzer and food. This is called *reacquisition*: old extinguished associations are reacquired faster than entirely new associations.

Why do associations show spontaneous recovery and reacquisition? Consider the ancient environment in which associative learning evolved.

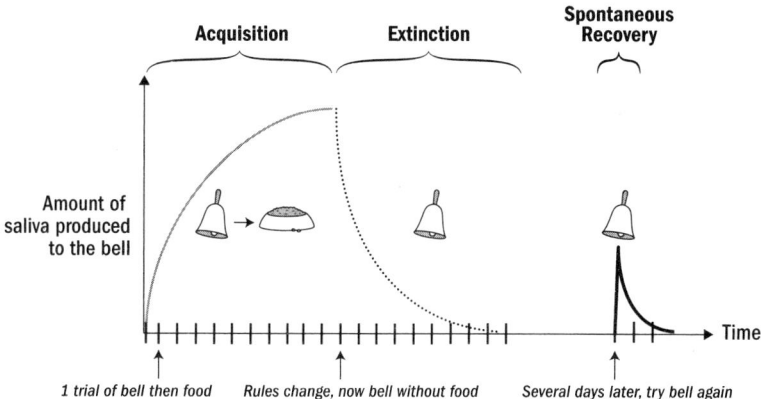

Figure 4.2: The time course of associative learning

Suppose a worm has many experiences of finding food alongside salt. And then one day, it detects salt, steers toward it, and finds no food. After the worm spends an hour sniffing around without finding food, the association becomes extinguished, and the worm begins steering toward other cues, no longer attracted to salt. If two days later it detects salt again, would it be smarter to steer toward or away from it? In all of the worm's past experiences, except the most recent one, when it smelled salt it also found food. And so the smarter choice would be to steer toward salt again—the most recent experience may have been a fluke. This is the benefit of spontaneous recovery—it enables a primitive form of long-term memory to persist through the tumult of short-term changes in the contingencies of the world. Of course, if the next *twenty* times the worm detects salt it fails to find food, the association may eventually be permanently extinguished.

The effect of reacquisition—the accelerated relearning of old previously broken associations—evolved in ancient worms for similar reasons. Suppose this same worm finds salt alongside food after the association was long ago extinguished. How quickly should the worm restrengthen the association between salt and food? It would make sense to relearn this association rapidly, given the long-term memory the worm has: *In some cases, salt leads to food, and it seems that right now is one of those situations!* Thus, old associations are primed to reemerge whenever the world provides hints that old contingencies are newly reestablished.

Spontaneous recovery and reacquisition enabled simple steering brains to navigate changing associations, temporarily suppress old associations that were currently inaccurate, and remember and relearn broken associations that became effective again.

The first bilaterians used these tricks of acquisition, extinction, spontaneous recovery, and reacquisition to navigate changing contingencies in their world. These solutions to continual learning are found across many reflexes in many animals, even the most ancient animals like the nematode, embedded in the simplest neural circuits, inherited from the first brain, originally crafted to make steering work in the ever-changing world of the ancient Ediacaran Sea.

The Credit Assignment Problem

Associative learning comes with another problem: When an animal gets food, there is never a single predictive cue beforehand but rather a whole swath of cues. If you pair a tap to the side of a slug with a shock, how does a slug's brain know to associate only the tap with the shock and not the many other sensory stimuli that were present, such as the surrounding temperature, the texture of the ground, or the diverse chemicals floating around the seawater? In machine learning, this is called the *credit assignment problem*: When something happens, what previous cue do you give credit for predicting it? The ancient bilaterian brain, which was capable of only the simplest forms of learning, employed four tricks to solve the credit assignment problem. These tricks were both crude and clever, and they became foundational mechanisms for how neurons make associations in all their bilaterian descendants.

The first trick used what are called *eligibility traces*. A slug will associate a tap with a subsequent shock only if the tap occurs one second before the shock. If the tap occurs two seconds or more before the shock, no association will be made. A stimulus like a tap creates a short eligibility trace that lasts for about a second. Only within this short time window can associations be made. This is clever, as it invokes a reasonable rule of thumb: stimuli that are useful for predicting things should occur *right before* the thing you are trying to predict.

The second trick was *overshadowing*. When animals have multiple predictive cues to use, their brains tend to pick the cues that are the strongest—strong cues *overshadow* weak cues. If a bright light and a weak odor are both present before an event, the bright light, not the weak odor, will be used as the predictive cue.

The third trick was *latent inhibition*—stimuli that animals regularly experienced in the past are inhibited from making future associations. In other words, frequent stimuli are flagged as irrelevant background noise. Latent inhibition is a clever way to ask, "What was different this time?" If a slug has experienced the current texture of the ground and the current temperature a thousand times but has never experienced a tap before, then the tap is far more likely to be used as a predictive cue.

The fourth and final trick for navigating the credit assignment problem was *blocking*. Once an animal has established an association between a predictive cue and a response, all further cues that overlap with the predictive cue are blocked from association with that response. If a slug has learned that a tap leads to shock, then a new texture, temperature, or chemical will be blocked from being associated with the shock. Blocking is a way to stick to one predictive cue and avoid redundant associations.

The Original Four Tricks for Tackling the Credit Assignment Problem

ELIGIBILITY TRACES	OVERSHADOWING	LATENT INHIBITION	BLOCKING
Pick the predictive cue that occurred between 0 to 1 second *before* the event.	Pick the predictive cue that was the strongest.	Pick the predictive cue that you haven't seen before.	Stick to predictive cues once you have them and ignore others.

Eligibility traces, overshadowing, latent inhibition, and blocking are ubiquitous across Bilateria. Pavlov identified these in the conditional reflexes of his salivating dogs; they are found in the involuntary reflexes of humans; and they are seen in the associative learning of flatworms, nematodes, slugs, fish, lizards, birds, rats, and most every bilaterian in the animal kingdom. These tricks for navigating the credit assignment problem evolved as far back as the very first brains to make associative learning work.

These tricks are hardly perfect. In some circumstances, the best predictive cue may occur one minute before the event, not one second before. In other circumstances, the best predictive cue may be the weak cue, not the strong one. Over time, brains evolved more sophisticated strategies for solving the credit assignment problem (stay tuned for breakthrough #2 and breakthrough #3). But the remnants of the first solutions—eligibility traces, overshadowing, latent inhibition, and blocking—still exist in modern brains. They are seen in our involuntary reflexes and in our most ancient brain circuits. Indeed, rats with their entire brains removed, with nothing left but their neural circuits in their spinal cord, still show latent inhibition, blocking, and overshadowing. Along with acquisition, extinction, spontaneous recovery, and reacquisition, this portfolio of tricks make up the foundation of the neural mechanisms of associative learning, mechanisms that are embedded deep into the inner workings of neurons, neural circuits, and brains themselves.

The Ancient Mechanisms of Learning

For thousands of years, two groups of philosophers have been debating the relationship between the brain and the mind. One group, the dualists, like Plato, Aquinas, and Descartes, argue that the mind exists separately from the brain. The entities might interact with each other, but they are distinct; the mind is something beyond the physical. The materialists, like Kanada, Democritus, Epicurus, and Hobbes, argued that whatever the mind is, it is located entirely in the physical structure of the brain. There is nothing beyond the physical. This debate still rages in philosophy departments around the world. If you have made it this far in the book, I will assume you lean on the side of materialism, that you—like me—tend to reject nonphysical explanations for things, even the mind. But by siding with the materialists, we introduce several issues that, at first, are hard to explain physically, the most obvious being learning.

You can read a sentence once and then immediately repeat it out loud. If we stick to a materialist view, this means that reading this sentence instantaneously *changed something physical in your brain*. Anything that leads to learning causes physical reorganization of *something* in the 86 billion neurons in each of our heads. Keeping track of a conversation, watching a movie, and learning to tie your shoes all must change the physicality of our brains.

People have been speculating on the physical mechanisms of learning for thousands of years, and even the dualists pontificated on materialist explanations for learning. Plato believed that the brain was like a wax tablet in which perceptions left lasting impressions; he believed that memories were these impressions. Descartes argued that memories were formed through creating new "folds" in the brain, "not unlike the folds which remain in this paper after it has once been folded." Others speculated that memories were persistent "vibrations." These ideas were all wrong, although through no fault of their originators; at the time, no one understood even the basic building blocks of the nervous system, so they could not even begin to conceive of how learning might work.

The flurry of discoveries about neurons in the early twentieth century provided a host of new building blocks. The discovery of the connections between neurons—synapses—was the most obvious new thing that could presumably change in the brain during learning. Indeed, it turns out that learning rises not from impressions, folds, or vibrations but from changes to these synaptic connections.

Learning occurs when synapses change their strength or when new synapses are formed or old synapses are removed. If the connection between two neurons is weak, the input neuron will have to fire many spikes to get the output neuron to spike. If the connection is strong, the input neuron will have to fire only a few spikes to get the output neuron to spike. Synapses can increase their strength by the input neuron releasing more neurotransmitter in response to a spike or the postsynaptic neuron increasing the number of protein receptors (hence more responsive to the same quantity of neurotransmitter).

Synapses have many mechanisms for choosing when to strengthen or weaken. These mechanisms are extremely old evolutionary innovations, originating from the associative learning of the first bilaterians. For example, there is clever protein machinery in the synapses of bilaterian neurons

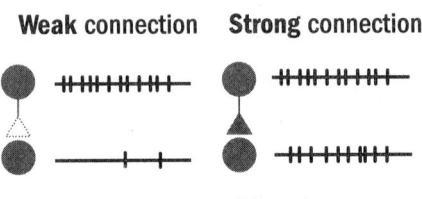

Figure 4.3

that detect whether the input neuron fired *within a similar time window* as the output neuron. In other words, individual connections can detect whether an input (like the sensory neuron activated by a tap) is activated at the same time as an output (like the motor neuron activated by a shock). And if these neurons are activated at the same time, this protein machinery triggers a process that strengthens the synapse.[†] Thus, next time the tap neuron activates, it activates the motor neuron on its own (because the connection between the neurons was strengthened), and you now have a conditional reflex. This learning mechanism is called *Hebbian learning* after the psychologist Donald Hebb, who eloquently hypothesized the existence of such a mechanism in the 1940s, several decades before the mechanism was discovered. Hebbian learning is often referred to as the rule that "neurons that fire together wire together."

But the logic of changing synaptic strengths gets more complex than this. There are molecular mechanisms in synapses to measure *timing*, whereby associations are built only if the input neuron fires *right before* the output neuron, thereby enabling the trick of eligibility traces. Neuromodulators like serotonin and dopamine can modify the learning rules of synapses; some synapses undergo Hebbian learning only when dopamine or serotonin receptors are *also* activated, thereby enabling neuromodulators to gate the ability of synapses to build new associations. A worm that sniffs a chemical and then finds food has its brain flooded with dopamine, which then can trigger the strengthening of specific synapses.

Although we don't yet fully understand all the mechanisms by which neurons rewire themselves, these mechanisms are remarkably similar

---

[†] The protein machinery for this is beautiful but beyond the scope of this book. If interested, look up coincidence detection using NMDA receptors.

among bilaterians; the neurons in the brain of a nematode change their synapses in largely the same way as the neurons in your brain. In contrast, when we examine the neurons and synapses of non-bilaterians like coral polyps, we do not find the same machinery; for example, they lack certain proteins known to be involved in Hebbian learning. But given our evolutionary past, this is to be expected: if our shared ancestor with coral polyps had no associative learning, then we should expect them to lack the mechanisms that underly such learning.

Learning had humble beginnings. While early bilaterians were the first to learn associations, they were still unable to learn most things. They could not learn to associate events separated by more than a few seconds; they could not learn to predict the exact timing of things; they could not learn to recognize objects; they could not recognize patterns in the world; and they could not learn to recognize locations or directions.

But still, the ability of the human brain to rewire itself, to make associations between things, is not a uniquely human superpower but one we inherited from this ancient bilaterian ancestor that lived over 550 million years ago. All the feats of learning that followed (the ability to learn spatial maps, language, object recognition, music, and everything else) were built on these same learning mechanisms. From the bilaterian brain onward, the evolution of learning was primarily a process of finding new applications of preexisting synaptic learning mechanisms, without changing the learning mechanisms themselves.

Learning was not the core function of the first brain; it was merely a feature, a trick to optimize steering decisions. Association, prediction, and learning emerged for tweaking the goodness and badness of things. In some sense, the evolutionary story that will follow is one of learning being transformed from a cute feature of the brain to its core function. Indeed, the next breakthrough in brain evolution was all about a brilliant new form of learning, one that was possible only because it was built on the foundation of valence, affect, and associative learning.

Summary of Breakthrough #1: Steering

Our ancestors from around 550 million years ago transitioned from a radially symmetric brainless animal, like a coral polyp, to a bilaterally symmetric brain-enabled animal, like a nematode. And while many neurological changes occurred across this transition, a surprisingly broad set of them can be understood through the lens of enabling a singular breakthrough: that of *navigating by steering*. These include:

- A bilateral body plan that reduces navigational choices to two simple options: go forward or turn
- A neural architecture for valence in which stimuli is evolutionarily hard-coded into good and bad
- Mechanisms for modulating valence responses based on internal states
- Circuits whereby different valence neurons can be integrated into a singular steering decision (hence a big cluster of neurons we identify as a brain)
- Affective states for making persistent decisions as to whether to leave or stay
- The stress response for energy management of movements in the presence of hardship
- Associative learning for changing steering decisions based on previous experience
- Spontaneous recovery and reacquisition for dealing with changing contingencies in the world (making continual learning work, even if imperfectly)
- Eligibility traces, overshadowing, latent inhibition, and blocking for (imperfectly) tackling the credit assignment problem

All of these changes made steering possible and solidified our ancestors' place as the first large multicellular animals who survived by *navigating*—moving not with microscopic cellular propellers but with muscles and neurons. And all these changes, along with the predatory ecosystem they begot, laid the foundation for breakthrough #2, which was when learning finally took its central role in the function of our brains.

BREAKTHROUGH #2

Reinforcing and the First Vertebrates

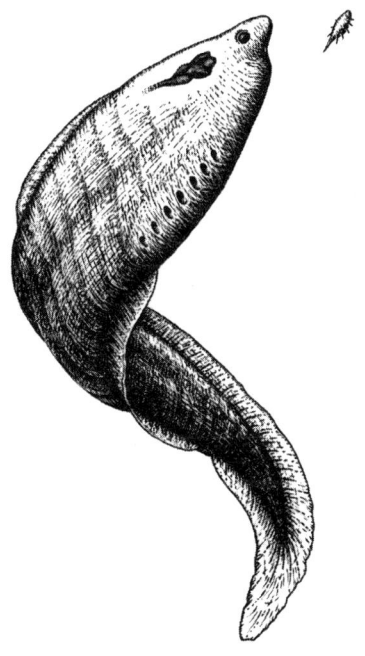

Your brain 500 million years ago

5

The Cambrian Explosion

TO GET TO the next milestone in brain evolution, we must leave the era when the first bilaterians were wiggling around and jump forward fifty million years. The ancient world this brings us to is the Cambrian period, an era that stretched from 540 to 485 million years ago.

If you peered around the Cambrian, you would see a world very different from the older Ediacaran. The gooey microbial mats of the Ediacaran that turned the ocean floor green would have long since faded and given way to a more familiar sandy underbelly. The sensile, slow, and small creatures of the Ediacaran would have been replaced by a bustling zoo of large mobile animals as varied in form as in size. This wouldn't resemble a zoo you would enjoy—this was a world ruled by *arthropods*, the ancestors of insects, spiders, and crustaceans. These arthropods were far more terrifying than their modern descendants; they were massive and armed with hauntingly oversize claws and armored shells. Some grew to over five feet long.

The discovery of steering in our nematode-like ancestor accelerated the evolutionary arms race of predation. This triggered what is now known as the Cambrian explosion, the most dramatic expansion in the diversity of animal life Earth has ever seen. Ediacaran fossils are rare and sought after, but Cambrian fossils, if you dig deep enough, are all over the place, and they encompass a mind-boggling diversity of creatures. During the Ediacaran period, animals with brains were humble inhabitants of the seafloor, smaller and less numerous than their brainless animal cousins like the coral and anemones. During the Cambrian period, however, animals with brains began their reign over the animal kingdom.

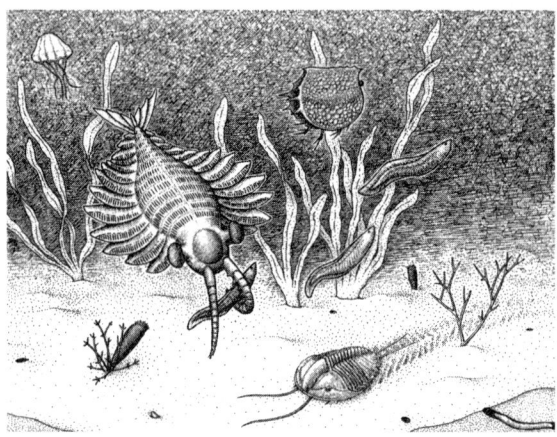

Figure 5.1: The Cambrian world

One lineage of Grandma Worm remained relatively unchanged and shrank in size, becoming the nematodes of today. Another lineage became the masters of this era, the arthropods. Lineages of these arthropods would independently develop their own brain structures with their own intellectual abilities. Some, such as the ants and honeybees, would go on to become impressively smart. But neither the arthropod nor the nematode lineage is ours. Our ancestors were likely not very conspicuous in the Cambrian cacophony of terrifying creatures; they were barely bigger than early bilaterians, only a few inches long, and not particularly numerous. But if you spotted them, they would have looked refreshingly familiar—they would have resembled a modern fish.

Fossil records of these ancient fish show several familiar features. They had fins, gills, a spinal cord, two eyes, nostrils, and a heart. The easiest-to-spot feature in fossils of these creatures is the *vertebral column*, the thick interlocking bones that encased and protected their spinal cord. Indeed, taxonomists refer to the descendants of this ancient fishlike ancestor as *vertebrates*. But of all the familiar changes that emerged in these early vertebrates, the most remarkable was surely the brain.

The Vertebrate Brain Template

The brains of invertebrates (nematodes, ants, bees, earthworms) have no recognizably similar structures to the brains of humans. The evolutionary

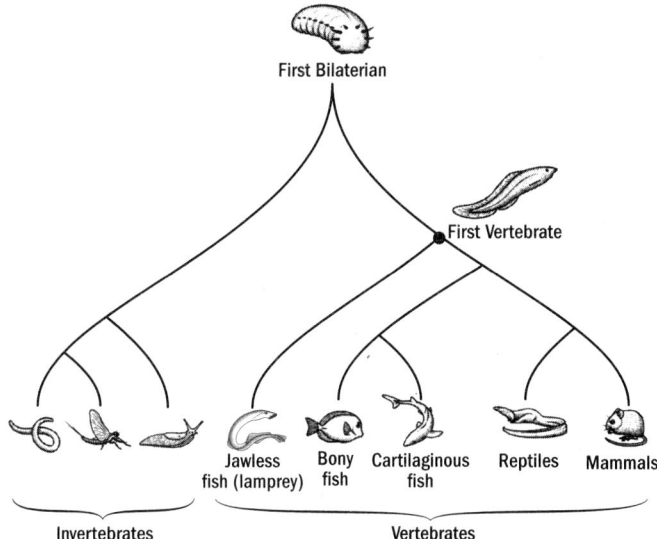

Figure 5.2: Our Cambrian ancestors

distance between humans and invertebrates is too distant; our brains are derived from too basic a template in our bilaterian ancestor to reveal any common structures. But when we peer into the brain of even the most distant vertebrates, such as the jawless lamprey fish—with whom our most recent common ancestor was the first vertebrate over five hundred million years ago—we see a brain that shares not only some of the same structures but *most* of them.

From the heat of the Cambrian explosion was forged the vertebrate brain template, one that, even today, is shared across all the descendants of these early fishlike creatures. If you want a crash course in how the *human* brain works, learning how the *fish* brain works will get you half of the way there.

The brains of all vertebrate embryos, from fish to humans, develop in the same initial steps. First, brains differentiate into three bulbs, making up the three primary structures that scaffold all vertebrate brains: a forebrain, midbrain, and hindbrain. Second, the forebrain unfolds into two subsystems. One of these goes on to become the cortex and the basal ganglia, and the other goes on to become the thalamus and the hypothalamus.

This results in the six main structures found in all vertebrate brains:

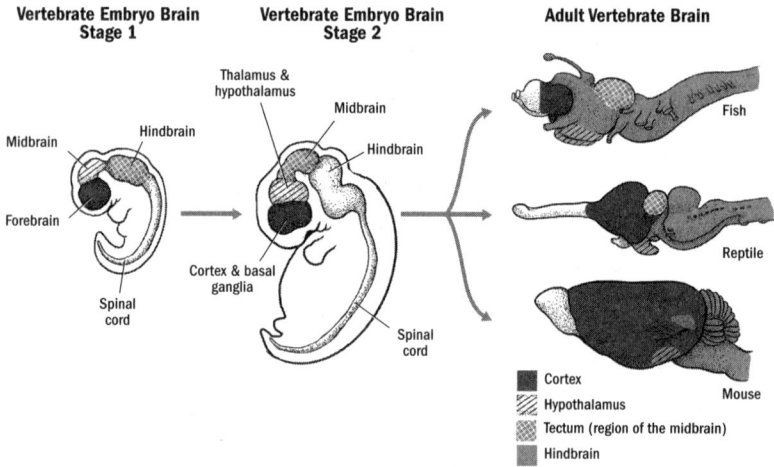

Figure 5.3: The shared embryonic development of vertebrates

the cortex, basal ganglia, thalamus, hypothalamus, midbrain, and hindbrain. Revealing their common ancestry, these structures are remarkably similar across modern vertebrates (except for the cortex, which has unique modifications in some vertebrates, such as mammals; stay tuned for breakthrough #3). The circuitry of the human basal ganglia, thalamus, hypothalamus, midbrain, and hindbrain and that of a fish are incredibly similar.

The first animals gifted us neurons. Then early bilaterians gifted us brains, clustering these neurons into centralized circuits, wiring up the first system for valence, affect, and association. But it was early vertebrates who transformed this simple proto-brain of early bilaterians into a true machine, one with subunits, layers, and processing systems.

The question is, of course, what did this early vertebrate brain *do*?

Thorndike's Chickens

Around the same time that Ivan Pavlov was unraveling the inner workings of conditional reflexes in Russia, an American psychologist by the name of Edward Thorndike was probing animal learning from a different perspective.

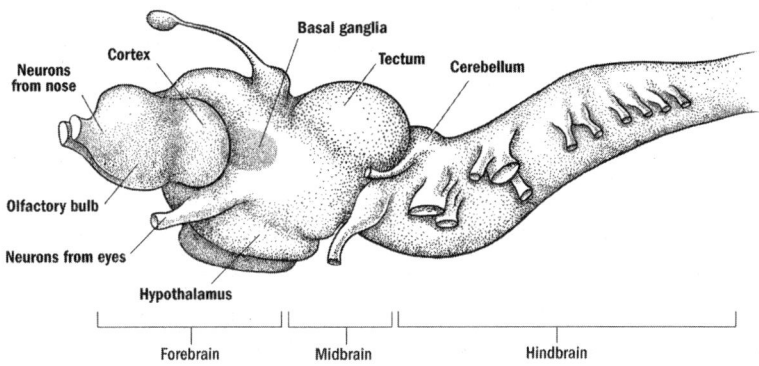

Figure 5.4: The brain of the first vertebrates

In 1896, Edward Thorndike found himself in a room full of chickens. Thorndike had recently enrolled in Harvard's master's program in psychology. His main interest was studying how children learn: How best can we teach children new things? He had numerous ideas for experiments, but to Thorndike's chagrin, Harvard would not allow him to conduct experiments on human children. So Thorndike had no choice but to focus on subjects that were easier to obtain: chickens, cats, and dogs.

To Thorndike, this wasn't all bad. A staunch Darwinist, he was unwavering in his view that there should be common principles in the learning of chickens, cats, dogs, and humans. If these animals shared a common ancestor, then they all should have inherited similar learning mechanisms. By probing how these other animals learned, he believed he might be able to also illuminate the principles of how humans learned.

Thorndike was both extremely shy and incredibly smart, so he was perhaps the perfect person to engage in the solitary, meticulously repetitive, and undeniably clever animal studies that he pioneered. Pavlov did his groundbreaking psychology work when he was middle-aged, after an already famed career as a physiologist, but Thorndike's most famous work was his first. It was his doctoral dissertation, published in 1898, when he was twenty-three, for which he is most well known. His dissertation: "Animal Intelligence: An Experimental Study of the Associative Processes in Animals."

Thorndike's genius, like Pavlov's, was in how he reduced hopelessly complex theoretical problems to simple measurable experiments. Pavlov explored learning by measuring the amount of saliva released in response to a buzzer. Thorndike explored learning by measuring the speed with which animals learned to escape from what he called *puzzle boxes*.

Thorndike constructed a multitude of cages, each with a different puzzle inside that, if solved correctly, would open an escape door. These puzzles weren't particularly complex—some had latches that when pushed would open the door; others had hidden buttons; others had hoops to pull. Sometimes the puzzle did not require a physical contraption, and Thorndike would just manually open the door whenever the animal did something specific, such as lick itself. He placed various animals in these cages, put food outside to motivate the animals to get out of the boxes, and measured exactly how long it took them to figure out the puzzle.

Once the animal escaped, he would record the animal's time, and then have the animal do it again, and again, and again. He would calculate the average time it took animals to solve a given puzzle on their first trial, compare that with the time for their second, and go all the way to how fast they solved it after as many as a hundred trials.

Thorndike originally wanted to probe the dynamics of *imitation*, a feature of learning he believed would exist across many animal species. He allowed untrained cats to watch trained cats escape from various puzzle boxes to see if it had any effect on their own learning. In other words,

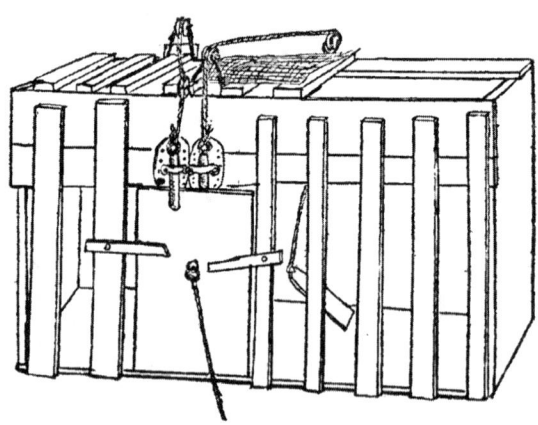

Figure 5.5: One of Thorndike's puzzle boxes

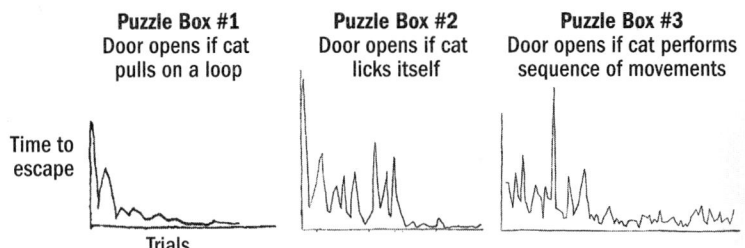

Figure 5.6: Animals learning through trial and error

could cats learn through imitation? It seemed at the time that the answer was no; they didn't get any better by watching (note that some animals *can* do this; stay tuned for breakthrough #4). But in this failure, he discovered something surprising. He found that these animals did all share a learning mechanism—it just wasn't the one he originally expected.

When first placed in a cage, the cat would try a whole host of behaviors: scratching at the bars, pushing at the ceiling, digging at the door, howling, trying to squeeze through the bars, pacing around the cage. Eventually the cat would accidently press the button or pull the hoop, and the door would open; the cat would exit and happily eat its prize. The animals became progressively faster at repeating the behaviors that got them out of the box. After many trials, cats stopped doing any of their original behaviors and immediately performed the actions required to escape. These cats were learning through *trial and error*. He could quantify this trial-and-error learning with the gradual decay in the time it took for animals to escape (figure 5.6).

What was most surprising was how much intelligent behavior emerged from something as simple as trial-and-error learning. After enough trials, these animals could effortlessly perform incredibly complex sequences of actions. It was originally believed that the only way to explain such intelligent behavior in animals was through some notion of insight or imitation or planning, but Thorndike showed how simple trial and error was all an animal really needed. Thorndike summarized his result in his now famous *law of effect*:

> Responses that produce a satisfying effect in a particular situation become more likely to occur again in that situation, and responses

that produce a discomforting effect become less likely to occur again in that situation.

Animals learn by first performing random exploratory actions and then adjusting future actions based on valence outcomes—positive valence reinforces recently performed actions, and negative valence un-reinforces previously performed actions. The terms *satisfying* and *discomforting* went out of favor over the decades following Thorndike's original research; they had an uncomfortable allusion to an actual internal sensation or feeling. Psychologists, including Thorndike, eventually replaced the terms *satisfying* and *discomforting* with *reinforcing* and *punishing*.

One of Thorndike's intellectual successors, B. F. Skinner, went so far as to suggest that *all* animal behavior, even in humans, was a consequence of nothing more than trial and error. As we will see with breakthroughs #3, #4, and #5 in this book, B. F. Skinner turned out to be wrong. But while trial and error does not explain all of animal learning, it undergirds a surprisingly large portion of it.

Thorndike's original research was on cats, dogs, and birds—animals that share a common ancestor around 350 million years ago. But what about more distant vertebrate cousins, those that we share an ancestor with as far back as 500 million years ago? Do they too learn through trial and error?

A year after his 1898 dissertation, Thorndike published an additional note showing the results of these same studies performed on a different animal: fish.

The Surprising Smarts of Fish

If there is any member of the vertebrate group that humans bear the most prejudice against, it is fish. The idea that fish are, well, *dumb* is embedded in many cultures. We have all heard the folklore that fish cannot retain memories for more than three seconds. Perhaps all this prejudice is to be expected; fish are the vertebrates that are the least like us. But this prejudice is unfounded; fish are far smarter than we give them credit for.

In Thorndike's original experiment, he put a fish in a tank with a series of transparent walls with hidden openings. He put the fish on one side of the tank (in a bright light, which fish dislike), and on the other side of the tank was a desirable location (the dark, which fish prefer). At first, the fish tried lots of random things to get across the tank, frequently banging into parts of the transparent wall. Eventually the fish found one of the gaps and made it through to the next wall. It then repeated the process until it found the next gap. Once the fish made it past all the walls to the other side, Thorndike picked it up, brought it back to the beginning, and had it start again, each time clocking how long it took the fish to get to the other side. Just as Thorndike's cats learned to escape puzzle boxes through trial and error, so did his fish learn to quickly zip through each of the hidden openings to escape the bright side of the tank.

This ability of fish to learn arbitrary sequences of actions through trial and error has been replicated many times. Fish can learn to find and push a specific button to get food; fish can learn to swim through a small escape hatch to avoid getting caught in a net; and fish can even learn to jump through hoops to get food. Fish can remember how to do these tasks for months or even *years* after being trained. The process of learning is the same in all these tests: fish try some relatively random actions and then progressively refine their behavior depending on what gets reinforced. Indeed, Thorndike's trial and error learning often goes by another name: reinforcement learning.

If you tried to teach a simple bilaterian like a nematode, flatworm, or slug to perform any of these tasks, it would fail. A nematode cannot be trained to perform arbitrary sequences of actions; it will never learn to navigate through hoops to get food.

Over the next four chapters we will explore the challenges of reinforcement learning and learn why ancestral bilaterians, like modern nematodes, were unable to learn this way. We will learn about how the first vertebrate brains worked, how they overcame these earlier challenges, and how these brains flowered into general reinforcement learning machines.

The second breakthrough was reinforcement learning: the ability to learn arbitrary sequences of actions through trial and error. Thorndike's idea of trial-and-error learning sounds so simple—reinforce behaviors that lead to good things and punish behaviors that lead to bad things. But this is an example where our intuitions about what is intellectually easy and what is hard are mistaken. It was only when scientists tried to get AI systems to learn through reinforcement that they realized that it wasn't as easy as Thorndike had thought.

6

The Evolution of Temporal Difference Learning

THE FIRST REINFORCEMENT learning computer algorithm was built in 1951 by a doctoral student at Princeton named Marvin Minsky. This was the beginning of the first wave of excitement around artificial intelligence. The prior decade had seen the development of the main building blocks for AI: Alan Turing had published his mathematical formulation of general purpose problem-solving machines; the global war effort in the 1940s led to the development of modern computers; an understanding of how neurons worked was beginning to provide clues to how biological brains worked on the *micro* level; and the study of animal psychology in the vein of Thorndike's law of effect had provided general principles for how animal intelligence worked on the *macro* level.

And so Marvin Minsky set out to build an algorithm that would learn like a Thorndikian animal. He named his algorithm the Stochastic Neural-Analog Reinforcement Calculator, or SNARC. He created an artificial neural network with forty connections and trained it to navigate through various mazes. The training process was simple: whenever his system successfully got out of the maze, he strengthened the recently activated synapses. Like Thorndike training a cat to escape a puzzle box with *food* reinforcements, Minsky was training an AI to escape mazes with *numerical* reinforcements.

Minsky's SNARC did not work well. The algorithm got better at navigating out of simple mazes over time, but whenever it faced even slightly more complex situations, it failed. Minsky was one of the first to realize that training algorithms the way that Thorndike believed animals learned—by directly reinforcing positive outcomes and punishing negative outcomes—was not going to work.

Here's why. Suppose we teach an AI to play checkers using Thorndike's version of trial-and-error learning. This AI would start by making random moves, and we would give it a reward whenever it won and a punishment whenever it lost. Presumably, if it played enough games of checkers, it should get better. But here's the problem: The reinforcements and punishments in a game of checkers—the outcome of winning or losing—occur only at the end of the game. A game can consist of hundreds of moves. If you win, which moves should get credit for being good? If you lose, which moves should get credit for being bad?

This is, of course, just another version of the credit assignment problem we saw in chapter 4. When a light and a sound both occur alongside food, which stimulus should get associated with food? We already reviewed tricks that simple bilaterians use to decide this: overshadowing (choosing the strongest stimulus), latent inhibition (choosing the novel stimulus), and blocking (choosing what has been associated before). While these solutions are useful when assigning credit between stimuli that *overlap* in time, they are useless when assigning credit between stimuli that are *separated* in time. Minsky realized that reinforcement learning would not work without a reasonable strategy for assigning credit across time; this is called the *temporal* credit assignment problem.

One solution is to reinforce or punish actions that occurred *recently* before winning or losing. The greater the time window between an action and a reward, the less it gets reinforced. This was how Minsky's SNARC worked. But this works only in situations with short time windows. Even in the game of checkers this is an untenable solution. If a checkers-playing AI assigned credit in this way, then the moves toward the end of the game would always get most of the credit and those toward the beginning very little. This would be dumb—the entire game might have been won on a single clever move in the beginning, long before the game was actually won or lost.

An alternative solution is to reinforce all the prior moves at the end of a winning game (or conversely, *punish* all the prior moves at the end of a *losing* game). Your opening blunder, the tide-turning move in the middle, and the inevitable finish will all get reinforced or punished equally depending on whether you won or lost. The argument goes like this: If the AI plays enough games, it will eventually be able to tell the difference between the specific moves that were good and those that were bad.

But this solution also does not work. There are too many configurations of games to learn which moves are good in any reasonable amount of time. There are over five hundred quintillion possible games of checkers. There are over 10^{120} possible games of chess (more than the number of atoms in the universe). Such a method would require an AI to play so many games that we would all be long dead before it became an even reasonably good player.

This leaves us stuck. When training an AI to play checkers, navigate a maze, or do any other task using reinforcement learning we cannot merely reinforce *recent* moves and we cannot merely reinforce *all* the moves. How, then, can AI ever learn through reinforcement?

Minsky identified the temporal credit assignment problem as far back as 1961, but it was left unsolved for decades. The problem was so severe that it rendered reinforcement learning algorithms impotent to solve real-world problems, let alone play a simple game of checkers.

And yet today, artificial reinforcement learning algorithms work far better. Reinforcement learning models are becoming progressively more common in technologies all around us; self-driving cars, personalized ads, and factory robots are frequently powered by them.

How did we get from the complete hopelessness of reinforcement learning in the 1960s to the boom of today?

A Magical Bootstrapping

In 1984, decades after Minsky, a man named Richard Sutton submitted his final PhD dissertation. Sutton proposed a new strategy for solving the temporal credit assignment problem. He had spent the prior six years as a graduate student at UMass Amherst under the supervision of the postdoc

Andrew Barto. Sutton and Barto dug up old ideas on reinforcement learning and attempted another stab at it. Six years of work culminated with Sutton's dissertation, in which he laid one of the intellectual cornerstones for the reinforcement learning revolution. Its title: "Temporal Credit Assignment in Reinforcement Learning."

Sutton—who had studied psychology, not computer science, as an undergraduate—tackled the problem from a uniquely biological perspective. He didn't want to understand the *best* way to tackle the temporal credit assignment problem; he wanted to understand the *actual* way that animals solved it. Sutton's undergraduate thesis was titled "A Unified Theory of Expectation." And Sutton had a hunch that expectation was what was missing from previous attempts to make reinforcement learning work.

Sutton proposed a simple but radical idea. Instead of reinforcing behaviors using *actual* rewards, what if you reinforced behaviors using *predicted* rewards? Put another way: Instead of rewarding an AI system when it wins, what if you reward it when the AI system *thinks* it is winning?

Sutton decomposed reinforcement learning into two separate components: an actor and a critic. The critic predicts the likelihood of winning at every moment during the game; it predicts which board configurations are great and which are bad. The actor, on the other hand, chooses what action to take and gets rewarded *not* at the end of the game but whenever the critic thinks that the actor's move *increased the likelihood of winning*. The signal on which the actor *learns* is not rewards, per se, but the temporal difference in the predicted reward from one moment in time to the next. Hence Sutton's name for his method: *temporal difference learning*.

Imagine you are playing checkers. For the first nine moves, it is mostly neck and neck between you and your opponent. And then on the tenth move you pull off some clever maneuver that turns the tide of the game; suddenly you realize you are in a far better position than your opponent. It is *that* moment where a temporal difference learning signal reinforces your action.

This, Sutton proposed, might solve the temporal credit assignment problem. This would enable an AI system to learn as it goes instead of having to wait until the end of each game. An AI system can reinforce some moves and punish others throughout a long game of checkers, whether

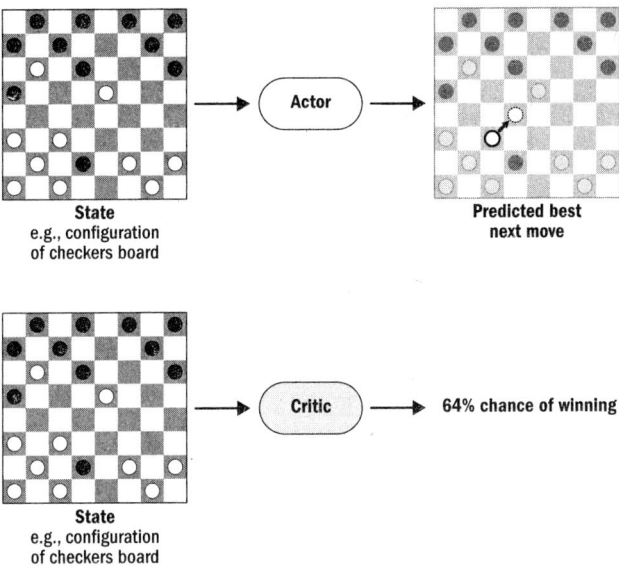

Figure 6.1

or not it won or lost the overall game. Indeed, sometimes a player makes many good moves in a game he or she ultimately loses, and sometimes a player makes many bad moves in a game he or she ultimately wins.

Despite the intuitive appeal of Sutton's approach, we should not expect it to work. Sutton's logic is circular. The critic's prediction of how likely you are to win given a board position depends on what future actions the actor will take (a good board position isn't good if the actor doesn't know how to take advantage of it). Similarly, the actor's decision of what action to take depends on how accurate the critic's temporal difference reinforcement signals have been at reinforcing and punishing past actions. In other words, the critic depends on the actor, and the actor depends on the critic. This strategy seems doomed from the start.

In his simulations, however, Sutton found that by training an actor and a critic simultaneously, a magical bootstrapping occurs between them. Sure, in the beginning the critic often rewards the wrong actions, and the actor often fails to take the necessary actions to fulfill the predictions of the critic. But over time, with enough games, each refines the other until they converge to produce an AI system capable of making remarkably

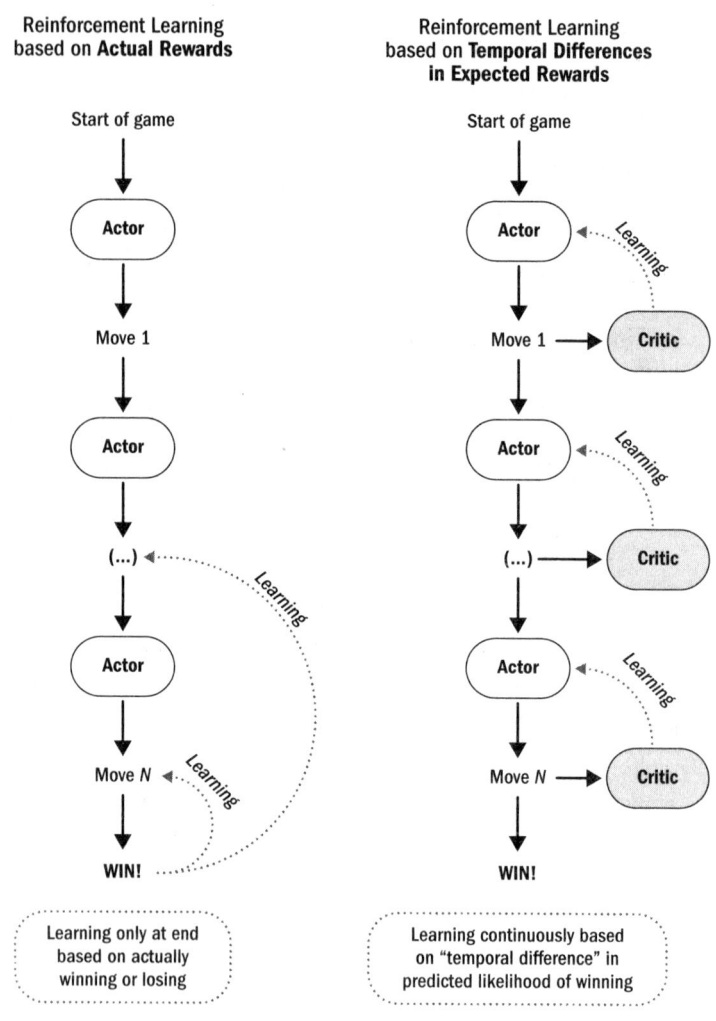

Figure 6.2

intelligent decisions. At least, that's what happened in Sutton's simulations. It wasn't clear whether this would work in practice.

At the same time that Sutton was working on TD learning, a young physicist by the name of Gerald Tesauro was working on getting AI systems to play backgammon. Tesauro was at IBM Research, the same group that would later build Deep Blue (the program that famously beat Garry

Kasparov in chess) and Watson (the program that famously beat Ken Jennings in *Jeopardy!*). But before Deep Blue or Watson, there was Neurogammon. Neurogammon was a backgammon-playing AI system that was trained on transcripts of hundreds of expertly played backgammon games. It learned not through trial and error but by attempting to replicate what it believed a human expert would do. By 1989, Neurogammon could beat every other backgammon-playing computer program, but it was lackluster compared to a human, unable to beat even an intermediate-level player.

By the time Tesauro stumbled on Sutton's work on TD learning, he had spent years trying every conceivable technique to get his computer to play backgammon as well as a human. His crowning achievement was Neurogammon, which was clever but stuck at an intermediate level. And so Tesauro was open to new ideas, even Sutton's radical idea of allowing a system to teach itself from its own predictions.

It was Tesauro who first put Sutton's idea to a practical test. In the early 1990s he began working on TD-Gammon, a system that learned to play backgammon using temporal difference learning.

Tesauro was skeptical. Neurogammon had been taught with examples from expert human players—it was shown the best moves—while TD-Gammon learned solely from trial and error, requiring it to discover the best moves on its own. And yet, by 1994, TD-Gammon achieved, in Tesauro's own words, a "truly staggering level of performance." It not only blew Neurogammon away but was as good as some of the best backgammon players in the world. While it was Sutton who proved that temporal difference learning worked in theory, it was Tesauro who proved it worked in practice. In the decades that followed, TD learning would be used to train AI systems to do many tasks with human-level skill, from playing Atari games to changing lanes in self-driving cars.

The real question, however, was whether TD learning was merely a clever technique that happened to work or a technique that captured something fundamental about the nature of intelligence. Was TD learning a technological invention, or was it, as Sutton had hoped, an ancient technique that evolution had stumbled upon and long ago weaved into animal brains to make reinforcement learning work?

The Grand Repurposing of Dopamine

While Sutton had hoped there was a connection between his idea and the brain, it was one of his colleagues, Peter Dayan, who found it. At the Salk Institute in San Diego, Dayan and his fellow postdoc Read Montague were convinced that brains implemented some form of TD learning. In the 1990s, emboldened by the success of Tesauro's TD-Gammon, they went hunting for evidence in the ever-growing mound of neuroscience data.

They knew where to start. Any attempt to understand how reinforcement learning works in vertebrate brains surely had to begin with a little neuromodulator we have already seen: dopamine.

Deep within the midbrain of all vertebrates is a small cluster of dopamine neurons. These neurons, while few in number, send their output to many regions of the brain. In the 1950s researchers discovered that if you put an electrode into the brain of a rat and stimulate these dopamine neurons, you can get a rat to do pretty much anything. If you stimulate these neurons every few times a rat pushes a lever, the rat will push this lever over five thousand times an hour for twenty-four hours straight. In fact, if given the choice between a dopamine-releasing lever and eating food, rats will *choose the lever*. Rats will ignore food and starve themselves in favor of dopamine stimulation.

This effect is also found in fish. A fish will return to places where it is given dopamine and continue doing so even if those areas are paired with unpleasant things it usually avoids (like being repeatedly removed from water).

Indeed, most drugs of abuse—alcohol, cocaine, nicotine—work by triggering the release of dopamine. All vertebrates, from fish to rats to monkeys to humans, are susceptible to becoming addicted to such dopamine-enhancing chemicals.

Dopamine was undeniably related to reinforcement, but how exactly was not so clear. The original interpretation was that dopamine was the brain's pleasure signal; animals repeated behaviors that activated dopamine neurons because it *felt good*. This made sense in Thorndike's original concept of trial-and-error learning as a process of repeating behaviors that led to satisfying outcomes. But we already saw in chapter 3 that dopamine *does not produce pleasure.* It is less about liking and more about wanting. So then why was dopamine so reinforcing?

The only way to know what dopamine is signaling is to, well, measure the signal. It wasn't until the 1980s that technology was advanced enough for scientists to do this. A German neuroscientist named Wolfram Schultz was the first to measure the activity of individual dopamine neurons.

Schultz devised a simple experiment to probe the relationship between dopamine and reinforcement. Schultz showed monkeys different cues (such as pictures of a geometric shape) and then a few seconds later delivered some sugar water into their mouths.

Sure enough, even in this simple reward-prediction task, it was immediately clear that dopamine was *not* a signal for Thorndike's satisfying outcomes—it was not a signal for pleasure or valence. At first, dopamine neurons did respond like a valence signal, getting uniquely excited whenever a hungry monkey got sugar water. But after a few trials, dopamine neurons *stopped responding to the reward itself* and instead responded *only to the predictive cue.*

When a picture popped up that monkeys knew would lead to sugar, their dopamine neurons got excited, but when these monkeys got sugar water a few moments later, their dopamine neurons did not deviate from their baseline level of activity. Perhaps, then, dopamine was actually a signal for *surprise*? Perhaps dopamine got excited only when events deviated from expectations, like a surprising picture popping up or a surprsing delivery of sugar water?

When Schultz performed additional experiments, it became clear that this "dopamine as surprise" idea was wrong. Once one of his monkeys had learned to expect sugar water after a specific picture was presented, Schultz again presented this reward-predicting picture but didn't give sugar. In this case, despite an equal amount of surprise, dopamine activity *dramatically declined.* While the presentation of an unexpected reward *increases* dopamine activity, the omission of an expected reward *decreases* dopamine activity.*

---

* Dopamine neurons always have a background static, firing at about one to two spikes per second. During these omissions, these neurons go silent (see figure 6.3).

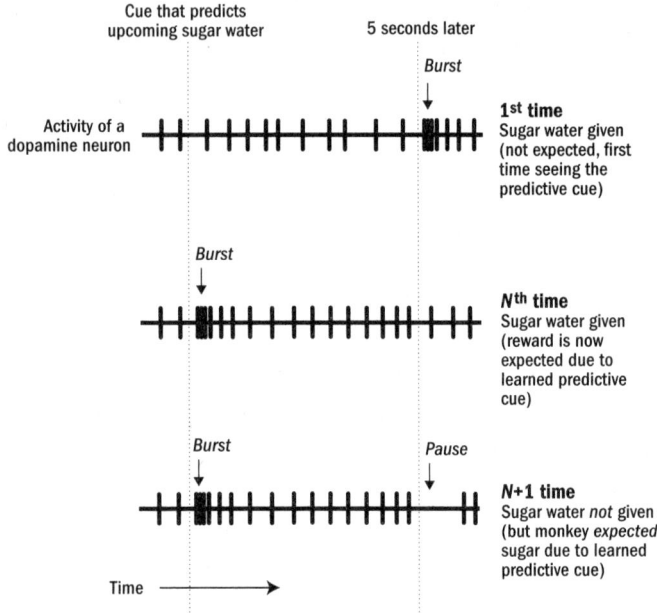

Figure 6.3: Responses of dopamine neurons to predictive cues, rewards, and omissions

Schultz was confused by these results. What was dopamine a signal for? If not for valence or pleasure or surprise, then what? Why did dopamine activity shift from rewards to the predictive cues for rewards? Why did dopamine activity decline when expected rewards were omitted?

For many years, the neuroscience community was unsure how to interpret Schultz's data, an oddity laid bare in the clicks and pauses of an ancient type of neuron.

It wasn't until a decade later that it was solved. Indeed, it was a decade later when Dayan and Montague began scouring the literature for clues that brains implement some form of TD learning. When they eventually came across Schultz's data, they immediately knew what they were seeing. The dopamine responses that Schultz found in monkeys aligned *exactly* with Sutton's temporal difference learning signal. Dopamine neurons in Schultz's monkeys got excited by predictive cues because these cues led to an increase in predicted future rewards (a *positive* temporal difference); dopamine neurons were unaffected by the delivery of an expected reward because there was no change in predicted future reward (no temporal difference); and dopamine-neuron activity decreased when expected rewards were omitted because there was a decrease in predicted future rewards (a *negative* temporal difference).

Even the subtleties of dopamine responses aligned exactly with a temporal difference signal. For example, Schultz found that a cue that predicts food in four seconds triggers more dopamine than a cue that predicts food in sixteen seconds. This is called *discounting*, something Sutton also incorporated into his TD-learning signal; discounting drives AI systems (or animals) to choose actions that lead to rewards sooner rather than later.

Even the way dopamine responds to probabilities aligned with a TD-learning signal—a cue that predicts food with a 75 percent probability triggers more dopamine than a cue that predicts food with a 25 percent probability.

Dopamine is not a signal for reward but for reinforcement. As Sutton found, reinforcement and reward must be decoupled for reinforcement learning to work. To solve the temporal credit assignment problem, brains must reinforce behaviors based on changes in *predicted* future rewards, not *actual* rewards. This is why animals get addicted to dopamine-releasing behaviors despite it not being pleasurable, and this is why dopamine responses quickly shift their activations to the moments when animals *predict* upcoming reward and away from rewards themselves.

In 1997 Dayan and Montague published a landmark paper, coauthored with Schultz, titled "A Neural Substrate of Prediction and Reward." To this day, this discovery represents one of the most famous and beautiful partnerships between AI and neuroscience. A strategy inspired by how Sutton thought the brain might work turned out to successfully overcome practical challenges in AI, and this in turn helped us interpret mysterious data about the brain. Neuroscience informing AI, and AI informing neuroscience.

Most studies that record the activity of dopamine neurons have been done in mammals, but there is every reason to believe these properties of dopamine extend to fish as well. The dopamine-system circuitry is largely the same in both fish and mammal brains, and the same TD-learning signals have been found in the brain structures of fish, rat, monkey, and human brains. In contrast, no TD-learning signals have been found in the dopamine neurons of nematodes or other simple bilaterians.*

In early bilaterians, dopamine was a signal for good things nearby—a primitive version of wanting.† In the transition to vertebrates, however, this good-things-are-nearby signal was elaborated to not only trigger a state of wanting but also to communicate a precisely computed temporal difference learning signal. Indeed, it makes sense that dopamine was the neuromodulator that evolution reshaped into a temporal difference learning signal; as the signal for nearby rewards, it was the closest thing to a measure of predicted future reward. And so, dopamine was transformed from a good-things-are-nearby signal to a there-is-a-35 percent-chance-of-something-awesome-happening-in-exactly-ten-seconds signal. Repurposed from a fuzzy average of recently detected food to an ever fluctuating, precisely measured, and meticulously computed predicted-future-reward signal.

---

* It is important to note that *some* invertebrates, specifically arthropods, do show such reward-prediction errors, but this is believed to have evolved independently given the fact that these reward-prediction errors are not found in other simple bilaterians, and the fact that, in arthropods, the brain structures these responses are found within are uniquely arthropod brain structures.

† In fact, recent studies show how elegantly evolution modified the function of dopamine while still retaining its earlier role of generating a state of wanting. The *amount* of dopamine in the input nuclei of the basal ganglia (called the "striatum") seems to measure the discounted predicted future reward, triggering the state of wanting based on how good things are likely to be and driving animals to focus on and pursue nearby rewards. As an animal approaches a reward, dopamine ramps up, peaking at the moment when an animal expects the reward to be delivered. During this ramping-up process, if predicted rewards *change* (some omission or new cue changes the probability of getting a reward), then dopamine levels rapidly increase or decrease to account for the new level of predicted future reward. These rapid fluctuations in dopamine levels are produced through the bursting and pausing of dopamine neurons that Schultz found; these rapid fluctuations in dopamine levels are the temporal difference learning signal. The *quantity* of dopamine floating around in the striatum modifies the excitability of neurons, which shifts behavior toward exploitation and wanting. In contrast, the *rapid changes* in dopamine levels trigger modifications in the strength of various connections, thereby reinforcing and punishing behaviors. In other words, dopamine in vertebrates is both a signal for wanting and a signal for reinforcement.

The Emergence of Relief, Disappointment, and Timing

From the ancient seed of TD learning sprouted several features of intelligence. Two of these—disappointment and relief—are so familiar that they almost disappear from view, so ubiquitous that it is easy to miss the unavoidable fact that they did not always exist. Both disappointment and relief are emergent properties of a brain designed to learn by predicting future rewards. Indeed, without an accurate prediction of a future reward, there can be no disappointment when it does not occur. And without an accurate prediction of future pain, there can be no relief when it does not occur.

Consider the following task of a fish learning through trial and error. If you turn on a light and then after five seconds gently zap the fish if it does not swim to the opposite side of a tank, it will learn to automatically swim to the opposite side of the tank whenever you turn the light on. Seems like straightforward trial-and-error learning right? Unfortunately not. The ability of vertebrates to perform this type of task—called an avoidance task—has long been a source of debate among animal psychologists.

How would Thorndike have explained a fish's ability to do this? When one of Thorndike's cats finally got out of a puzzle box, it was the *presence* of food rewards that reinforced the cat's actions. But when our fish swam to the safe location, it was the *omission* of a predicted shock that reinforced the fish's actions. How can the absence of something be reinforcing?

The answer is that the omission of an expected punishment is itself reinforcing; it is *relieving*. And the omission of an expected reward is itself punishing; it is *disappointing*. This is why the activity of Schultz's dopamine neurons decreased when food was omitted. He was observing the biological manifestation of disappointment—the brain's punishment signal for a failed prediction of future reward.

Indeed, you can train vertebrates, even fish, to perform arbitrary actions not only with rewards and punishments but also with the *omission* of expected rewards or punishments. To some, a surprise piece of dessert (a reward) is as reinforcing as a surprise day off from school (the omission of something expected but disliked).

A nematode, on the other hand, cannot learn to perform arbitrary behaviors through the omission of rewards. Even crabs and honeybees, who independently evolved many intellectual faculties, are not able to learn from omission of things.*

In this intellectual divide between vertebrates and invertebrates we find another familiar feature of intelligence, one that also emerges from TD learning and its counterparts of disappointment and relief. If we looked closely at our fish learning to swim to specific locations to avoid a zap, we would observe something remarkable. When the light turns on, the fish does not immediately dash to safety. Instead, it leisurely ignores the light until *just before* the end of the five-second interval and then rapidly dashes to safety. In this simple task, fish learn not only *what* to do but *when* to do it; fish know that the shock occurs precisely five seconds after light.

Many forms of life have mechanisms to track the passage of time. Bacteria, animals, and plants all have circadian clocks to track the cycle of the day. But vertebrates are unique in the precision with which they can measure time. A vertebrate can remember that one event occurs precisely five seconds after another event. In contrast, simple bilaterians like slugs and flatworms are entirely unable to learn the precise time intervals between events. Indeed, simple bilaterians like slugs cannot even learn to *associate* events separated by more than two seconds, let alone learn that one thing happens exactly five seconds after another. Even the advanced invertebrates like crabs and bees are unable to learn precise time intervals between events.

---

* It requires some experimental cleverness to distinguish between an association merely fading because a contingency no longer applies (e.g., a light no longer leads to a zap) and learning from the omission of something. In one study with fish, the distinction was shown by adding a new cue specifically in trials where rewards were omitted. If an association were merely fading, then this new cue would not become rewarding (nothing was reinforced in the omission trial), but if instead an animal's brain treats an omitted zap as rewarding in and of itself, then this new cue (which showed up uniquely when zaps were omitted) should be learned to be as rewarding. Researchers have shown that in such experiments fish do, in fact, treat this new cue as rewarding and will approach it in the future. In contrast, we know a nematode cannot do this because they cannot even associate events separated in time, and there is evidence (although it is still unsettled) that even smart invertebrates such as honeybees and crabs do not learn from omission in this way.

TD learning, disappointment, relief, and the perception of time are all related. The precise perception of time is a necessary ingredient to learn from omission, to know when to trigger disappointment or relief, and thereby to make TD learning work. Without time perception, a brain cannot know whether something was omitted or simply hasn't happened yet; our fish would know that the light was *associated* with a zap but not *when* it should occur. Our fish would cower in fear in the presence of the light long after the risk of the zap had passed, blind to its own safety. It is only with an inner clock that fish can predict the exact moment the zap *would* occur and thus, if omitted, the exact moment it deserves a relieving dopamine burst.

The Basal Ganglia

My favorite part of the brain is a structure called the basal ganglia.

For most brain structures, the more one learns about them, the less one understands them—simplified frameworks crumble under the weight of messy complexity, the hallmark of biological systems. But the basal ganglia is different. Its inner wiring reveals a mesmerizing and beautiful design, exposing an orderly computation and function. As one might feel awe that evolution could construct an eye, with such symmetry and elegance, one could equally feel awe that evolution could construct the basal ganglia, also endowed with its own symmetry and elegance.

The basal ganglia is wedged between the cortex and the thalamus (see the figure in the first pages of this book). The input to the basal ganglia comes from the cortex, thalamus, and midbrain, enabling the basal ganglia to monitor an animal's actions and external environment. Information then flows through a labyrinth of substructures within the basal ganglia, branching and merging, transforming and permuting until it reaches the basal ganglia's output nucleus, which contains thousands to millions of inhibitory neurons that send massive and powerful connections to motor centers in the brainstem. This output nucleus of the basal ganglia is, by default, activated. The motor circuits of the brainstem are constantly being suppressed and gated by the basal ganglia. It is only when specific neurons in the basal ganglia turn off that specific motor circuits in the brainstem are *ungated* from activation. The basal ganglia is thereby in a perpetual state of gating and ungating specific actions, operating as a global puppeteer of an animal's behavior.

The functioning of the basal ganglia is essential to our lives. The canonical symptom of Parkinson's disease is the inability to initiate movement. Patients will sit in a chair for many minutes before they can muster the will to even sit up. This symptom of Parkinson's disease primarily emerges due to disruption of the basal ganglia, leaving it in a perpetual state of gating all actions, thereby depriving patients of the ability to initiate even the simplest of movements.

What is the computation performed by the basal ganglia? How does it use incoming information about an animal's actions and external environment to decide which actions to gate (prevent from occuring) and which actions to ungate (allow to occur)?

In addition to receiving information about an animal's actions and external environment, the basal ganglia also receives input from a cluster of dopamine neurons. Whenever these dopamine neurons get excited, the basal ganglia is rapidly flooded with dopamine; whenever these dopamine neurons are inhibited, the basal ganglia is rapidly starved of dopamine. The synapses within the basal ganglia have different dopamine receptors, each responding in unique ways; these fluctuating levels of dopamine strengthen and weaken specific synapses, modifying how the basal ganglia processes input.

As neuroscientists traced the circuitry of the basal ganglia, its function became quite clear. The basal ganglia learns to repeat actions that maximize dopamine release. Through the basal ganglia, actions that lead to dopamine release become more likely to occur (the basal ganglia *ungates* those actions), and actions that lead to dopmaine inhibition become less likely to occur (the basal ganglia *gates* those actions). Sound familiar? The basal ganglia is, in part, Sutton's "actor"—a system designed to repeat behaviors that lead to reinforcement and inhibit behaviors that lead to punishment.

Remarkably, the circuitry of the basal ganglia is practically identical between a human brain and a lamprey fish brain, two species whose shared ancestors were the first vertebrates over 500 million years ago. The various subclusters, the types of neurons, and the overall function seem to be the same. In the brain of early vertebrates emerged the basal ganglia, the biological locus of reinforcement learning.

Reinforcement learning emerged not from the basal ganglia acting alone, but from an ancient interplay between the basal ganglia and another uniquely vertebrate structure called the hypothalamus, which is a small structure at the base of the forebrain.

In vertebrate brains, dopamine release is initially controlled by the hypothalamus. It is the hypothalamus that houses valence neurons inherited from the valence sensory apparatus of ancestral bilaterians. When you are cold, it is your hypothalamus that triggers shivering and that makes you enjoy warmth; just as when you are hot, it is your hypothalamus that triggers sweating and that makes you enjoy the cold. When your body needs calories, it is your hypothalamus that detects hunger signals in your bloodstream and that makes you hungry. The positive valence food-sensitive neurons in early bilaterians functioned just as the positive valence food-sensitive neurons in your hypothalamus do, becoming highly responsive to food when you are hungry and less responsive to food when you are full. This is why you will be salivating over pizza one moment, but then after engorging yourself want *absolutely nothing to do* with pizza just ten minutes later.

In other words, the hypothalamus is, in principle, just a more sophisticated version of the steering brain of early bilaterians; it reduces external stimuli to good and bad and triggers reflexive responses to each. The valence neurons of the hypothalamus connect to the same cluster of dopamine neurons that propogates dopamine throughout the basal ganglia. When the hypothalamus is happy, it floods the basal ganglia with dopamine, and when it is upset, it deprives the basal ganglia of dopamine. And so, in some ways, the basal ganglia is a student, always trying to satisfy its vague but stern hypothalamic judge.

The hypothalamus doesn't get excited by *predictive* cues; it gets excited only when it actually gets what it wants—food when hungry, warmth when cold. The hypothalamus is the decider of *actual* rewards; in our AI-playing-backgammon metaphor, the hypothalamus tells the brain whether it won or lost the game but not how well it is doing as the game is unfolding.

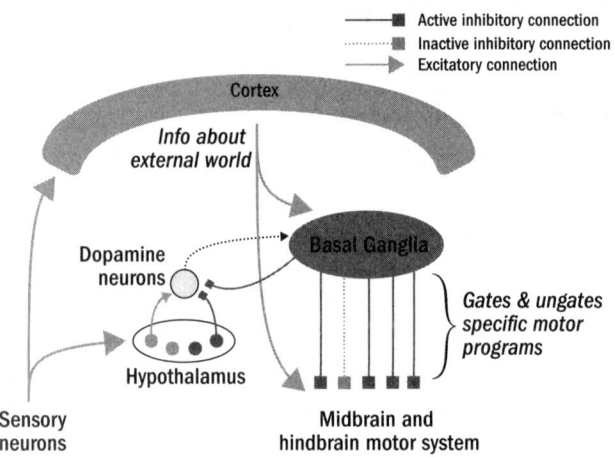

Figure 6.4: A simplified framework for the design of the first vertebrate brain

But as Minsky found with his attempts to make reinforcement learning algorithms in the 1950s, if brains learned only from *actual* rewards, they would never be able to do anything all that intelligent. They would suffer from the problem of temporal credit assignment. So then how is dopamine transformed from a valence signal for *actual rewards* to a temporal difference signal for *changes in predicted future reward*?

In all vertebrates, there is a mysterious mosaic of parallel circuits within the basal ganglia, one that flows down to motor circuits and gates movement, and another that flows back toward dopamine neurons directly. One leading theory of basal ganglia function is that these parallel circuits are literally Sutton's actor-critic system for implementing temporal difference learning. One circuit is the "actor," learning to repeat the behaviors that trigger dopamine release; the other circuit is the "critic," learning to predict future rewards and trigger its own dopamine activation.

In our metaphor, the basal ganglian student initially learns solely from the hypothalamic judge, but over time learns to judge itself, knowing when it makes a mistake *before* the hypothalamus gives any feedback. This is why dopamine neurons initially respond when rewards are delivered, but over time shift their activation toward predictive cues. This is also why receiving a reward that you knew you were going to receive doesn't trigger dopamine release; predictions from the basal ganglia cancel out the excitement from the hypothalamus.

The beautifully conserved circuitry of the basal ganglia, first emerging in the minuscule brain of early vertebrates and maintained for five hundred million years, seems to be the biological manifestation of Sutton's actor-critic system. Sutton discovered a trick that evolution had already stumbled upon over five hundred million years ago.

TD learning, the wiring of vertebrate basal ganglia, the properties of dopamine responses, the ability to learn precise time intervals, and the ability to learn from omissions are all interwoven into the same mechanisms for making trial-and-error learning work.

7

The Problems of Pattern Recognition

FIVE HUNDRED MILLION years ago, the fish-like ancestor of every vertebrate alive today—the inch-long grandmother of every pigeon, shark, mouse, dog, and, yes, human—swam unknowingly toward danger. She swam through the translucent underwater plants of the Cambrian, gently weaving between their thick seaweed-like stalks. She was hunting for coral larvae, the protein-rich offspring of the brainless animals populating the sea. Unbeknownst to her, she too was being hunted.

An *Anomalocaris*—a foot-long arthropod with two spiked claws sprouting from its head—lay hidden in the sand. *Anomalocaris* was the apex predator of the Cambrian, and it was waiting patiently for an unlucky creature to come within lunging distance.

Our vertebrate ancestor would have noticed the unfamiliar smell and the irregular-shaped mound of sand in the distance. But there were always unfamiliar smells in the Cambrian ocean; it was a zoo of microbes, plants, fungi, and animals, each releasing their own unique portfolio of scents. And there was always a backdrop of unfamiliar shapes, an ever-moving portrait of countless objects, both living and inanimate. And so she thought nothing of it.

As she emerged from the safety of the Cambrian plants, the arthropod spotted her and lurched forward. Within milliseconds, her reflexive escape response kicked in. Grandma Fish's eyes detected a fast-moving object in her periphery, triggering a hardwired reflexive turn and dash in the opposite direction. The activation of this escape response flooded her brain with norepinephrine, triggering a state of high arousal, making sensory responses more sensitive, pausing all restorative functions, and reallocating energy to her muscles. In the nick of time, she escaped the clasping talons and swam away.

This has unfolded billions of times, a never-ending cycle of hunting and escaping, of anticipation and fear. But this time was different—our vertebrate ancestor would *remember the smell* of that dangerous arthropod; she would *remember the sight* of its eyes peeking through the sand. She wouldn't make the same mistake again. Sometime around five hundred million years ago, our ancestor evolved pattern recognition.

The Harder-Than-You-Would-Think Problem of Recognizing a Smell

Early bilaterians could not perceive what humans experience as smell. Despite how little effort it takes for you to distinguish the scent of a sunflower from that of a salmon, it is, in fact, a remarkably complicated intellectual feat, one inherited from the first vertebrates.

Just as you have in your nose today, within the nostrils of early vertebrates were thousands of olfactory neurons. In the lamprey fish, there are about fifty different types of olfactory neurons, each type containing a unique olfactory receptor that responds to specific types of molecules. Most smells are made up not of a single molecule but of multiple molecules. When you come home and recognize the smell of your family's best pulled pork, your brain isn't recognizing the pulled-pork molecule (there is no such thing). Rather, it is recognizing a particular soup of many molecules that activates a symphony of olfactory neurons. Any given smell is represented by a pattern of activated olfactory neurons. In summary, smell recognition is nothing more than pattern recognition.

Our nematode-like ancestor's ability to recognize the world was constrained to only the sensory machinery of individual neurons. It could recognize the presence of light by the activation of a single photosensitive neuron or the presence of touch from the activation of a single mechanosensory neuron. Although useful for steering, this rendered a painfully

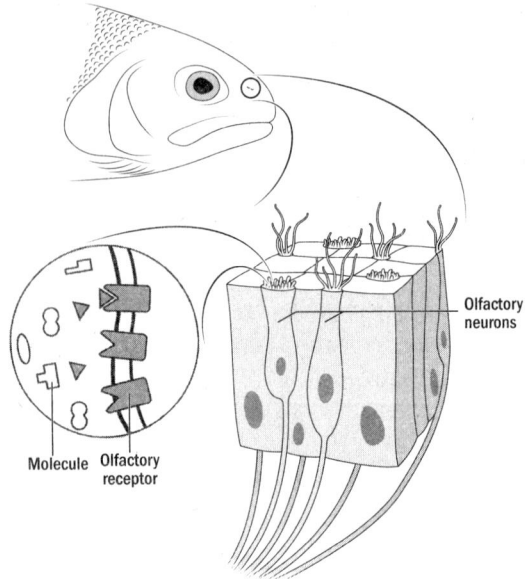

Figure 7.1: Inside the nose of a vertebrate

opaque picture of the outside world. Indeed, the brilliance of steering was that it enabled the first bilaterians to find food and avoid predators without perceiving much of anything about the world.

However, most of the information about the world around you can't be found in a single activated neuron but only in the pattern of activated neurons. You can distinguish a car from a house based on the pattern of photons hitting your retina. You can distinguish the ramblings of a person from the roar of a panther based on the pattern of sound waves hitting your inner ear. And, yes, you can distinguish the smell of a rose from the smell of chicken based on the pattern of olfactory neurons activated in your nose. For hundreds of millions of years, animals were deprived of this skill, stuck in a perceptual prison.

When you recognize that a plate is too hot or a needle too sharp, you are recognizing attributes of the world the way early bilaterians did, with the activations of individual neurons. However, when you recognize a smell, a face, or a sound, you are recognizing things in the world in a way that was beyond early bilaterians; you are using a skill that emerged later in early vertebrates.

Early vertebrates could recognize things using brain structures that decoded *patterns* of neurons. This dramatically expanded the scope of what animals could perceive. Within the small mosaic of only fifty types of olfactory neurons lived a universe of different patterns that could be recognized. Fifty cells can represent over *one hundred trillion patterns.*[*]

HOW EARLY BILATERIANS RECOGNIZED THINGS IN THE WORLD	HOW EARLY VERTEBRATES RECOGNIZED THINGS IN THE WORLD
A single neuron detects a specific thing	Brain decodes the pattern of activated neurons to recognize a specific thing
Small number of things can be recognized	Large number of things can be recognized
New things can be recognized only through evolutionary tinkering (new sensory machinery needed)	New things can be recognized without evolutionary tinkering but through learning to recognize a new pattern (no new sensory machinery needed)

Pattern recognition is hard. Many animals alive today, even after another half billion years of evolution, never acquired this ability—the nematodes and flatworms of today show no evidence of pattern recognition.

There were two computational challenges the vertebrate brain needed to solve to recognize patterns. In figure 7.2, you can see an example of three fictional smell patterns: one for a dangerous predator, one for yummy food, and one for an attractive mate. Perhaps you can see from this figure why pattern recognition won't be easy—these patterns overlap with each other despite having different meanings. One should trigger escape and the others approach. This was the first problem of pattern recognition, that of *discrimination*: how to recognize overlapping patterns as distinct.

The first time a fish experiences fear in the presence of a novel

---

[*] There are 2^{50} possible combinations of 50 elements that can be either on or off: $2^{50} = \sim 1.1 \times 10^{15}$

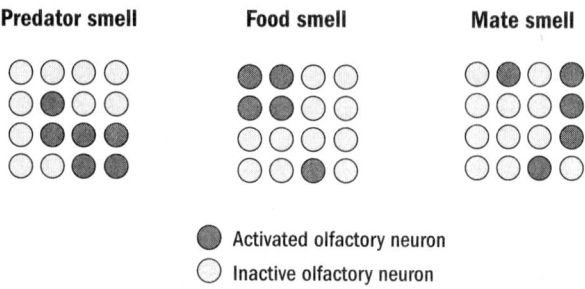

Figure 7.2: The discrimination problem

predator smell, it will remember that specific smell pattern. But the next time the fish encounters that same predator smell, it won't activate the exact same pattern of olfactory neurons. The balance of molecules will never be identical—the age of the new arthropod, or its sex, or its diet, or many other things might be different that could slightly alter its scent. Even the background smells from the surrounding environment might be different, interfering in slightly different ways. The result of all these minor perturbations is that the next encounter will be *similar* but *not the same*. In figure 7.3 you can see three examples of the olfactory patterns that the next encounter with the predator smell might activate. This is the second challenge of pattern recognition: how to *generalize* a previous pattern to recognize novel patterns that are similar but not the same.

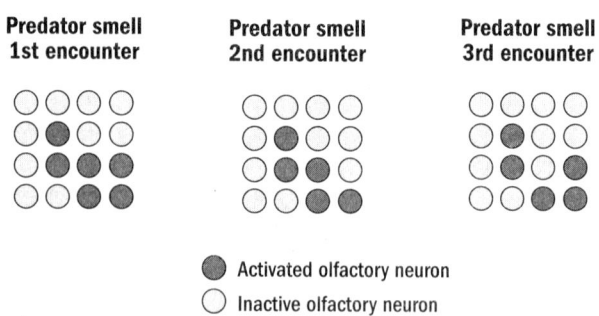

Figure 7.3: The generalization problem

How Computers Recognize Patterns

You can unlock your iPhone with your face. Doing this requires your phone to solve the generalization and discrimination problems. Your iPhone needs to be able to tell the difference between your face and other people's faces, despite the fact that faces have overlapping features (discrimination). And your iPhone needs to identify your face despite changes in shading, angle, facial hair, and more (generalization). Clearly, modern AI systems successfully navigate these two challenges of pattern recognition. How?

The standard approach is the following: Create a network of neurons like in figure 7.4 where you provide an input pattern on one side that flows through layers of neurons until they are transformed into an output on the other end of the network. By adjusting the weights of the connections between neurons, you can make the network perform a variety of operations on its input. If you can edit the weights to be *just right*, you can get an algorithm to take an input pattern and recognize it correctly at the end of the network. If you edit the weights one way, it can recognize faces. If you edit the weights a different way, it can recognize smells.

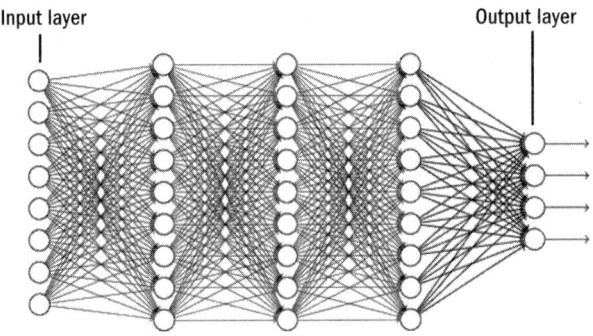

Figure 7.4: An artificial neural network

The hard part is teaching the network how to learn the right weights. The state-of-the-art mechanism for doing this was popularized by Geoffrey Hinton, David Rumelhart, and Ronald Williams in the 1980s.

Their method is as follows: If you were training a neural network to categorize smell patterns into egg smells or flower smells, you would show it a bunch of smell patterns and simultaneously tell the network whether each pattern is from an egg or a flower (as measured by the activation of a specific neuron at the end of the network). In other words, you *tell* the network the correct answer. You then compare the actual output with the desired output and nudge the weights across the entire network in the direction that makes the actual output closer to the desired output. If you do this many times (like, *millions* of times), the network eventually learns to accurately recognize patterns—it can identify smells of eggs and flowers. They called this learning mechanism *backpropagation*: they propagate the error at the end back throughout the entire network, calculate the exact error contribution of each synapse, and nudge that synapse accordingly.

The above type of learning, in which a network is trained by providing examples alongside the correct answer, is called supervised learning (a human has *supervised* the learning process by providing the network with the correct answers). Many supervised learning methods are more complex than this, but the principle is the same: the correct answers are provided, and networks are tweaked using backpropagation to update weights until the categorization of input patterns is sufficiently accurate. This design has proven to work so generally that it is now applied to image recognition, natural language processing, speech recognition, and self-driving cars.

But even one of the inventors of backpropagation, Geoffrey Hinton, realized that his creation, although effective, was a poor model of how the brain actually works. First, the brain does not do supervised learning—you are not given labeled data when you learn that one smell is an egg and another is a strawberry. Even before children learn the words *egg* and *strawberry*, they can clearly recognize that they are different. Second, backpropagation is biologically implausible. Backpropagation works by magically nudging millions of synapses simultaneously and in exactly the right amount to move the output of the network in the right direction. There is no conceivable way the brain could do this. So then how does the brain recognize patterns?

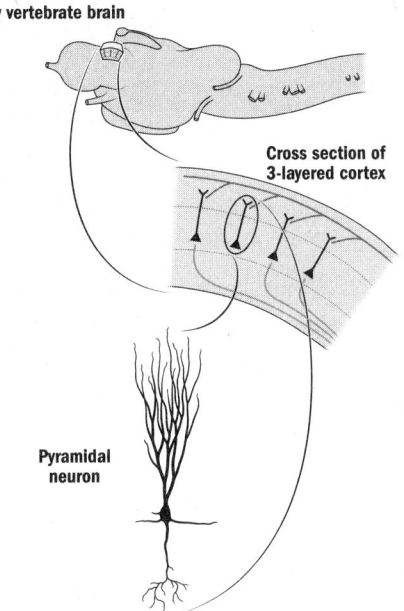

Figure 7.5: The cortex of early vertebrates

The Cortex

Olfactory neurons of fish send their output to a structure at the top of the brain called the cortex. The cortex of simpler vertebrates, like the lamprey fish and reptiles, is made up of a thin sheet of three layers of neurons.

In the first cortex evolved a new morphology of neuron, the *pyramidal* neuron, named after their pyramid-like shape. These pyramidal neurons have *hundreds* of dendrites and receive inputs across *thousands* of synapses. These were the first neurons designed for the purpose of recognizing patterns.

Olfactory neurons send their signals to the pyramidal neurons of the cortex. This network of olfactory input to the cortex has two interesting properties. First, there is a large dimensionality expansion—a small number of olfactory neurons connect to a much larger number of cortical neurons. Second, they connected *sparsely*; a given olfactory cell will connect to only a subset of these cortical cells. These two seemingly innocuous features of wiring may solve the discrimination problem.

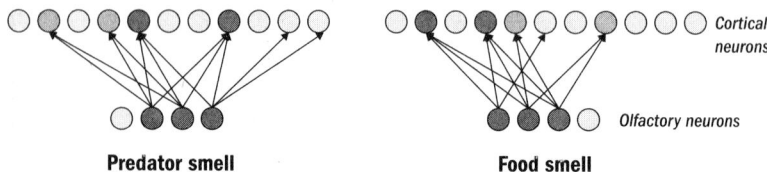

Predator smell   **Food smell**

Figure 7.6: Expansion and sparsity (also called expansion recoding) can solve the discrimination problem

Using figure 7.6 you can intuit why expansion and sparsity achieve this. Even though the predator-smell and food-smell patterns are overlapping, the cortical neurons that get input from *all* the activated neurons will be different. As such, the pattern that gets activated in the cortex will be different despite the fact that the input is overlapping. This operation is sometimes called *pattern separation, decorrelation,* or *orthogonalization.*

Neuroscientists have also found hints of how the cortex might solve the problem of generalization. Pyramidal cells of the cortex send their axons back *onto themselves*, synapsing on hundreds to thousands of other nearby pyramidal cells. This means that when a smell pattern activates a pattern of pyramidal neurons, this ensemble of cells gets automatically wired together through Hebbian plasticity.* The next time a pattern shows up, even if it is incomplete, the full pattern can be reactivated in the cortex. This trick is called *auto-association*; neurons in the cortex automatically learn associations with themselves. This offers a solution to the generalization problem—the cortex can recognize a pattern that is similar but not the same.

Auto-association reveals an important way in which vertebrate memory differs from computer memory. Auto-association suggests that vertebrate brains use *content-addressable memory*—memories are recalled by providing subsets of the original experience, which reactivate the original pattern. If I tell you the beginning of a story you've heard before, you can recall the rest; if I show you half a picture of your car, you can draw the rest. However, computers use *register-addressable memory*—memories

---

* The trick we saw in chapter 4, shorthanded as "neurons that fire together wire together."

that can be recalled only if you have the unique *memory address* for them. If you lose the address, you lose the memory.

Auto-associative memory does not have this challenge of losing memory addresses, but it does struggle with a different form of forgetfulness. Register-addressable memory enables computers to segregate where information is stored, ensuring that new information does not overwrite old information. In contrast, auto-associative information is stored in a shared population of neurons, which exposes it to the risk of accidentally overwriting old memories. Indeed, as we will see, this is an essential challenge with pattern recognition using networks of neurons.

Catastrophic Forgetting (or The Continual Learning Problem, Part 2)

In 1989, Neal Cohen and Michael McCloskey were trying to teach artificial neural networks to do math. Not complicated math, just addition. They were neuroscientists at Johns Hopkins, and they were both interested in how neural networks stored and maintained memories. This was before artificial neural networks had entered the mainstream, before they had proven their many practical uses—neural networks were still something to be probed for missing capabilities and unseen limitations.

Cohen and McCloskey converted numbers into patterns of neurons, then trained a neural network to do addition by transforming two input numbers (e.g., 1 and 3) into the correct output number (in this case, 4). They first taught the network to add ones (1+2, 1+3, 1+4, and so on) until it got good at it. Then they taught the same network to add twos (2+1, 2+2, 2+3, and so on) until it got good at this as well.

But then they noticed a problem. After they taught the network to add twos, it *forgot how to add ones*. When they propagated errors back through the network and updated the weights to teach it to add twos, the network had simply overridden the memories of how to add ones. It successfully learned the new task at the expense of the previous task.

Cohen and McCloskey referred to this property of artificial neural networks as *the problem of catastrophic forgetting*. This was not an esoteric finding but a ubiquitous and devastating limitation of neural networks:

when you train a neural network to recognize a new pattern or perform a new task, you risk interfering with the network's previously learned patterns.

How do modern AI systems overcome this problem? Well, they don't yet. Programmers merely avoid the problem by freezing their AI systems after they are trained. We don't let AI systems learn things sequentially; they learn things all at once and then stop learning.

The artificial neural networks that recognize faces, drive cars, or detect cancer in radiology images do not learn continually from new experiences. As of this book going to print, even ChatGPT, the famous chatbot released by OpenAI, does not continually learn from the millions of people who speak to it. It too stopped learning the moment it was released into the world. These systems are not allowed to learn new things because of the risk that they will forget old things (or learn the wrong things). So modern AI systems are frozen in time, their parameters locked down; they are allowed to be updated only when retrained from scratch with humans meticulously monitoring their performance on all the relevant tasks.

The humanlike artificial intelligences we strive to create are, of course, not like this. Rosey from *The Jetsons* learned as you spoke to her—you could show her how to play a game and she could then play it without forgetting how to play other games.

While we are only just beginning to explore how to make continual learning work, animal brains have been doing so for a long time.

We saw in chapter 4 that even early bilaterians learned continually; the connections between neurons were strengthened and weakened with each new experience. But these early bilaterians never faced the problem of catastrophic forgetting because they never learned patterns in the first place. If things are recognized in the world using only individual sensory neurons, then the connection between these sensory neurons and motor neurons can be strengthened and weakened without interfering with each other. It is only when knowledge is represented in a *pattern of neurons*, like in artificial neural networks or in the cortex of vertebrates, that learning new things risks interfering with the memory of old things.

As soon as pattern recognition evolved, so too did a solution to the problem of catastrophic forgetting. Indeed, even fish avoid catastrophic forgetting fantastically well. Train a fish to escape from a net through a small escape hatch, leave the fish alone for an entire year, and then test it again. During this long stretch of time, its brain will have received a constant stream of patterns, learning continually to recognize new smells, sights, and sounds. And yet, when you place the fish back in the same net an *entire year* later, it will remember how to get out with almost the same speed and accuracy as it did the year before.

There are several theories about how vertebrate brains do this. One theory is that the cortex's ability to perform pattern separation shields it from the problem of catastrophic forgetting; by separating incoming patterns in the cortex, patterns are inherently unlikely to interfere with each other.

Another theory is that learning in the cortex selectively occurs only during moments of surprise; only when the cortex sees a pattern that passes some threshold of novelty are the weights of synapses allowed to change. This enables learned patterns to remain stable for long periods of time, as learning occurs only selectively. There is some evidence that the wiring between the cortex and the thalamus—both structures that emerged alongside each other in early vertebrates—are always measuring the level of novelty between incoming sensory data through the thalamus and the patterns represented in the cortex. If there is a match, then no learning is allowed, hence noisy inputs don't interfere with existing learned patterns. However, if there is a mismatch—if an incoming pattern is sufficiently new—then this triggers a process of neuromodulator release, which triggers changes in synaptic connections in the cortex, enabling it to now learn this new pattern.

We do not yet understand exactly how simple vertebrate brains, like those of fish, reptiles, and amphibians, are capable of overcoming the challenges of catastrophic forgetting. But the next time you spot a fish, you will be in the presence of the answer, hidden in its small cartilaginous head.

The Invariance Problem

Look at the two objects below.

As you view each object, a specific pattern of neurons in the backs of your eyes light up. The minuscule half-millimeter-thick membrane in the

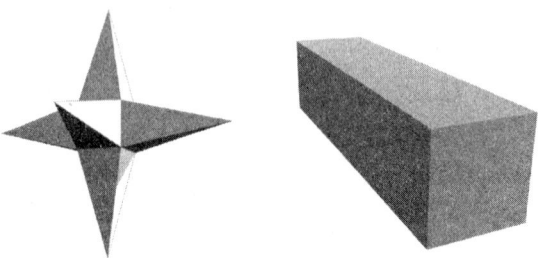

Figure 7.7

back of the eye—the retina—contains over one hundred million neurons of five different types. Each region of the retina receives input from a different location of the visual field, and each type of neuron is sensitive to different colors and contrasts. As you view each object, a unique pattern of neurons activates a symphony of spikes. Like the olfactory neurons that make up a smell pattern, the neurons in the retina make up a visual pattern; your ability to see exists only in your ability to recognize these visual patterns.

The activated neurons in the retina send their signals to the thalamus, which then sends these signals to the part of the cortex that processes visual input (the visual cortex). The visual cortex decodes and memorizes the visual pattern the same way the olfactory cortex decodes and memorizes smell patterns. This is, however, where the similarity between sight and smell ends.

Look at the objects below. Can you identify which shapes are the same as the ones in the first picture?

Figure 7.8

The fact that it is so effortlessly obvious to you that the objects in figure 7.8 are the same as those in figure 7.7 is mind-blowing. Depending on where you focus, the activated neurons in your retina could be completely non-overlapping, with not a single shared neuron, and yet you could still identify them as the same object.

The pattern of olfactory neurons activated by the smell of an egg is the same no matter the rotation, distance, or location of the egg. The same molecules diffuse through the air and activate the same olfactory neurons. But this is not the case for other senses such as vision.

The same visual object can activate different patterns depending on its rotation, distance, or location in your visual field. This creates what is called the invariance problem: how to recognize a pattern as the same despite large variances in its inputs.

Nothing we have reviewed about auto-association in the cortex provides a satisfactory explanation for how the brain so effortlessly did this. The auto-associative networks we described cannot identify an object you have never seen before from completely different angles. An auto-associative network would treat these as different objects because the input neurons are completely different.

This is not only a problem with vision. When you recognize the same set of words spoken by the high-pitched voice of a child and the low-pitched voice of an adult, you are solving the invariance problem. The neurons activated in your inner ear are different because the pitch of the sound is completely different, and yet you can still tell they are the same words. Your brain is somehow recognizing a common pattern despite huge variances in the sensory input.

In 1958, decades before Cohen and McCloskey discovered the problem of catastrophic forgetting, a different team of neuroscientists, also at Johns Hopkins, were exploring a different aspect of pattern recognition.

David Hubel and Torsten Wiesel anesthetized cats, put electrodes into their cortices, and recorded the activity of neurons as they presented the cats with different visual stimuli. They presented dots, lines, and various shapes in different locations in the cats' visual field. They wanted to know how the cortex encoded visual input.

In mammal brains (cats, rats, monkeys, humans, et cetera), the part of the cortex that first receives input from the eye is called *V1* (the first visual

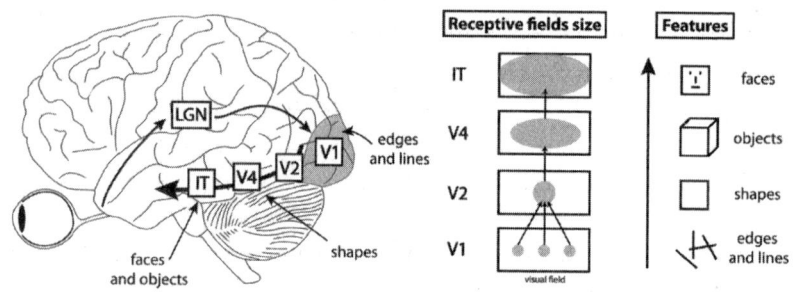

Figure 7.9

area). Hubel and Wiesel discovered that individual neurons in V1 were surprisingly selective with what they respond to. Some neurons were activated only by vertical lines at a specific location in a cat's visual field. Other neurons were activated only by horizonal lines at some other location, and still others by 45-degree lines at a different location. The entire surface area of V1 makes up a map of the cat's full field of view, with individual neurons selective for lines of specific orientations at each location.

V1 decomposes the complex patterns of visual input into simpler features, like lines and edges. From here, the visual system creates a hierarchy: V1 sends its output to a nearby region of cortex called V2, which then sends information to an area called V4, which then sends information to an area called IT.

Neurons at progressively higher levels of this cortical hierarchy become sensitive to progressively more sophisticated features of visual stimuli—neurons in V1 are primarily activated by basic edges and lines, neurons in V2 and V4 are sensitive to more complex shapes and objects, and neurons in IT are sensitive to complex whole objects such as specific faces. A neuron in V1 responds only to input in a specific region of one's visual field; in contrast, a neuron in IT can detect objects across any region of the eye. While V1 decomposes pictures into simple features, as visual information flows up the hierarchy, it is pieced back together into whole objects.

In the late 1970s, well over twenty years after Hubel and Wiesel's initial work, a computer scientist by the name of Kunihiko Fukushima was trying to get computers to recognize objects in pictures. Despite his best attempts, he couldn't get standard neural networks, like those depicted earlier in the chapter, to successfully do it; even small changes in the location, rotation, or size of an object activated entirely different sets of neurons, which blinded networks to generalizing different patterns to the same object—a square over here would be incorrectly perceived as different from the same square over there. He had stumbled on the invariance problem. And he knew that somehow, brains solved it.

Fukushima had spent the prior four years working in a research group that included several neurophysiologists, and so he was familiar with the work of Hubel and Wiesel. Hubel and Wiesel had discovered two things. First, visual processing in mammals was *hierarchical*, with lower levels having smaller receptive fields and recognizing simpler features, and higher levels having larger receptive fields and recognizing more complex objects. Second, at a given level of the hierarchy, neurons were all sensitive to similar features, just in different places. For example, one area of V1 would look for lines at one location, and another area would look for lines for another location, but *they were all looking for lines.*

Fukushima had a hunch that these two findings were clues as to how brains solved the invariance problem, and so Fukushima invented a new architecture of artificial neural networks, one designed to capture these two ideas discovered by Hubel and Wiesel. His architecture departed from the standard approach of taking a picture and throwing it into a fully connected neural network. His architecture first decomposed input pictures into multiple feature maps, like V1 seemed to do. Each feature map was a grid that signaled the location of a feature—such as vertical or horizontal lines—within the input picture. This process is called a *convolution*, hence the name applied to the type of network that Fukushima had invented: *convolutional neural networks.*[*]

After these feature maps identified certain features, their output was compressed and passed to another set of feature maps that could combine

---

[*] Note that he did not use the word *convolution*, but he is credited with coming up with the approach and architecture. Also note that it was Yann LeCun who updated this architecture to use backpropagation, which is what catalyzed the widespread adoption of convolutional neural networks in practical applications.

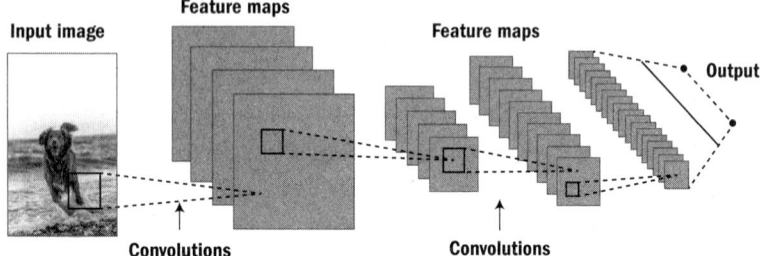

Figure 7.10: A convolutional neural network

them into higher-level features across a wider area of the picture, merging lines and edges into more complex objects. All this was designed to be analogous to the visual processing of the mammalian cortex. And, amazingly, *it worked*.

Most modern AI systems that use computer vision, from your self-driving car to the algorithms that detect tumors in radiology images use Fukushima's convolutional neural networks. AI was blind, but now can see, a gift that can be traced all the way back to probing cat neurons over fifty years ago.

The brilliance of Fukushima's convolutional neural network is that it imposes a clever "inductive bias." An inductive bias is an assumption made by an AI system by virtue of how it is designed. Convolutional neural networks are designed with the *assumption* of translational invariance, that a given feature in one location should be treated the same as that same feature but in a different location. This is an impregnable fact of our visual world: the same thing can exist in different places without the thing being different. And so, instead of trying to get an arbitrary web of neurons to learn this fact about the visual world, which would require too much time and data, Fukushima simply encoded this rule directly into the architecture of the network.

Despite being inspired by the brain, convolutional neural networks (CNNs) are, in fact, a poor approximation of how brains recognize visual patterns. First, visual processing isn't as hierarchical as originally thought; input frequently skips levels and branches out to multiple levels simultaneously. Second, CNNs impose the constraint of *translation*, but they don't inherently understand rotations of 3D objects, and thus don't do a great job recognizing objects when rotated.* Third, modern CNNs are still founded on supervision and backpropagation—with its magical simultaneous updating of many connections—while the cortex seems to recognize objects without supervision and without backpropagation.

And fourth, and perhaps most important, CNNs were inspired by the *mammal* visual cortex, which is much more complex than the simpler visual cortex of fish; and yet the fish brain—lacking any obvious hierarchy or the other bells and whistles of the mammalian cortex—is still eminently capable of solving the invariance problem.

In 2022, the comparative psychologist Caroline DeLong at Rochester Institute of Technology trained goldfish to tap pictures to get food. She presented the goldfish with two pictures. Whenever the fish tapped specifically a picture of a frog, she gave the fish food. Fish quickly learned to swim right up to the frog picture whenever it was presented. DeLong then changed the experiment. She presented the picture of the same frog but from new angles that the fish had never seen before. If fish were unable to recognize the same object from different angles, they would have treated this like any other photo. And yet, amazingly, the fish swam right up to the new frog picture, clearly able to immediately recognize the frog despite the new angle, just as you recognized the 3D objects a few pages ago.

How the fish brain does this is not understood. While auto-association captures some principles of how pattern recognition works in the cortex, clearly even the cortex of fish is doing something far more sophisticated. Some theorize that the vertebrate brain's ability to solve the invariance problem derives not from the unique cortical structures in mammals, but from the complex interactions between the cortex and the thalamus, interactions that have been present since the first vertebrates. Perhaps the thalamus—a ball-shaped structure at the center of the brain—operates like a three-dimensional blackboard, with the cortex providing initial sensory

---

* The way modern CNNs get around this rotation problem is by augmenting the training data to include huge numbers of examples of the same object rotated.

input and the thalamus intelligently routing this sensory information around other areas of the cortex, together rendering full 3D objects from 2D input, thereby flexibly able to recognize rotated and translated objects.

Perhaps the best lesson from CNNs is not the success of the specific assumptions they attempt to emulate—such as translational invariance—but the success of assumptions themselves. Indeed, while CNNs may not capture exactly how the brain works, they reveal the power of a good inductive bias. In pattern recognition, it is good assumptions that make learning fast and efficient. The vertebrate cortex surely has such an inductive bias, we just don't know what it is.

In some ways, the tiny fish brain surpasses some of our best computer-vision systems. CNNs require incredible amounts of data to understand changes in rotations and 3D objects, but a fish seems to recognize new angles of a 3D object in one shot.

In the predatory arms race of the Cambrian, evolution shifted from arming animals with new sensory neurons for detecting specific things to arming animals with general mechanisms for recognizing anything.

With this new ability of pattern recognition, vertebrate sensory organs exploded with complexity, quickly flowering into their modern form. Noses evolved to detect chemicals; inner ears evolved to detect frequencies of sound; eyes evolved to detect sights. The coevolution of the familiar sensory organs and the familiar brain of vertebrates is not a coincidence—they each facilitated the other's growth and complexity. Each incremental improvement to the brain's pattern recognition expanded the benefits to be gained by having more detailed sensory organs; and each incremental improvement in the detail of sensory organs expanded the benefits to be gained by more sophisticated pattern recognition.

In the brain, the result was the vertebrate cortex, which somehow recognizes patterns without supervision, somehow accurately discriminates overlapping patterns and generalizes patterns to new experiences, somehow continually learns patterns without suffering from catastrophic forgetting, and somehow recognizes patterns despite large variances in its input.

The elaboration of pattern recognition and sensory organs, in turn, also found themselves in a feedback loop with reinforcement learning itself. It is also not a coincidence that pattern recognition and reinforcement learning evolved simultaneously in evolution. The greater the brain's ability to learn arbitrary actions in response to things in the world, the greater the benefit to be gained from *recognizing more things in the world*. The more unique objects and places a brain can recognize, the more unique actions it can learn to take. And so the cortex, basal ganglia, and sensory organs evolved together, all emerging from the same machinations of reinforcement learning.

8

Why Life Got Curious

IN THE AFTERMATH of the success of TD-Gammon, researchers began applying Sutton's temporal difference learning to all kinds of different games. And one by one, games that had previously been "unsolvable" were successfully beaten by these algorithms; TD learning algorithms eventually surpassed human-level performance in video games like Pinball, Star Gunner, Robotank, Road Runner, Pong, and Space Invaders. And yet there was one Atari game that was perplexingly out of reach: Montezuma's Revenge.

In Montezuma's Revenge, you start in a room filled with obstacles. In each direction is another room, each with its own obstacles. There is no sign or clue as to which direction is the right way to go. The first reward is earned when you find your way to a hidden door in a faraway hidden room. This makes the game particularly hard for reinforcement learning systems: the first reward occurs so late in the game that there is no early nudging of what behavior should be reinforced or punished. And yet somehow, of course, humans beat this game.

It wasn't until 2018 when an algorithm was developed that finally completed level one of Montezuma's Revenge. This new algorithm, developed by Google's DeepMind, accomplished this feat by adding something familiar that was missing from Sutton's original TD learning algorithm: curiosity.

Sutton had always known that a problem with any reinforcement learning system is something called the *exploitation-exploration dilemma*. For trial-and-error learning to work, agents need to, well, have lots of trials from which to learn. This means that reinforcement learning can't work by just *exploiting* behaviors they predict lead to rewards; it must also *explore* new behaviors.

In other words, reinforcement learning requires two opponent processes—one for behaviors that were previously reinforced (exploitation) and the other for behaviors that are new (exploration). These choices are, by definition, opposing each other. Exploitation will always drive behavior toward known rewards, and exploration will always drive toward what is unknown.

In early TD learning algorithms, this trade-off was implemented in a crude way: these AI systems spontaneously—say, 5 percent of the time—did something totally random. This worked okay if you were playing a constrained game with only so many next moves, but it worked terribly in a game like Montezuma's Revenge, where there were practically an infinite number of directions and places you could go.

There is an alternative approach to tackling the exploitation-exploration dilemma, one that is both beautifully simple and refreshingly familiar. The approach is to make AI systems explicitly *curious*, to reward them for exploring new places and doing new things, to make *surprise itself reinforcing*. The greater the novelty, the larger the compulsion to explore it. When AI systems playing Montezuma's Revenge were given this intrinsic motivation to explore new things, they behaved very differently—indeed, more like a human player. They became motivated to explore areas, go to new rooms, and expand throughout the map. But instead of exploring through random actions, they explored deliberately; they specifically wanted to go to new places and to do new things.

Even though there are no explicit rewards until you get past all the rooms in level one, these AI systems didn't need any external rewards to explore. They were motivated on their own. Simply finding their way to a new room was valuable in and of itself. Armed with curiosity, suddenly these models started making progress, and they eventually beat level one.

The importance of curiosity in reinforcement learning algorithms suggests that a brain designed to learn through reinforcement, such as the brain of early vertebrates, should also exhibit curiosity. And indeed, evidence suggests that it was early vertebrates who first became curious. Curiosity is seen across all vertebrates, from fish to mice to monkeys to human infants. In vertebrates, surprise itself triggers the release of dopamine, even if there is no "real" reward. And yet, most invertebrates do not exhibit curiosity; only the most advanced invertebrates, such as insects and cephalopods, show curiosity, a trick that evolved independently and wasn't present in early bilaterians.

The emergence and mechanisms of curiosity help explain gambling, which is an irrational oddity of vertebrate behavior. Gamblers violate Thorndike's law of effect—they continue to gamble their money away despite the fact that the expected reward is negative.

B. F. Skinner was the first to realize that rats will gamble. The best way to get a rat to obsessively push a lever for food is *not* to have the lever release food pellets every time it is pressed; instead, it is to have the lever *randomly* release food pellets. Such variable-ratio reinforcement makes rats go crazy; they endlessly push the lever, seemingly obsessed with seeing if just *one more push* might be the one to produce a pellet. Even when the total number of pellets released are identical overall, such variable-ratio reinforcement leads to far more lever pushes than fixed-ratio reinforcement. Fish exhibit this effect as well.

One explanation for this is that vertebrates get an extra boost of reinforcement when something is *surprising*. To make animals curious, we evolved to find surprising and novel things reinforcing, which drives us to pursue and explore them. This means that even if the reward of an activity is negative, if it is novel, we might pursue it anyway.

Games of gambling are carefully designed to exploit this. In games of gambling, you don't have a 0 percent chance of winning (which would lead you not to play); you have a 48 percent chance of winning, high enough to make it possible, uncertain enough to make it surprising when you win (giving you a dopamine boost), and low enough so that the casino will, in the long run, suck you dry.

Our Facebook and Instagram feeds exploit this as well. With each scroll, there is a new post, and randomly, after some number of scrolls, something interesting shows up. Even though you might not want to use Instagram, the same way gamblers don't want to gamble or drug addicts don't want to use anymore, the behavior is subconsciously reinforced, making it harder and harder to stop.

Gambling and social feeds work by hacking into our five-hundred-million-year-old preference for surprise, producing a maladaptive edge case that evolution has not had time to account for.

Curiosity and reinforcement learning coevolved because curiosity is a requirement for reinforcement learning to work. With the newfound ability to recognize patterns, remember places, and flexibly change behavior based on past rewards and punishments, the first vertebrates were presented with a new opportunity: for the first time, *learning* became, in and of itself, an extremely valuable activity. The more patterns a vertebrate recognized and the more places she remembered, the better she would survive. And the more new things she tried, the more likely she was to learn the correct contingencies between her actions and their corresponding outcomes. And so it was 500 million years ago in the tiny brain of our fishlike ancestors when curiosity first emerged.

9

The First Model of the World

HAVE YOU EVER tried to navigate through your home in the dark? I'm guessing not on purpose, but perhaps during a power outage or a midnight stroll to the bathroom. If you have ever tried this, you probably had the (not very surprising) realization that it is hard to do. As you step out of your bedroom and walk toward the end of the hall, you are prone to mispredicting the length of the hallway or the exact location of the bathroom door. You might stub a toe.

But you would also notice that, despite your blindness, you have a reasonable hunch about where the end of the hallway is, some *intuition* about where you are in the labyrinth of your home. You might be off by a step or two, but your intuition nonetheless proves an effective guide. What is remarkable about this is not that it is hard, but that it is achievable *at all*.

The reason you can do this is that your brain has built a spatial map of your home. Your brain has an internal *model* of your home, and thus, as you move, your brain can update your position in this map on its own. This trick, the ability to construct an internal model of the external world, was inherited from the brains of first vertebrates.

The Maps of Fish

This same find-your-way-to-the-bathroom-in-the-dark test can be done in fish. Well, not the bathroom part, but the general test of remembering a location without a visual guide. Put a fish in an empty tank with a grid of twenty-five identical containers throughout the tank. Hide food in one of the containers. The fish will explore the tank, randomly inspecting each container until it stumbles on the food. Now take the fish out of the tank, put food back in the same container, and put the fish back into the tank. Do this a few times, and the fish will learn to quickly dart directly to the container with the food.

Fish are not learning some fixed rule of *Always turn left when I see this object*—they navigate to the correct location no matter where in the tank they are initially placed. And they are not learning some fixed rule of *Swim toward this image or smell of food*; fish will go back to the correct container even if you don't put *any* food back in the container. In other words, even if every container is exactly identical because none of them have any food at all, fish still correctly identify which container is currently placed at the exact location that previously contained food.

The only clue as to which container previously held the food was the walls of the tank itself, which had markings to designate specific sides. Thus, fish somehow identified the correct container based solely on the container's location relative to the landmarks on the side of the tank. The only way fish could have accomplished this is by building a spatial map—an internal model of the world—in their minds.

The ability to learn a spatial map is seen across vertebrates. Fish, reptiles, mice, monkeys, and humans all do this. And yet simple bilaterians like nematodes are incapable of learning such a spatial map—they cannot remember the location of one thing relative to another thing.

Even many advanced invertebrates such as bees and ants are unable to solve spatial tasks. Consider the following ant study. Ants form a path from a nest to food and another path from the food source to their nest. They return to the nest with scavenged food, then leave again empty-handed to acquire more food. Suppose you took one of the ants on its way *back* to the nest and placed it on the path that is *leaving* the nest. This ant clearly wants to go back to the nest, but is now placed in a location from which it has only ever navigated *away* from the nest. If the ant had an internal model of space, it would realize that the fastest way home is to just turn around and go in the opposite direction. That is what a fish would do. But if instead the ant learned only a series of movements (turn right at cue X, turn left at cue Y), then it would just dutifully begin the loop all over again. Indeed, ants go through the entire loop again. Ants navigate by following a set of rules of when to turn where, not by constructing a map of space.

Your Inner Compass

Here's another test you can do on yourself: Sit in one of those swivel chairs, close your eyes, ask someone to turn the chair, and then guess what direction of the room you are facing before opening your eyes. You will be amazingly accurate. How did your brain do this?

Deep in your inner ear are semicircular canals, small tubes filled with fluid. These canals are lined with sensory neurons that float in this fluid and activate whenever they detect movement. The semicircular canals are organized in three loops, one for facing forward, one for facing sideways, and one for facing upward. The fluid in each of these canals moves only when you move in that specific dimension. Thus, the ensemble of activated sensory cells signal the direction of head movement. This creates a unique sense—the *vestibular* sense. This is why you get dizzy if you are spun in a

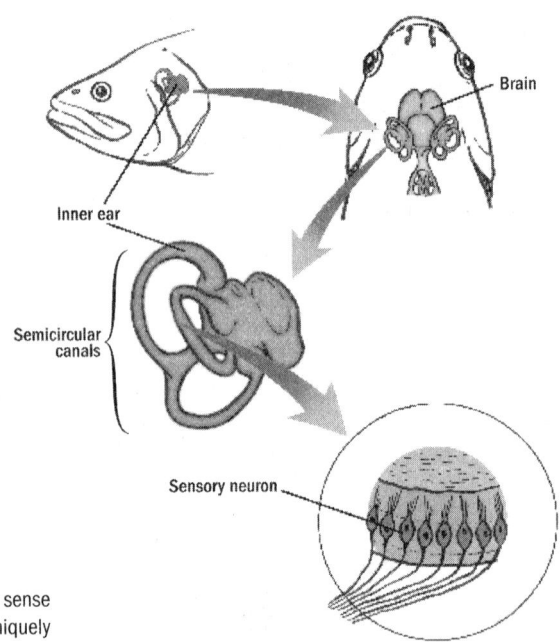

Figure 9.1: The vestibular sense of fish emerges from the uniquely vertebrate semicircular canals

chair continuously—eventually, this overactivates these sensory cells and when you stop turning, they are still active, incorrectly signaling rotation even when you aren't rotating.

The evolutionary origin of semicircular canals was in early vertebrates, emerging at the same time as reinforcement learning and the ability to construct spatial maps. Modern fish have the same structure in their inner ears, and it enables them to identify when and by how much they are moving.

The vestibular sense is a necessary feature of building a spatial map. An animal needs to be able to tell the difference between something swimming toward it and it swimming toward something. In each case, the visual cues are the same (both show an object coming closer), but each means *very* different things in terms of movement through space. The vestibular system helps the fish tell the difference: If it starts swimming toward an object, the vestibular system will detect this acceleration. In contrast, if an object starts moving toward *it*, no such activation will occur.

In the hindbrain of vertebrates, in species as diverse as fish and rats, are what are called "head-direction neurons" that fire only when an animal is facing a certain direction. These cells integrate visual and vestibular input to create a neural compass. The vertebrate brain evolved, from its very beginnings, to model and navigate three dimensional space.

But if the hindbrain of fish constructs a compass of an animal's own direction, where is the model of external space constructed? Where does the vertebrate brain store information about the locations of things relative to other things?

The Medial Cortex (aka the Hippocampus)

The cortex of early vertebrates had three subareas: lateral cortex, ventral cortex, and medial cortex. The lateral cortex is the area where early vertebrates recognized smells and that would later evolve into the olfactory cortex in early mammals. The ventral cortex is the area where early vertebrates learned patterns of sights and sounds and that would later evolve into the amygdala in early mammals. But folded into the middle of the brain was the third area, the medial cortex.

Early vertebrate cortex

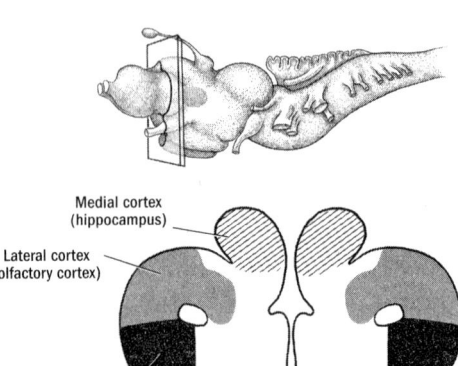

Figure 9.2: The cortex of early vertebrates

The medial cortex is the part of cortex that later became the hippocampus in mammals. If you record neurons in the hippocampus of fish as they navigate around, you will find some neurons that activate only when the fish are at a specific location in space, others only when the fish are at a border of a tank, and others only when the fish are facing specific directions. Visual, vestibular, and head-direction signals propagate to the medial cortex, where they are all mixed together and converted into a spatial map.

Indeed, if you damage the hippocampus of a fish, it can learn to swim toward or away from cues, but it loses the ability to remember locations. These fish fail to use distant landmarks to figure out the right direction to turn in a maze; fail to navigate to specific locations in an open space to get food; and fail to figure out how to escape a simple room when given different starting locations.

The function and structure of the hippocampus has been conserved across many lineages of vertebrates. In humans and rats, the hippocampus contains place cells, which are neurons that activate only when an animal is in a specific location in an open maze. Damage to the hippocampus in lizards, rats, and humans similarly impairs spatial navigation.

Clearly the three-layered cortex of early vertebrates performed computations far beyond simple auto-association. Not only is it also seemingly capable of recognizing objects despite large changes in rotation and scale (solving the invariance problem), but it is also seemingly capable of constructing an internal model of space. To speculate: Perhaps the ability of the cortex to recognize objects despite changes in rotation and its ability to model space are related. Perhaps the cortex is tuned to model 3D *things*—whether those things are objects or spatial maps.

The evolution of spatial maps in the minds of early vertebrates marked numerous firsts. It was the first time in the billion-year history of life that an organism could recognize *where* it was. It is not hard to envision the advantage this would have offered. While most invertebrates steered around and executed reflexive motor responses, early vertebrates could remember the places where arthropods tended to hide, how to get back to safety, and the locations of nooks and crannies filled with food.

It was also the first time a brain differentiated the self from the world. To track one's location in a map of space, an animal needs to be able to tell the difference between "something swimming toward me" and "me swimming toward something."

And most important, it was the first time that a brain constructed an *internal model*—a representation of the external world. The initial use of this model was, in all likelihood, pedestrian: it enabled brains to recognize arbitrary locations in space and to compute the correct direction to a given target location from any starting location. But the construction of this internal model laid the foundation for the next breakthrough in brain evolution. What began as a trick for remembering locations would go on to become much more.

Summary of Breakthrough #2: Reinforcing

Our ancestors from around five hundred million years ago transitioned from simple wormlike bilaterians to fishlike vertebrates. Many new brain structures and abilities emerged in these early vertebrate brains, most of which can be understood as enabling and emerging from breakthrough #2: reinforcement learning. These include

- Dopamine became a temporal difference learning signal, which helped solve the temporal credit assignment problem and enabled animals to learn through trial and error.
- The basal ganglia emerged as an actor-critic system, enabling animals to generate this dopamine signal by predicting future rewards and to use this dopamine signal to reinforce and punish behaviors.
- Curiosity emerged as a necessary part of making reinforcement learning work (solving the exploration-exploitation dilemma).
- The cortex emerged as an auto-associative network, making pattern recognition possible.
- The perception of precise timing emerged, enabling animals to learn, through trial and error, not only *what* to do but *when* to do it.
- The perception of three-dimensional space emerged (in the hippocampus and other structures), enabling animals to recognize where they were and remember the location of things relative to other things.

Reinforcement learning in early vertebrates was possible only because the mechanisms of valence and associative learning had already evolved in early bilaterians. Reinforcement learning is bootstrapped on simpler valence signals of good and bad. Conceptually, the vertebrate brain is built on top of the more ancient steering system of bilaterians. Without steering, there is no starting point for trial and error, no foundation on which to measure what to reinforce or un-reinforce.

SUMMARY OF BREAKTHROUGH #2: REINFORCING

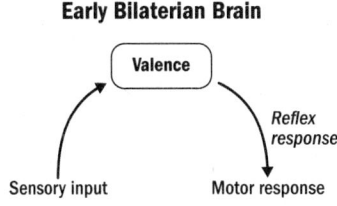

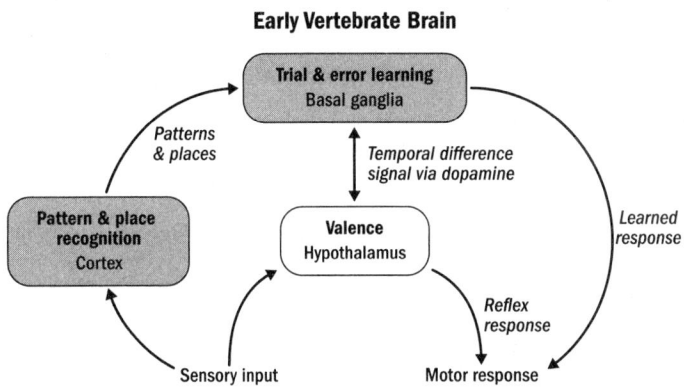

Figure 9.3

Steering bilaterians made it possible for later vertebrates to learn through trial and error. And trial and error in vertebrates, in turn, made it possible for the even more perplexing and monumental breakthrough that would follow. It was early mammals who first figured out how to engage in a different flavor of trial and error: learning not by *doing* but by *imagining*.

BREAKTHROUGH #3

Simulating and the First Mammals

Your brain 200 million years ago

10

The Neural Dark Ages

FROM 420 TO 375 million years ago, oceans became filled with progressively more diverse predatory fish of many shapes and sizes. What would have resembled the sharks and stingrays of today were common sightings. Twenty-foot-long placoderms, fish with armored head plates and thick bone-crushing teeth, found themselves at the top of this food chain.

Arthropods and other invertebrates were relegated to various niches. Some got smaller. Some evolved thicker shells. Some even took a cue from early vertebrates and survived by getting smarter—it was during this period that the cephalopods emerged, the ancestors of today's squids and octopuses. Under severe pressure to survive their mass hunting by fish, cephalopods became impressively intelligent down an independent lineage with brains that work very differently than our own.

The most radical invertebrate survival strategy was to escape the sea altogether. The arthropods, driven from their homeland by relentless predation, were the first animals to walk out of the oceans and populate the land. They found respite among the small leafless land-faring plants that had sparsely sprouted along the seashores.

The period between 420 to 375 million years ago is called the Devonian period, and it was here when land plants first evolved *leaves* for better absorption of sunlight and *seeds* for spreading, both of which enabled plants to propagate to previously inhospitable areas. Plants that resembled today's trees first developed, growing thick roots and creating stable soils for nearby arthropods to live. In the early Devonian period, plants on land were no more than thirty centimeters tall, but by the end of it, they were thirty *meters* tall. It was only at this point that our planet began to appear green from above, as land plants spread across Earth's surface.

While life for arthropods was horrifying in the sea, it was heavenly on land. Arthropods developed new tricks to meet the needs of life on land, diversifying into what resembled today's spiders and insects. Unfortunately, as we have seen with today's problem of climate change, Earth's biosphere is unforgiving to those who proliferate rapidly and unsustainably. What began as a small oasis for arthropod refugees eventually became an overextended orgy of plant life, triggering a global extinction event that would eradicate close to half of all life.

A Story of Two Great Deaths

History repeats itself.

One and a half billion years ago, the explosion of cyanobacteria suffocated the Earth with carbon dioxide and polluted it with oxygen. Over a billion years later, the explosion of plants on land seems to have committed a similar crime.

The inland march of plants was too rapid for evolution to accommodate and rebalance carbon dioxide levels through the expansion of more CO_2-producing animals. Carbon dioxide levels plummeted, which caused the climate to cool. The oceans froze over and gradually became inhospitable to life. This was the Late Devonian Extinction, the first great death of this era. There are competing theories of what caused it; some argue that it was not overproliferation of plants but some other natural disaster. In any case, it was from the icy graves of this tragedy that our ancestors emerged from the sea.

Extinction events create opportunities for small niches to transform into dominant strategies. Before the Late Devonian Extinction, our ancestors had found such a niche. Most fish stayed far away from the shore to avoid deadly beaching, a situation where a fish becomes stuck on land as tides recede. Although the risk of beaching made it dangerous to pursue, there was a big nutritional prize to be found close to the shore: the warm earthy puddles were full of small insects and vegetation.

Our ancestors were the first fish to evolve the ability to survive out of water. They developed a pair of lungs that augmented their gills, enabling them to extract oxygen from both water and air. And so our ancestors would use their fins both for swimming in water and for wading themselves short distances on land, traveling from puddle to puddle in search of insects.

When the Late Devonian Extinction Event began to freeze over the oceans, our air-breathing and land-walking ancestors were one of the few warm-water fish to survive. As the food supply in warm waters began to die, our ancestors spent more of their time living in the inland puddles. They lost their gills (and thus their ability to breath underwater), and their webbed fins gave way to fingered hands and feet. They became the first *tetrapods* (*tetra* for "four" and *pods* for "feet"), most closely resembling a modern amphibian such as a salamander.

One evolutionary lineage of tetrapods, who were lucky enough to live in parts of the Earth that still supported these warmer puddles, would maintain this lifestyle for hundreds of millions of years—they would become the amphibians of today. Another lineage abandoned the dying shores and wandered farther inland in search of food. This was the lineage of amniotes—the creatures that developed the ability to lay leathery eggs that could survive out of the water.

The first amniotes probably best resembled a lizard of today. Amniotes found an inland ecosystem abundant with food—insects and plants were everywhere for the feasting. Eventually, the Devonian ice age faded and amniotes spread and diversified to all corners of the Earth. The Carboniferous and Permian eras, which collectively lasted from 350 million years ago to 250 million years ago, saw an explosion of amniotes on land.

Living on land presented unique challenges to the amniotes that their fish cousins never faced. One such challenge was temperature fluctuations. Cycles of the day and season create only muted temperature changes deep in the oceans. In contrast, temperatures can fluctuate dramatically on the surface. Amniotes, like fish, were cold-blooded—their only strategy for regulating their body temperature was to physically relocate to warmer places.

One amniote lineage were the reptiles, who would eventually diversify into dinosaurs, lizards, snakes, and turtles. Most of these reptiles dealt with daily temperature fluctuations by becoming immobile at night. Temperatures were too low for their muscles and metabolisms to function properly, so they simply shut down. The fact that reptiles were shut down for a third of their lives presented an opportunity—creatures that could hunt in the night would reap an incredible feast of motionless lizards.

The other lineage of amniotes were our ancestors: the therapsids. The therapsids differed from reptiles at the time in one important way: they developed *warm-bloodedness*. Therapsids were the first vertebrates to evolve the ability to use energy to generate their own internal heat.* This was a gamble. They would require far more food to survive, but in return they had the ability to hunt at any time, including the cold nights when their reptile cousins lay immobile—an easy feast offered on a Permian platter.

In the Permian era, when the land was full of edible reptiles and arthropods, this gamble paid off. During the period from 300 to 250 million years ago, therapsids became the most successful land animals. They grew to the size of a modern tiger and began to grow hair to further maintain their heat. These therapsids would have looked like large hairy lizards.

Perhaps you can already see a trend emerging from the evolutionary history of life on Earth: all reigns come to an end. The therapsid reign on Earth was no different: the Permian-Triassic mass extinction event, which occurred around 250 million years ago, was the deadliest of all extinction events in Earth's history. It was the second great death in this era. This extinction event was the most severe and perhaps the most enigmatic. Within five to ten million years, 96 percent of all marine life died, and 70 percent of all land life died. There is still controversy over what caused this—theories include asteroids, volcanic explosions, and methane-producing microbes. Some suggest that it was no single reason but rather a perfect storm of multiple unlucky occurrences. Regardless of the cause, we know the effects.

The large therapsids went almost entirely extinct. The gamble of warm-bloodedness that originally facilitated their rise was also the cause of their

---

* Note that some dinosaurs, later in their evolutionary story, are believed to have evolved warm-bloodedness, as evidenced by chemical analyses of their fossils and the fact that birds are warm-blooded.

Figure 10.1: The first therapsid

downfall. During a period of reduced access to food, the therapsids, with their need for huge amounts of calories, died first. The reptiles and their comparatively scant diets were much better suited to weather this storm.

For about five million years, life survived only in tiny pockets of the world. The only therapsids that survived were the small plant-eating ones, such as the burrowing cynodonts. The cynodonts originally evolved into the niche of burrowing underground to hide from the larger and more predatory therapsids that dominated the world. As food supply went away and all those bigger animals died off, these small cynodonts were among the few surviving therapsids to emerge on the other side of the Permian-Triassic extinction.

Although the therapsid lineage was just barely preserved by the small cynodont, the world they found themselves in was different. On the other side of this extinction event, with 70 percent of land life extinguished, reptiles emerged numerous, diverse, and big. The eradication of the large therapsids handed the animal kingdom to their scaly reptilian cousins. From the end of this extinction event and for the next one hundred fifty million years, reptiles would rule.

Small lizards of the Permian evolved into twenty-foot-long predatory archosaurs with massive teeth and claws, resembling a smaller Tyrannosaurus. It was also during this period that vertebrates took to the skies—the pterosaur, a flying archosaur, was the first to grow wings and hunt from above.

In order to survive this ravenous era of predatory dinosaurs, pterosaurs, and other massive reptilian beasts, cynodonts got smaller and smaller until they were no more than four inches long. Equipped with

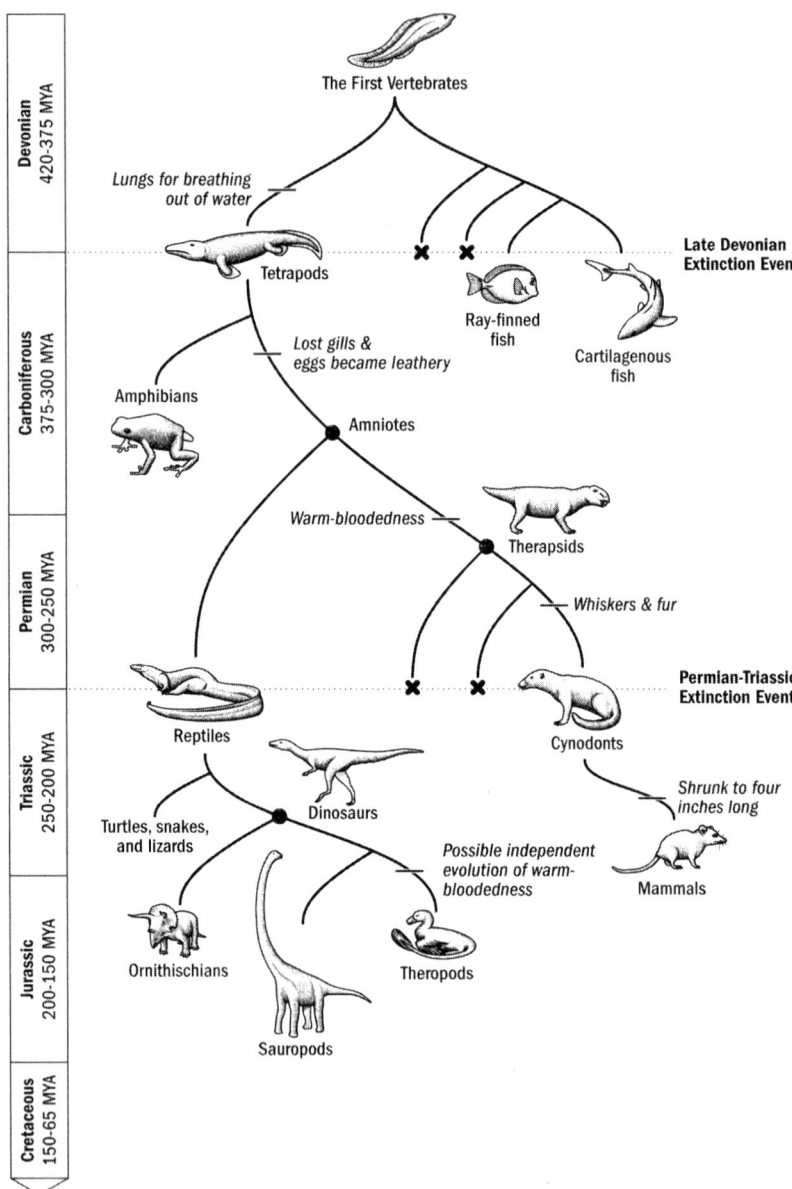

Figure 10.2: The evolutionary tree from the first vertebrates to the first mammals.
MYA = million years ago.

warm-bloodedness and miniaturization, they survived by hiding in burrows during the day and emerging during the cold night when archosaurs were relatively blind and immobile. They made their homes in dug-out burrowed mazes or in the thick bark of trees. They hunted by quietly wandering the twilight forest floors and tree branches in search of insects. They became the first mammals.

At some point in this hundred-million-year reign of dinosaurs, as these small mammals survived tucked away in nooks and crannies of the world, they added one more survival trick to their repertoire. They evolved a new cognitive ability, the biggest neural innovation since the Cambrian fish.

Surviving by Simulating

This early four-inch-long mammal, likely resembling a mouse or squirrel of today, was not stronger than dinosaurs or birds and surely unable to fight its way out of a predatory assault. It was probably also slower, or at least no faster, than an archosaur or a pterosaur swooping down from the sky. But the burrowing and arboreal lifestyle did indeed give early mammals a singular advantage: they got to make the *first move*. From an underground burrow or from behind a tree branch, they got to look around, spot a faraway bird and a tasty insect, and decide whether to make a run for it. This gift of the first move was left unexploited for hundreds of millions of years. But eventually a neural innovation emerged to exploit it: a region of the cortex transformed, through a currently unknown series of events, into a new region called the *neocortex* (*neo* for "new").

The neocortex gave this small mouse a superpower—the ability to simulate actions before they occurred. It could look out at a web of branches leading from its hole to a tasty insect. It could see the faraway eyes of a nearby predatory bird. The mouse could simulate going down different paths, simulate the bird chasing it and the insects hopping away, then pick the best path—the one that, in its simulation, it found itself both alive and well fed. If the reinforcement-learning early vertebrates got the power of learning *by doing*, then early mammals got the even more impressive power of learning *before doing*—of learning by imagining.

Many creatures had previously found themselves in positions of having the first move—crabs hide under sand and small fish weave between the leaves of coral plants. So then why was it only with mammals that simulating emerged?

It has been speculated that there were two requirements for simulating to evolve. First, you need far-ranging vision—you need to be able to see a lot of your surroundings in order for simulating paths to be fruitful. On land, even at night, you can see up to one hundred times farther than you can underwater. Thus, fish opted not to simulate and plan their movements but instead to respond quickly whenever something came at them (hence their large midbrain and hindbrain, and comparatively smaller cortex).

The second speculated requirement is warm-bloodedness. For reasons we will see in the next few chapters, simulating actions is astronomically more computationally expensive and time-consuming than the reinforcement-learning mechanisms in the cortex-basal-ganglia system. The electrical signaling of neurons is highly sensitive to temperature—at lower temperatures, neurons fire much more slowly than at warmer temperatures. This meant that a side effect of warm-bloodedness was that mammal brains could operate much faster than fish or reptile brains. This made it possible to perform substantially more complex computations. This is why reptiles, despite their long-range vision on land, were never endowed with the gift of simulating. The only nonmammals that have shown evidence of the ability to simulate actions and plan are birds. And birds are, conspicuously, the only nonmammal species alive today that independently evolved warm-bloodedness.

Inside the Brain of the First Mammals

Throughout this several-hundred-million-year-long story, from the emergence of fish onto land to the rise of dinosaurs, there was an expansive diversification of animal shapes, sizes, and organs. And yet, there was one thing that was surprisingly unchanged: brains.

From the early vertebrates to the first tetrapods to reptiles and therapsids, brains were largely stuck in a neural dark age. Evolution settled for, or at least was resigned to, the reinforcement-learning brain of the early vertebrates, and it shifted its focus toward tweaking other biological structures—creating jaws, armor, lungs, more ergonomic bodies, warm-bloodedness, scales, fur, and other such morphological modifications. This is why the brain of a modern fish and that of a modern reptile, despite hundreds of millions of years of evolutionary separation, are remarkably similar.*

It was only in early mammals that a spark of innovation emerged from the eternity of neural stagnation. The fish cortex split into four separate structures in early mammals, three of which were effectively the same as the subregions that had come before, only one of which, the neocortex, could truly be considered *new*. The ventral cortex of early vertebrates became the associative amygdala in mammals, containing similar circuitry and serving largely the same purpose: learning to recognize patterns across various modalities, especially those that were predictive of valence outcomes (e.g., predicting that sound A leads to good things and sound B leads to bad things). The smell-pattern detectors in the lateral cortex of early vertebrates became the olfactory cortex in mammals, working the same way—detecting smell patterns through auto-associative networks. The medial cortex of early vertebrates, where spatial maps were learned, became the hippocampus of mammals, performing a similar function using similar circuitry. But a fourth region of the cortex underwent a more meaningful change—it transformed into the *neo*cortex, which contained completely different circuitry.

Other than the emergence of the neocortex, the brain of early mammals was largely the same as that of early vertebrates. The basal ganglia integrated input about the world from the olfactory cortex, hippocampus, amygdala, and now also the neocortex to learn to take actions that maximized dopamine release. The hypothalamus still triggered direct valence responses and modulated other structures through neuromodulators such as dopamine. Midbrain and hindbrain structures still implemented reflexive movement patterns, albeit now specialized for walking as opposed to swimming.

---

* Although, to be fair, there are differences between fish and reptile brains. Some argue amniotes evolved a dorsal cortex, a possible precursor to the neocortex (although newer evidence suggests the dorsal cortex was not present in early amniotes).

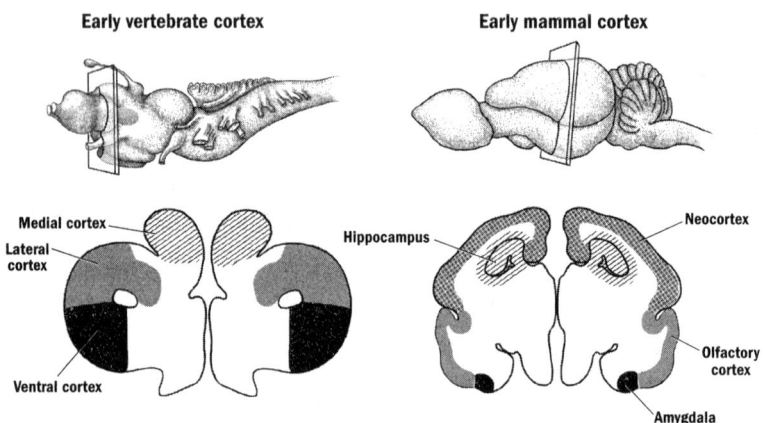

Figure 10.3: How the cortex changed in the transition from early vertebrates to early mammals

The neocortex of this early mammal was small and took up only a small fraction of the brain. Most volume was given to the olfactory cortex (early mammals, like many modern mammals, had an incredible sense of smell). But despite the small size of the neocortex in early mammals, it was still the kernel from which human intelligence would arise. In the human brain, the neocortex takes up 70 percent of brain volume. In the breakthroughs that followed, this originally small structure would progressively expand from a clever trick to the epicenter of intelligence.

11

Generative Models and the Neocortical Mystery

WHEN YOU LOOK at a human brain, almost everything you see is neocortex. The neocortex is a sheet about two to four millimeters thick. As the neocortex got bigger, the surface area of this sheet expanded. To fit in the skull, it became folded, the way you would bunch up a towel to fit it in a suitcase. If you unfolded a human neocortical sheet, it would be almost three square feet in surface area—about the size of a small desk.

Early experimentation led to the conclusion that the neocortex didn't serve any one function and instead subserved a multitude of different functions. For example, the back of the neocortex processes visual input and hence is called the visual cortex.* If you removed your visual cortex, you would become blind. If you record the activity of neurons in the visual cortex, they respond to specific visual features at specific locations, such as certain colors or line orientations. If you stimulate neurons within the visual cortex, people will report seeing flashes of lights.

In a nearby region called the auditory cortex, the same thing occurs with auditory perception. Damage to one's auditory cortex impairs one's ability to perceive and understand sounds. If you record the activity of neurons in the auditory cortex, you'll find they are responsive to specific

---

* Note that when referring to regions of neocortex in the brains of mammals, it is common to drop the *neo*—for instance, *visual cortex* rather than visual neocortex.

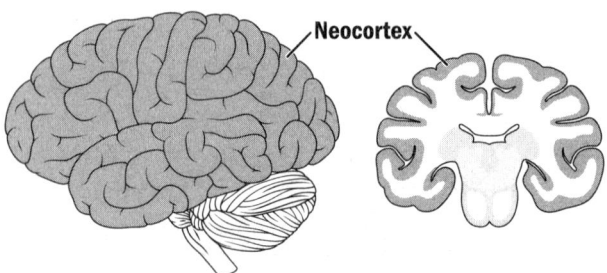

Figure 11.1: The human neocortex

frequencies of sound. If you stimulate certain neurons within the auditory cortex, people will report hearing noises.

There are other neocortical regions for touch, pain, and taste. And there are other areas of the neocortex that seem to serve even more disparate functions—there are areas for movement, language, and music.

At first glance, this makes no sense. How can one structure do so many different things?

Mountcastle's Crazy Idea

In the mid-twentieth century, the neuroscientist Vernon Mountcastle was pioneering what was, at the time, a new research paradigm: recording the activity of individual neurons in the neocortex of awake-behaving animals. This new approach offered a novel view of the inner workings of brains as animals went about their life. He used electrodes to record the neurons in the somatosensory cortices (the neocortical area that processes touch input) of monkeys to see what types of touch stimuli elicited what responses.

One of the first observations Mountcastle made was that neurons within a vertical column (about five hundred microns in diameter) of the neocortical sheet seemed to all respond similarly to sensory stimuli, while neurons horizontally farther away did not. For example, an individual column within the visual cortex might contain neurons that all similarly responded to bars of light at specific orientations at a specific location in the visual field. However, neurons within nearby columns responded only

GENERATIVE MODELS AND THE NEOCORTICAL MYSTERY 169

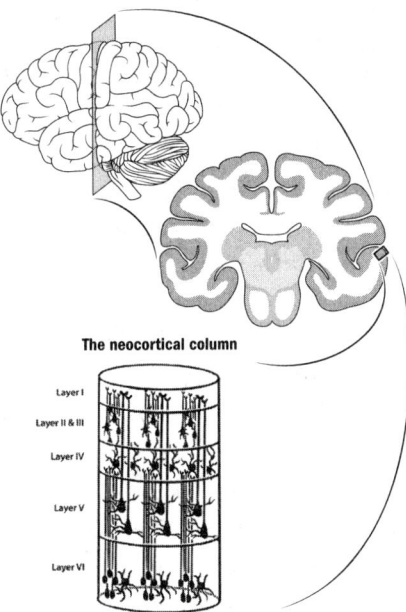

Figure 11.2: The neocortical column

to bars of light at different orientations or locations. This same finding has been confirmed within multiple modalities. In rats, there are columns of neocortex that respond only to the touch of a specific single whisker, with each nearby column responding to a completely different whisker. In the auditory neocortex, there are individual columns that are selective for specific frequencies of sound.

The second observation that Mountcastle made was that there were many connections vertically within a column and comparatively fewer connections between columns.

The third and final observation Mountcastle made was that under a microscope, the neocortex looked largely identical everywhere. The auditory neocortex, somatosensory neocortex, and visual neocortex all contain the same types of neurons organized in the same way. And this is true across species of mammals—the neocortex of a rat, a monkey, and a human all look relatively the same under a microscope.

These three facts—vertically aligned activity, vertically aligned connectivity, and observed similarity between all areas of neocortex—led

Mountcastle to a remarkable conclusion: the neocortex was made up of a repeating and duplicated microcircuit, what he called the *neocortical column*. The cortical sheet was just a bunch of neocortical columns packed densely together.

This provided a surprising answer to the question of how one structure can do so many different things. According to Mountcastle, the neocortex does *not* do different things; each neocortical column does *exactly the same* thing. The only difference between regions of neocortex is the input they receive and where they send their output; the actual computations of the neocortex itself are identical. The only difference between, for example, the visual cortex and the auditory cortex is that the visual cortex gets input from the retina, and the auditory cortex gets input from the ear.

In the year 2000, decades after Mountcastle first published his theory, three neuroscientists at MIT performed a brilliant test of Mountcastle's hypothesis. If the neocortex is the same everywhere, if there is nothing uniquely visual about the visual cortex or auditory about the auditory cortex, then you would expect these areas to be interchangeable. Experimenting on young ferrets, the scientists cut off input from the ears and rewired input from the retina to the auditory cortex instead of the visual cortex. If Mountcastle was wrong, the ferrets would end up blind or visually impaired—input from the eye in the auditory cortex would not be processed correctly. If the neocortex was indeed the same everywhere, then the auditory cortex receiving visual input should work the same way as the visual cortex.

Remarkably, the ferrets could see just fine. And when researchers recorded the area of the neocortex that was typically auditory but was now receiving input from the eyes, they found the area responded to visual stimuli just as the visual cortex would. The auditory and visual cortices are interchangeable.

This was further reinforced by studies of congenitally blind patients whose retinas had never sent any signals to their brains. In these patients, the visual cortex never received input from the eyes. However, if you record the activity of neurons in the visual cortex of congenitally blind humans, you find that the visual cortex has not been rendered a functionally useless region. Instead, it becomes responsive to a multitude of *other* sensory input, such as sounds and touch. This puts meat on the bone of the idea that people who are blind do, in fact, have superior hearing—the visual cortex becomes repurposed to aid in audition. Again, areas of neocortex seem interchangeable.

Consider stroke patients. When patients have damage to a specific area of neocortex, they immediately lose the function in that area. If the motor cortex is damaged, patients can become paralyzed. If the visual cortex is damaged, patients become partially blind. But over time, function can return. This is usually not the consequence of the damaged area of neocortex recovering; typically, that area of neocortex remains dead forever. Instead, nearby areas of neocortex become repurposed to fulfill the functions of the now-damaged area of neocortex. This too suggests that areas of neocortex are interchangeable.

To those in the AI community, Mountcastle's hypothesis is a scientific gift like no other. The human neocortex is made up of over ten billion neurons and trillions of connections; it is a hopeless endeavor to try and decode the algorithms and computations performed by such an astronomically massive hairball of neurons. So hopeless that many neuroscientists believe that attempting to decode how the neocortex works is a fruitless endeavor, doomed to fail. But Mountcastle's theory offers a more hopeful research agenda—instead of trying to understand the entire human neocortex, perhaps we only have to understand the function of the microcircuit that is repeated a million or so times. Instead of understanding the trillions of connections in the entire neocortex, perhaps we only have to understand the million or so connections within the neocortical column. Further, if Mountcastle's theory is correct, it suggests that the neocortical column implements some algorithm that is so general and universal that it can be applied to extremely diverse functions such as movement, language, and perception across every sensory modality.

The basics of this microcircuit can be seen under a microscope. The neocortex contains six layers of neurons (unlike the three-layered cortex seen in earlier vertebrates). These six layers of neurons are connected in a complicated but beautifully consistent way. There is a specific type of

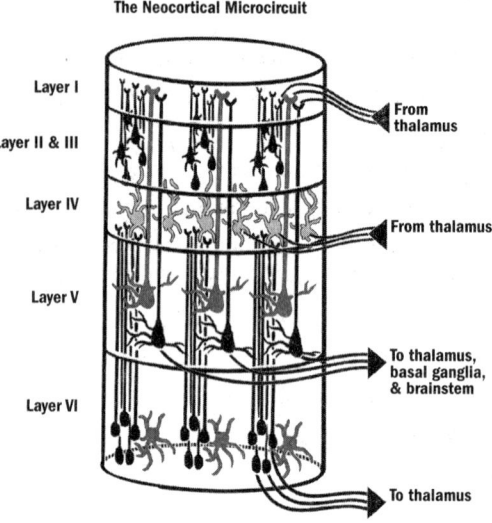

Figure 11.3: The microcircuitry of the neocortical column

neuron in layer five that always projects to the basal ganglia, the thalamus, and the motor areas. In layer four, there are neurons that always get input directly from the thalamus. In layer six, there are neurons that always project to the thalamus. It is not just a soup of randomly connected neurons; the microcircuit is prewired in a specific way to perform some specific computation.

The question is, of course: What is the computation?

Peculiar Properties of Perception

In the nineteenth century, the scientific study of human perception began in full force. Scientists around the world started probing the mind. How does vision work? How does audition work?

The inquiry into perception began with the use of illusions; by manipulating people's visual perceptions, scientists uncovered three peculiar properties of perception. And because much of perception, in humans at least, occurs in the neocortex, these properties of perception teach us about how the neocortex works.

Property #1: Filling In

The first thing that became clear to these nineteenth-century scientists was that the human mind automatically and unconsciously fills in missing things. Consider the images in figure 11.4. You immediately perceive the word *editor*. But this is not what your eye is actually seeing—most of the lines of the letters are missing. In the other images too, your mind perceives something that is not there: a triangle, a sphere, and a bar with something wrapped around it.

This filling in is not a property unique to vision; it is seen across most of our sensory modalities. This is how you can still understand what someone is saying through a garbled phone connection and how you can identify an object through touch even with your eyes closed.

Figure 11.4: Filling-in property of perception

Property #2: One at a Time

If your mind fills in what it thinks is there based on sensory evidence, what happens if there are multiple ways to fill in what you see? All three of the images in figure 11.5 are examples of visual illusions devised in the 1800s to probe this question. Each of these images can be interpreted in two different ways. On the left side of figure 11.5, you can see it as a staircase, but you can also see it as a protrusion from *under* a staircase.* In the middle of figure 11.5, the cube could be one where the bottom right square is the

---

* If you don't see this, look at the staircase and, while maintaining your gaze, rotate the page 180 degrees.

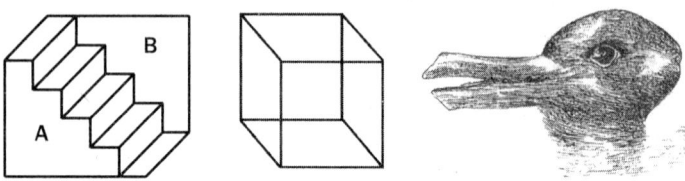

Figure 11.5: One-at-a-time property of perception

front or where the top left square is the front. On the right of figure 11.5, the picture could be a rabbit or a duck.

What is interesting about all these ambiguous pictures is that your brain can see only one interpretation at a time. You cannot see a duck and a rabbit simultaneously, even though the sensory evidence is equally suggestive of both. The mechanisms of perception in the brain, for some reason, require it to pick only one.

This also applies to audition. Consider the "cocktail-party effect." If you are at a noisy cocktail party, you can tune in to the conversation of the person you are speaking to or the conversation of a nearby group. But you cannot listen to both conversations at the same time. No matter which conversation you tune in to, the auditory input into your ear is identical; the only difference is what your brain infers from that input. You can perceive only a single conversation at a time.

Property #3: Can't Unsee

What happens when sensory evidence is *vague*—when it isn't clear that it can be interpreted as anything meaningful at all? Consider the image in 11.6. If you haven't seen these before, they will look like nothing—just blobs. If I give you a reasonable interpretation of these blobs, all of a sudden, your perception of them will change.

Figure 11.6 can be interpreted as a frog (see the next page if you don't see this). Once your mind perceives this interpretation, you will never be able to unsee it. This is what might be called the can't-unsee property of perception. Your mind likes to have an interpretation that explains sensory input. Once I give you a good explanation, your mind sticks to it. You now perceive a frog.

Figure 11.6: The can't-unsee property of perception

In the nineteenth century, a German physicist and physician named Hermann von Helmholtz proposed a novel theory to explain these properties of perception. He suggested that a person doesn't perceive what is experienced; instead, he or she perceives what the brain *thinks* is there—a process Helmholtz called *inference*. Put another way: you don't perceive what you actually see, you perceive a simulated reality that you have *inferred* from what you see.

This idea explains all three of these peculiar properties of perception. Your brain fills in missing parts of objects because it is trying to decipher the truth that your vision is *suggesting* ("Is there actually a sphere there?"). You can see only one thing at a time because your brain must pick a single reality to simulate—in reality, the animal can't be both a rabbit and a duck. And once you see that an image is best explained as a frog, your brain maintains this reality when observing it.

While many psychologists came to agree, in principle, with Helmholtz's theory, it would take another century before anyone proposed how Helmholtz's perception by inference might actually work.

Generative Models: Recognizing by Simulating

In the 1990s, Geoffrey Hinton and some of his students (including the same Peter Dayan that had helped discover that dopamine responses are

temporal difference learning signals) set their sights on building an AI system that learned in the way that Helmholtz suggested.

We reviewed in chapter 7 how most modern neural networks are trained with supervision: a picture is given to a network (e.g., a picture of a dog) along with the correct answer (e.g., "This is a dog"), and the connections in the network are nudged in the right direction to get it to give the right answer. It is unlikely the brain recognizes objects and patterns using supervision in this way. Brains must somehow recognize aspects of the world *without* being told the right answer; they must engage in *unsupervised* learning.

One class of unsupervised-learning methods are auto-associative networks, like those we speculated emerged in the cortex of early vertebrates. Based on correlations in input patterns, these networks cluster common patterns of input into ensembles of neurons, offering a way in which overlapping patterns can be recognized as distinct, and noisy and obstructed patterns can be completed.

But Helmholtz suggested that human perception was doing something more than this. He suggested that instead of simply clustering incoming input patterns based on their correlations, human perception might optimize for the accuracy with which the inner *simulated reality* predicts the current external sensory input.

In 1995, Hinton and Dayan came up with a proof of concept for

Helmholtz's idea of perception by inference; they named it the Helmholtz machine. The Helmholtz machine was, in principle, similar to other neural networks; it received inputs that flowed from one end to the other. But unlike other neural networks, it also had *backward* connections that flowed the *opposite* way—from the end to the beginning.

Hinton tested this network with images of handwritten numbers between 0 and 9. A picture of a handwritten number can be given at the bottom of the network (one neuron for each pixel) and will flow upward and activate a random set of neurons at the top. These activated neurons at the top can then flow back down and activate a set of neurons at the bottom to produce a picture of its own. Learning was designed to get the network to stabilize to a state where input that flows up the network is accurately re-created when it flows back down.

At first, there will be big discrepancies between the values in neurons from the image flowing in and the result flowing out. Hinton designed this network to learn with two separate modes: recognition mode and generative mode. When in *recognition mode*, information flows up the network (starting from an input picture of a 7 to some neurons at the top), and the backward weights are nudged to make the neurons activated at the top of the network better reproduce the input sensory data (make a good simulated 7). In contrast, when in *generative mode*, information flows *down* the network (starting from the goal to produce an imagined picture of a 7), and the forward weights are nudged so that the neurons activated at the bottom of the network are correctly recognized at the top ("I recognize what I just made as a 7").

Nowhere was this network told the right answer; it was never told what properties make up a 2 or even which pictures were 2s or 7s or any other number. The only data the network had to learn from was pictures of numbers. The question was, of course, would this work? Would this toggling back and forth between recognition and generation enable the network to both recognize handwritten numbers and generate its own unique pictures of handwritten numbers without ever having been told the right answer?

Amazingly, it did; it learned on its own. When these two processes toggle back and forth, the network magically stabilizes. When you give it a picture of the number 7, it will be able to, for the most part, create a similar

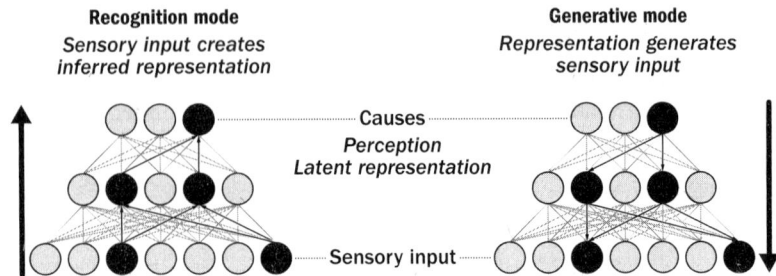

Figure 11.7: The Helmholtz Machine

image of a 7 on the way down. If instead you give it an image of an 8, it will be able to regenerate an input image of an 8.

This might not seem particularly impressive. You gave a network a picture of a number, and it spit out a picture of that same number—what's the big deal? There are three attributes of this network that are groundbreaking. First, the top of this network now reliably "recognizes" imperfectly handwritten letters without any supervision. Second, it generalizes impressively well; it can tell that two differently handwritten pictures of 7s are both a 7—they will activate a similar set of neurons at the top of the

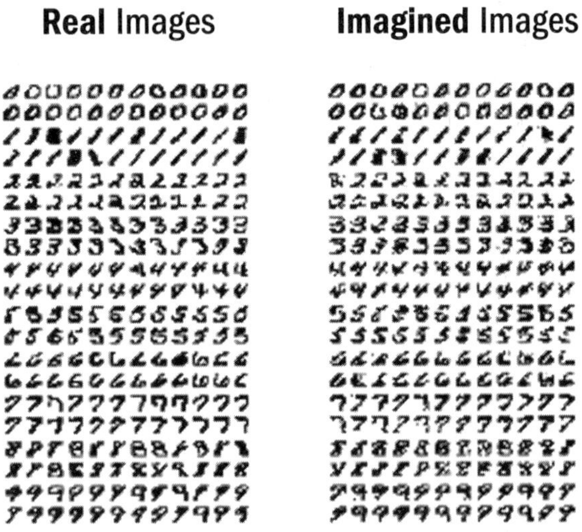

Real Images **Imagined** Images

Figure 11.8

network. And third, and most important, this network can now *generate novel pictures* of handwritten numbers. By manipulating neurons at the top of this network, you can create lots of handwritten 7s or handwritten 4s or any number it has learned. This network has learned to *recognize* by *generating* its own data.

The Helmholtz machine was an early proof of concept of a much broader class of models called generative models. Most modern generative models are more complicated than the Helmholtz machine, but they share the essential property that they learn to recognize things in the world by *generating* their own data and comparing the generated data to the actual data.

If you aren't impressed with the generation of small pixelated handwritten numbers, consider how far these generative models have come since 1995. As this book is going to print, there is an active website called thispersondoesnotexist.com. Every time you refresh the page, you will see a picture of a different person. The reality is more shocking: every time you reload the page, a generative model creates a completely new, never before seen, made-up face. The faces you see *do not exist*.

What is so amazing about these generative models is that they learn to capture the essential features of the input they are given without any

Figure 11.9: StyleGAN2 from thispersondoesnotexist.com

supervision. The ability to generate realistic novel faces requires the model to understand the essence of what constitutes a face and the many ways it can vary. Just as activating various neurons at the top of the Helmholtz machine can generate images of different handwritten numbers, if you activate various neurons at the top of this generative model of faces, you can control what types of faces it generates. If you change the value of one set of neurons, the network will spit out the same face but rotated. If you change the value of a different set of neurons, it will add a beard or change the age or alter the color of the hair (see figure 11.10).

While most AI advancements that occurred in the early 2000s involved applications of supervised-learning models, many of the recent advancements have been applications of generative models. Deepfakes, AI-generated art, and language models like GPT-3 are all examples of generative models at work.

Helmholtz suggested that much of human perception is a process of inference—a process of using a generative model to match an inner simulation of the world to the sensory evidence presented. The success of modern generative models gives weight to his idea; these models reveal that

Figure 11.10: Changing images by changing latent representations in generative models

something like this can work, at least in principle. It turns out that there is, in fact, an abundance of evidence that the neocortical microcircuit is implementing such a generative model.

Evidence for this is seen in filling-in, one-at-a-time, can't-unsee visual illusions; evidence is seen in the wiring of the neocortex itself, which has been shown to have many properties consistent with a generative model; and evidence is seen in the surprising symmetry—the ironclad inseparability—between perception and *imagination* that is found in both generative models and the neocortex. Indeed, the neocortex as a generative model explains more than just visual illusions—it also explains why humans succumb to hallucinations, why we dream and sleep, and even the inner workings of imagination itself.

Hallucinating, Dreaming, and Imagining: The Neocortex as a Generative Model

People whose eyes stop sending signals to their neocortex, whether due to optic-nerve damage or retinal damage, often get something called Charles Bonnet syndrome. You would think that when someone's eyes are disconnected from their brain, they would no longer see. But the opposite happens—for several months after going blind, people start seeing *a lot*. They begin to hallucinate. This phenomenon is consistent with a generative model: cutting off sensory input to the neocortex makes it unstable. It gets stuck in a drifting generative process in which visual scenes are simulated without being constrained to actual sensory input—thus you hallucinate.

Some neuroscientists refer to perception, even when it is functioning properly, as a "constrained hallucination." Without sensory input, this hallucination becomes *un*constrained. In our example of the Helmholtz machine, this is like randomly activating neurons at the top of the network and producing hallucinated images of numbers without ever grounding these hallucinations in real sensory input.

This idea of perception as a constrained hallucination is, of course, exactly what Helmholtz meant by inference and exactly what a generative model is doing. We match our inner hallucination of reality to the sensory data we are seeing. When the visual data suggests there is a triangle in a picture (even if there is not actually a triangle there), we hallucinate a triangle, hence the filling-in effect.

Generative models may also explain why we dream and why we need sleep. Most animals sleep, and it has numerous benefits, such as saving energy; but only mammals and birds show unequivocal evidence of *dreaming* as measured by the presence of REM sleep. And it is only mammals and birds who exhibit hallucinations and disordered perception if deprived of sleep. Indeed, birds seem to have independently evolved their own neocortex-like structure.

The neocortex (and presumably the bird equivalent) is always in an unstable balance between recognition and generation, and during our waking life, humans spend an unbalanced amount of time recognizing and comparatively less time generating. Perhaps dreams are a counterbalance to this, a way to stabilize the generative model through a process of forced generation. If we are deprived of sleep, this imbalance of too much recognition and not enough generation eventually becomes so severe that the generative model in the neocortex becomes unstable. Hence, mammals start hallucinating, recognition becomes distorted, and the difference between generation and recognition gets blurred. Fittingly, Hinton even called the learning algorithm to train his Helmholtz machine a "wake-sleep algorithm." Recognition step was when the model was "awake"; the generation step was when the model was "asleep."

Many features of imagination in mammals are consistent with what we would expect from a generative model. It is easy, even natural, for humans to imagine things that they are not currently experiencing. You can imagine the dinner you ate last night or imagine what you will be doing later today. What are you doing when you are imagining something? This is just your neocortex in generation mode. You are invoking a simulated reality in your neocortex.

The most obvious feature of imagination is that you cannot imagine things and recognize things simultaneously. You cannot read a book and imagine yourself having breakfast at the same time—the process of imagining is inherently at odds with the process of experiencing actual sensory data. In fact, you can tell when someone is imagining something by looking at that person's pupils—when people are imagining things, their pupils dilate as their brains stop processing actual visual data. People become pseudo-blind. As in a generative model, generation and recognition cannot be performed simultaneously.

Further, if you record neocortical neurons that become active during recognition (say, neurons that respond to faces or houses), those *exact same neurons* become active when you simply imagine the same thing. When you imagine moving certain body parts, the same area activates as if you were actually moving the body parts. When you imagine certain shapes, the same areas of visual cortex become activated as when you see those shapes. In fact, this is so consistent that neuroscientists can decode what people are imagining simply by recording their neocortical activity (and as evidence that dreaming and imagination are the same general process, scientists can also accurately decode people's dreams by recording their brains). People with neocortical damage that impairs certain sensor data (such as being unable to recognize objects on the left side of the visual field) become equally impaired at simply *imagining* features of that same sensory data (they struggle even to imagine things in the left visual field).

None of this is an obvious result. Imagination could have been performed by a system separate from recognition. But in the neocortex, this is not the case—they are performed in the exact same area. This is exactly what we would expect from a generative model: perception and imagination are not separate systems but two sides of the same coin.

Predicting Everything

One way to think about the generative model in the neocortex is that it renders a simulation of your environment so that it can predict things before they happen. The neocortex is continuously comparing the actual sensory data with the data predicted by its simulation. This is how you can immediately identify anything surprising that occurs in your surroundings.

As you walk down the street, you are not paying attention to the feelings of your feet. But with every movement you make, your neocortex is passively predicting what sensory outcome it expects. If you placed your left foot down and didn't feel the ground, you would immediately look to see if you were about to fall down a pothole. Your neocortex is running a simulation of you walking, and if the simulation is consistent with sensor data, you don't notice it, but if its predictions are wrong, you do.

Brains have been making predictions since early bilaterians, but over evolutionary time, these predictions became more sophisticated. Early bilaterians could learn that the activation of one neuron tended to precede the activation of another neuron and could thereby use the first neuron to *predict* the second. This was the simplest form of prediction. Early vertebrates could use patterns in the world to predict future rewards. This was a more sophisticated form of prediction. Early mammals, with the neocortex, learned to predict more than just the activation of reflexes or future rewards; they learned to predict *everything*.

The neocortex seems to be in a continuous state of predicting all its sensory data. If reflex circuits are reflex-prediction machines, and the critic in the basal ganglia is a reward-prediction machine, then the neocortex is a *world*-prediction machine—designed to reconstruct the entire three-dimensional world around an animal to predict exactly what will happen next as animals and things in their surrounding world move.

Somehow the neocortical microcircuit implements such a general system that it can render a simulation of many types of input. Give it visual input and it will learn to render a simulation of the visual aspects of the world; give it auditory input and it will learn to render a simulation of auditory aspects of the world. This is why the neocortex looks the same everywhere. Different subregions of neocortex simulate different aspects of the external world based on the input they receive. Put all these neocortical columns together, and they make a symphony of simulations that render a rich three-dimensional world filled with objects that can be seen, touched, and heard.

How the neocortex does this is still a mystery. At least one possibility is that it is prewired to make a set of clever assumptions. Modern AI models are often viewed as narrow—that is, they're able to work in a narrow set of situations they are specifically trained for. The human brain is considered general—it is able to work in a broad set of situations. The research agenda

The Evolution of Prediction

PREDICTION IN EARLY BILATERIANS	PREDICTION IN EARLY VERTEBRATES	PREDICTION IN EARLY MAMMALS
Predict reflex activation	Predict future rewards	Predict all sensory data
Reflex circuits	Cortex and basal ganglia	Neocortex

has therefore been to try and make AI more general. However, we might have it backward. One of the reasons why the neocortex is so good at what it does may be that, in some ways, it is far *less* general than our current artificial neural networks. The neocortex may make explicit narrow assumptions about the world, and it may be exactly these assumptions that enable it to be so general.

For example, the neocortex may be prewired to *assume* that incoming sensor data, whether visual, auditory, or somatosensory, represent three-dimensional objects that exist separately from ourselves and can move on their own. Therefore, it does not have to learn about space, time, and the difference between the self and others. Instead, it tries to explain all incoming sensory information it receives by *assuming* it must have been derived from a 3D world that unfolds over time.

This provides some intuition about what Helmholtz meant by inference—the generative model in the neocortex tries to *infer* the causes of its sensory input. *Causes* are just the inner simulated 3D world that the neocortex believes best matches the sensory input it is being given. This is also why generative models are said to try to *explain* their input—your neocortex attempts to render a state of the world that could produce the picture that you are seeing (e.g., if a frog was there, it would "explain" why those shadows look the way they do).

But why do this? What is the point of rendering an inner simulation of the external world? What value did the neocortex offer these ancient mammals?

There are many ongoing debates about what is missing in modern AI systems and what it will take to get AI systems to exhibit human-level intelligence. Some believe the key missing pieces are language and logic. But others, like Yann LeCun, head of AI at Meta, believe they are something else, something more primitive, something that evolved much earlier. In LeCun's words:

> We humans give way too much importance to language and symbols as the substrate of intelligence. Primates, dogs, cats, crows, parrots, octopi, and many other animals don't have human-like languages, yet exhibit intelligent behavior beyond that of our best AI systems. What they do have is an ability to learn powerful "world models" that allow them to predict the consequences of their actions and to search for and plan actions to achieve a goal. The ability to learn such world models is what's missing from AI systems today.

The simulation rendered in the neocortices of mammals (and perhaps in similar structures of birds or even octopuses) is exactly this missing "world model." The reason the neocortex is so powerful is not only that it can match its inner simulation to sensory evidence (Helmholtz's perception by inference) but, more important, that its simulation can be independently explored. If you have a rich enough inner model of the external world, you can explore that world in your mind and predict the consequences of actions you have never taken. Yes, your neocortex enables you to open your eyes and recognize the chair in front of you, but it also enables you to close your eyes and still see that chair in your mind's eye. You can rotate and modify the chair in your mind, change its colors, change its materials. It is when the simulation in your neocortex becomes *decoupled* from the real external world around you—when it imagines things that are not there—that its power becomes most evident.

This was the gift the neocortex gave to early mammals. It was imagination—the ability to render future possibilities and relive past events—that was the third breakthrough in the evolution of human intelligence. From it emerged many familiar features of intelligence, some of which we have re-created and surpassed in AI systems, others of which are still beyond our grasp. But all of them evolved in the minuscule brains of the first mammals.

In the coming chapters, we will learn how the neocortex enabled early mammals to perform feats like planning, episodic memory, and causal reasoning. We will learn how these tricks were repurposed to enable fine motor skills. We will learn about how the neocortex implements attention, working memory, and self-control. We will see that it is in the neocortex of early mammals where we will find many of the secrets to human-like intelligence, those that are missing from even our smartest AI systems.

12

Mice in the Imaginarium

THE EMERGENCE OF the neocortex was a watershed moment in the evolutionary history of human intelligence. The original function of the neocortex was surely not as broad as its modern applications—it wasn't for pondering the nature of existence, planning careers, or writing poetry. Instead, the first neocortex gifted early mammals something more foundational: the ability to imagine the world as it is not.

Most research on the neocortex has focused on its impressive ability to recognize objects: to see a single picture of a face and easily identify it at many scales, translations, and rotations. In the context of early generative models, the generative mode—the process of simulation—is often viewed as the *means* to achieving the benefit of recognition. In other words, recognition is what is useful; imagination is a byproduct. But the cortex that came before the *neo*cortex could also recognize objects quite well; even fish can recognize objects when rotated, rescaled, or perturbed.

The core evolutionary function of the neocortex might have been the opposite—recognition might have been the means that unlocked the adaptive benefit of *simulating*. This would suggest that the original evolutionary function of the neocortex was not recognizing the world—an ability the older vertebrate cortex already had—but instead imagining and simulating the world, an ability the older cortex was lacking.

There were three new abilities that neocortical simulating provided early mammals, all three of which were essential for surviving the one-hundred-and-fifty-million-year predatory onslaught of sharp-toothed dinosaurs.

New Ability #1: Vicarious Trial and Error

In the 1930s, the psychologist Edward Tolman, working at UC Berkeley, was putting rats in mazes to see how they learned. Normal psychology-type work at the time—this was the generation that followed Thorndike. The research paradigm of Thorndike's law of effect, in which animals repeated behavior that had pleasant consequences, was in full force.

Tolman noticed something odd. When rats reached forks in his mazes where the choice was not obvious—where it wasn't clear whether they should go left or right—rats would pause and look back and forth for a few seconds before choosing a direction. This made no sense in the standard Thorndikian view that all learning occurred through trial and error—why would the behavior of pausing and toggling one's head back and forth have been reinforced?

Tolman made a speculation: The rat was "playing out" each option before taking it. Tolman called this "*vicarious* trial and error."

Rats expressed this head toggling behavior only when decisions were hard. One way to make decisions hard is for the costs to be close to the benefits. Suppose you put a rat in a tunnel where it passes various doors, each of which leads to food. And suppose that when rats pass these doors, a specific sound is made that signals how long a rat will have to wait for food if it chooses to walk through the door. One sound signals that rats will have to wait only a few seconds; another sound signals that the rat will have to wait up to half a minute. Once rats learn how all this works, they do not perform their head toggling at the doors with a short delay (they just immediately run in to get the food, presumably thinking, *Obviously this is worth the wait*) or at doors with a long delay (they just immediately walk past the door: *No way this is worth the wait, I'll just check out the next door*). But they *do* perform the head toggling behavior at the doors with a medium delay (*Is this worth the wait or should I try the next option?*).

Another way to make decisions difficult is to change the rules. If suddenly food is no longer where a rat expects it in the maze, the next time the rat is placed in the maze it will substantially increase its head toggling behavior, seeming to consider alternative paths. Similarly, suppose a rat enters a maze with two types of food, and suppose recently the rat has eaten a lot of one of those types of food (and hence no longer wants that food), then this head toggling behavior emerges (*Do I want to turn left to get X instead of turning right and getting Y?*).

Of course, the fact that you can observe a rat pause and turn its head back and forth doesn't prove that it is, in fact, imagining going down different paths. And due to this lack of evidence, the idea of vicarious trial and error faded out of favor in the decades that followed Tolman's observation. It was only recently, in the 2000s, that technology reached the point where ensembles of neurons in the brain of a rat could be recorded in real time as rats navigated their environment. This was the first time that neuroscientists could literally watch what was happening in the brains of rats when they paused and toggled their heads back and forth.

It was David Redish and his student Adam Johnson, neuroscientists at the University of Minnesota, who first probed what was happening in the brain of rats during these choices. At the time, it was well known that when a rat navigates a maze, specific place cells in its hippocampus get activated. This was similar to the spatial map of a fish—specific hippocampal neurons encode specific locations. In a fish, these neurons become active only when the fish is physically present at the encoded location—but when Redish and Johnson recorded these neurons in rats, they found something different: when the rat stopped at the decision point and turned its head back and forth, its hippocampus ceased to encode the actual location of the rat and instead went back and forth rapidly playing out the sequence of place codes that made up both possible future paths from the choice point. Redish could literally see the rat imagining future paths.

How groundbreaking this was cannot be overstated—neuroscientists were peering directly into the brain of a rat, and directly observing the rat considering alternative futures. Tolman was right: the head toggling behavior he observed was indeed rats planning their future actions.

In contrast, the first vertebrates did not plan their actions ahead of time. We can see this by examining their cold-blooded descendants—modern fish and reptiles—who show no evidence of learning by vicarious trial and error.

Consider the detour task. Take a fish and put it in a tank with a transparent barrier in the middle of the tank. Put a small hole in one corner of the barrier so that the fish can get from one side to the other. Let the fish explore the tank, find the hole, and spend some time swimming back and forth. After several days, do something new: Put the fish on one side of the tank and put a treat on the opposite side of the transparent barrier. What happens?

The smart thing to do, if you wanted food, would be to immediately swim *away* from the treat to the corner of the barrier, go through the opening, and then turn back to get the food on the other side. But this is not what the fish does. The fish runs directly into the transparent barrier trying to get the food. After bopping against the wall enough times, it gives up and continues wandering its environment. Eventually, while wandering the environment, the fish happens to pass through the hole again, but even here it shows no understanding that it can now access the food; the fish does not turn toward the food. Instead, the fish just continues onward through the other side of the tank. Only when it happens to turn back and see the food again does it excitedly dart toward it. Indeed, it takes the same amount of time to find the food for a fish who has had many experiences navigating between each side of the barrier, as it does for a fish that has never experienced getting to the other side of the barrier at all.

Why is this? While the fish had swum through the hole to get to the other side of the tank before, it had never learned that the path through the hole provided dopamine. Trial-and-error learning had never trained the fish's basal ganglia that when it saw food across the transparent barrier, it should take the action of swimming through the hole to get the food.

This is a critical problem of learning only by doing: Although the fish had learned the path through the hole, it had never taken that path to get food before. So when it saw food, all it could do was generate an "approach" signal directly toward the food. Rats, however, are much savvier. In such detour tasks, they dramatically outperform fish. Both rats and fish will initially run up to the transparent barrier to try and get food. But rats are *much* better at figuring out how to navigate around a barrier. And a rat who has explored a map well—who knows how to get to the other side of a transparent barrier (even if doing so was never rewarded)—will get to the other side much faster than will a rat who has never navigated around the barrier before. This reveals one of the benefits of *vicarious* trial and error: once a rat has a world model of their environment, they can rapidly mentally explore it until they find a way to get around obstacles to get what they want.

Another problem with the older strategy of learning by doing is that sometimes past rewards are not predictive of current rewards because an animal's internal state has changed. For example, put a rat in a maze where one side provides overly salty food and the other side provides normal food. Let the rat navigate this maze normally and try the overly salty food (which it will hate and then avoid) and the normal food (which the rat will enjoy). Now suppose you put the rat back in that situation but with a twist: You make it severely salt-deprived. What does the rat do?

It *immediately* runs toward the salt. This is remarkable because the rat is now running toward a part of the maze that was previously *negatively* reinforced. This is possible only because the rat "simulated" each path and realized—vicariously—that the overly salty food would now be hugely rewarding. In other words, the path toward salt was *vicariously* reinforced before the rat even acted. I am unaware of any studies showing that fish or reptiles can perform such a task.

New Ability #2: Counterfactual Learning

Humans spend a painful amount of time wallowing in regret. Questions you might hear from an average human conversation: "What would life have been like if I had said yes when Ramirez offered to throw our lives away and move to Chile to work on his farm?" "What if I had followed my dream and pursued baseball instead of this desk job?" "Why did I say that stupid thing at work today? What would have happened had I said something cleverer?"

Buddhists and psychologists alike realize that ruminating about what could have been is a source of great misery for humanity. We cannot change the past, so why torture ourselves with it? The evolutionary roots of this go back to early mammals. In the ancient world, and in much of the world that followed, such ruminating was useful because often the same situation *would* recur and a better choice could be made.

The type of reinforcement learning we saw in early vertebrates has a flaw: It can only reinforce the specific action *actually* taken. The problem with this strategy is that the paths that were actually taken are a small subset of all the possible paths that *could have* been taken. What are the chances that an animal's first attempt picked the best path?

If a fish swims into a shoal to hunt some invertebrates and comes away with one, and a nearby fellow fish took a different path and came away with four, the first fish won't learn from that mistake; it will merely reinforce the path taken with the mediocre reward. What fish are missing is the ability to *learn from counterfactuals*. A counterfactual is what the world would be now if you had made a different choice in the past.

David Redish, having discovered that rats can imagine alternative futures, wanted to see if rats could also imagine alternative past choices. Redish and one of his students Adam Steiner put rats in a circular maze they called "restaurant row." Rats ran around counterclockwise and kept passing the same four corridors. At the end of each corridor was a different "restaurant," containing a different flavor of food (chocolate, cherry, banana). As the rat passed each corridor, a random tone was sounded that signaled the delay before the food would be released if the rat waited at this corridor instead of continuing on to the next. Tone A meant if the rat waited at the current corridor, it would get the food within forty-five seconds; tone B

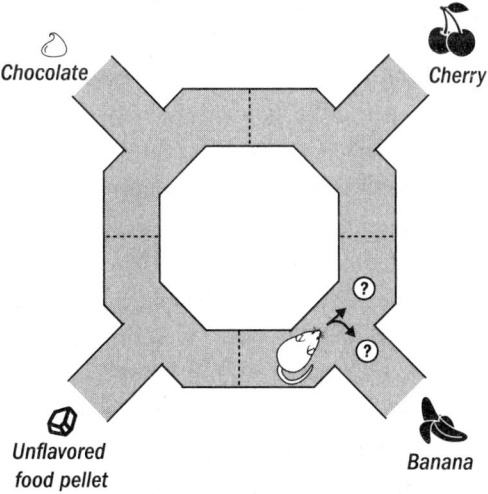

Figure 12.1: Redish restaurant-row test of regret in rats

meant it would get the food in five seconds. If a rat chose not to wait, then once it got to the next corridor, it could no longer go back; the food would not be released unless it went all the way around the circle again. This presented rats with a continuous set of *irreversible* choices. Rats were given an hour to keep trying to get as much of their favorite foods as they could.

Consider the choice presented to the rats at a given corridor: *Do I wait here for mediocre banana that the tone just signaled would be released in five seconds, or do I run to the next door, which contains my favorite food, cherry, and gamble that it will also be released quickly?* When rats chose to forgo quick access to a banana treat to try the cherry door and the next tone signaled a *long* wait of forty-five seconds, rats showed all the signs of regretting their choice. They paused and *looked back* toward the corridor that they had passed and could no longer go back to. And the neurons in the taste area of the neocortex reactivated the representation of banana, showing that rats were literally imagining a world where they had made a different choice and got to eat the banana.

The rats that turned back and reactivated the representation of the forgone choice also ended up changing their future choices. They waited longer the next time and ate other food more hastily to try and get back around the maze to the cherry to try again.

Primates also reason about counterfactuals. Consider an experiment where monkeys were taught to play rock, paper, scissors. Whenever monkeys lost, their next move was always biased toward the move that *would* have won in the prior game. If a monkey lost because it chose paper while its opponent chose scissors, next time it is most likely to pick rock (which would have won against scissors). This cannot be explained with the temporal difference learning of early vertebrates. According to the evolutionary framework presented here, if a fish could play rock, paper, scissors, it would not show this effect. If a fish lost after playing paper against scissors, it would become less likely to play paper next time (that action was punished because it led to losing), but the fish would be *equally likely* to play rock or scissors next time (neither action was punished or reinforced). In contrast, because a monkey can replay what *could have been* after losing by playing paper, it will realize that if it *had* played rock it would have won. Thus, monkeys are changing their actions based on counterfactual learning.

The perception of causation may be intricately tied to the notion of counterfactual learning. What we mean when we say "X caused Y" is that in the counterfactual case where X did not occur, then Y did not occur either. This is how we tease out the difference between correlation and causation. If you saw lightning strike a dry forest, and a fire immediately started, you would say that it was the lightning that *caused* the fire, not the fire that caused the lightning. You say this because when you imagine the counterfactual case in which lightning did not strike, the fire did not appear. Without counterfactuals, there is no way to distinguish between causation and correlation. You can never know what caused what; you can know only that "X always happens before Y" or "Whenever X happens, Y happens" or "Y has never happened without X happening," and so on.

Counterfactual learning represented a major advancement in how ancestral brains solved the credit assignment problem. As a reminder, the credit assignment problem is this: When some important event occurs that you want to be able to predict ahead of time, how do you choose what previous actions or events to *give credit* for having been predictive of the event? In simple terms: A bunch of things happen (a bird chirps, a gust of wind blows, a leaf moves, and lightning strikes), and then a fire appears—what do you give credit for being a good predictor of fire? In early bilaterians, simple tricks like blocking, latent inhibition, and overshadowing drove the logic by which simple predictions and associations were made. In early vertebrates, the evolution of temporal difference learning enabled the basal ganglia to assign credit using changes in future predicted reward; when the critic *thinks* the situation just got better or worse is when cues or actions are given credit. But in early mammals, with their ability to simulate alternative pasts, credit could also be assigned with the counterfactual. By asking "Would I have lost the game had I not made this move?," mammals can determine whether a move truly deserves credit for winning the game.

Causation itself may live more in psychology than in physics. There is no experiment that can definitively prove the presence of causality; it is entirely immeasurable. Controlled experiments we run may suggest

The Evolution of Credit Assignment

CREDIT ASSIGNMENT IN EARLY BILATERIANS	CREDIT ASSIGNMENT IN EARLY VERTEBRATES	CREDIT ASSIGNMENT IN EARLY MAMMALS
Credit assigned based on basic rules of blocking, latent inhibition, and overshadowing	Credit assigned based on when the critic predicts changes in future rewards	Credit assigned based on the counterfactual—which previous events or actions, if they had not occurred, would have prevented the subsequent event from occurring (i.e., what truly caused the event?)

causation, but they always fall short of proof because you can, in fact, never run a perfectly controlled experiment. Causation, even if real, is always empirically out of reach. In fact, modern experiments in the field of quantum mechanics suggest that causation may not even exist, at least not everywhere. The laws of physics may contain rules of how features of reality progress from one time step to the next without any real causal relationships between things at all. Ultimately, whether causality is real or not, the evolution of our intuitive perception of causality does not derive from its reality but from its usefulness. Causation is constructed by our brains to enable us to learn vicariously from alternative past choices.

While such counterfactual learning and causal reasoning is indeed seen across many mammals, even rats, there is no convincing evidence that fish or reptiles are capable of learning from counterfactuals or reasoning about causation (however, there is evidence of this type of reasoning in birds). This suggests that, at least in our lineage, this ability first emerged in our mammalian ancestors.

New Ability #3: Episodic Memory

In September 1953, a twenty-seven year old man named Henry Molaison underwent an experimental procedure that removed his entire hippocampus—the source of his debilitating seizures. The surgery was, in one sense, a success: severity of his seizures were markedly reduced and he retained his personality and intellect. But his doctors quickly discovered that the surgery had deprived their patient of something precious: upon waking up, Molaison was entirely unable to produce new memories. He could hold a conversation for a minute or two, but shortly thereafter would forget everything that just happened. Even forty years later, he could fill out crosswords with facts from before 1953, but not with facts that occurred after. Molaison was stuck in 1953.

We don't review the past only for the purpose of considering alternative past choices; we also review the past to remember previous life events. People can easily recall what they did five minutes ago or what they majored in at college or that funny joke made during a wedding speech. This form of memory, in which we recall specific past episodes of our lives, is called "episodic memory." This is distinct from, say, procedural memory, where we remember how to do various movements, such as speaking, typing, or throwing a baseball.

But here is the weird thing—we don't truly remember episodic events. The process of episodic remembering is one of simulating an approximate re-creation of the past. When imagining future events, you are simulating a *future* reality; when remembering past events, you are simulating a *past* reality. Both are simulations.

We know this based on two pieces of evidence. First, remembering past events and imagining future events use similar if not the same neural circuitry. The same networks get activated when you're imagining the future as when you're remembering the past. Remembering specific things (faces, houses) reactivates the same neurons in the sensory neocortex as when you actually perceive those things (just as we saw with imagining things).

The second piece of evidence that episodic memory is just a simulation comes from phenomena associated with such memories. For example, it turns out people's episodic memories are "filled in" during the remembering process (much like the filling in of visual shapes). This is why episodic memories feel so real but are much less accurate than we think. The clearest demonstrations of episodic memory's flaws are in eyewitness testimonies. Someone looking at a lineup of potential criminals is often 100 percent sure which member of the lineup committed the crime. However, contrary to our perceptions of how accurate our memories are, it has been shown that eyewitness testimony is terrible: 77 percent of the wrongfully convicted individuals exonerated by the Innocence Project were originally convicted because of mistaken eyewitness testimony. The distinction between a made-up imagined scene and an actual episodic memory is thin in the neocortex; studies show that repeatedly imagining a past event that did *not* occur falsely increases a person's confidence that the event *did* occur.

One can test the presence of episodic memory in animals by asking them *unexpected* questions about recent events. For example, a rat may learn that when presented with a certain maze, it will find food down one path if and only if it encountered food in the prior few minutes; otherwise, it should go down a different path to get food. The fact that this question is asked randomly (by randomly presenting the rat with this maze) makes it hard to see how the TD-learning mechanisms of the first vertebrates would learn this contingency; the maze is equally paired with each direction. And yet rats easily learn, when placed in this maze, to recall whether they recently encountered food and then choose the right corresponding path to get more food. The only nonmammals that have been demonstrated to have such episodic memory are birds and cephalopods, the two groups of species that seem to have independently evolved their own brain structures for rendering simulations.

After Molaison's surgery, he became the most studied neuroscience patient in history: why was the hippocampus required for creating new episodic memories, but not for retrieving old ones? This is an example of evolution repurposing old structures for new purposes. In mammal brains, episodic memory emerges from a partnership between the older hippocampus and the newer neocortex. The hippocampus can quickly learn patterns, but cannot render a simulation of the world; the neocortex can simulate detailed aspects of the world, but cannot learn new patterns quickly. Episodic memories must be stored rapidly, and thus the hippocampus, designed for the rapid pattern recognition of places, was repurposed to also aid in the rapid encoding of episodic memories. Distributed neural activations of sensory neocortex (i.e., simulations) can be "retrieved" by reactivating the corresponding pattern in the hippocampus. Just as rats reactivated place cells in the hippocampus to simulate going down different

MODEL-FREE REINFORCEMENT LEARNING	MODEL-BASED REINFORCEMENT LEARNING
Learns direct associations between a current state and the best actions	Learns a model of how actions affect the world and uses this to simulate different actions before choosing
Faster decisions but less flexible	Slower decisions but more flexible
Emerged in early vertebrates	Emerged in early mammals
Neocortex is not required	Neocortex is required
Example: Habitually going to work by just responding to each cue (traffic light, landmark) as it comes up	Example: Considering different ways to get to work and picking the one that got you there the fastest in your mind.

paths, rats can reactivate these "memory codes" in the hippocampus to rerender simulations of recent events.

This dynamic provided a new solution to the catastrophic forgetting problem, whereby neural networks forget old patterns when they learn new ones. By retrieving and replaying recent memories alongside old memories, the hippocampus aids the neocortex in incorporating new memories without disrupting old ones. In AI, this process is called "generative replay" or "experience replay" and has been shown to be an effective solution to catastrophic forgetting. This is why the hippocampus is necessary for creating new memories, but not for retrieving old ones; the neocortex can retrieve memories on its own after a sufficient amount of replay.

All this simulating of futures and pasts has a larger analog in machine learning. The type of reinforcement learning—temporal difference learning—that we saw with breakthrough #2 is a form of *model-free* reinforcement learning. In this type of reinforcement learning, AI systems learn by making *direct* associations between stimuli, actions, and rewards. These systems are called "model-free" because they do not require a model to play out possible future actions before making a decision. While this makes TD learning systems efficient, it also makes them less flexible.

There is another category of reinforcement learning called *model-based* reinforcement learning. These systems must learn something more complicated: a model of how their actions affect the world. Once such a model is constructed, these systems then play out sequences of possible actions before making choices. These systems are more flexible but are burdened with the difficult task of building and exploring an inner world model when making decisions.

Most of the reinforcement learning models employed in modern technology are model-free. The famous algorithms that mastered various Atari games and many self-driving-car algorithms are model-free. These systems don't pause and consider their choices; they immediately act in response to the sensory data they are given.

Model-based reinforcement learning has proven to be more difficult to implement for two reasons.

First, building a model of the world is hard—the world is complex and the information we get about it is noisy and incomplete. This is LeCun's missing world model that the neocortex somehow renders. Without a world model, it is impossible to simulate actions and predict their consequences.

The second reason model-based reinforcement learning is hard is that *choosing what to simulate* is hard. In the same paper that Marvin Minsky identified the temporal credit assignment problem as an impediment to artificial intelligence, he also identified what he called the "search problem": In most real-world situations, it is impossible to search through all possible options. Consider chess. Building a world model of the game of chess is relatively trivial (the rules are deterministic, you know all the pieces, all their moves, and all the squares of the board). But in chess, you cannot search through all the possible future moves; the tree of possibilities in chess has more branching paths than there are atoms in the universe. So the problem is not just constructing an inner model of the external world but also figuring out how to explore it.

And yet clearly, somehow, the brains of early mammals solved the search problem. Let's see how.

13

Model-Based Reinforcement Learning

AFTER THE SUCCESS of TD-Gammon, people tried to apply Sutton's temporal difference learning (a type of model-free reinforcement learning) to more complex board games like chess. The results were disappointing.

While model-free approaches like temporal difference learning can do well in backgammon and certain video games, they do not perform well in more complex games like chess. The problem is that in complex situations, model-free learning—which contains no planning or playing out of possible futures—is not good at finding the moves that don't look great right now but set you up well for the future.

In 2017, Google's DeepMind released an AI system called AlphaZero that achieved superhuman performance in not only the game of chess but also the game of Go, beating the world Go champion Lee Sedol. Go is an ancient Chinese board game that is even more complex than chess; there are a trillion upon trillion times more possible board positions in Go than in chess.

How did AlphaZero achieve superhuman performance at Go and chess? How did AlphaZero succeed where temporal difference learning could not? The key difference was that AlphaZero *simulated future possibilities*. Like TD-Gammon, AlphaZero was a reinforcement learning system—its strategies were not programmed into it with expert rules but learned through trial and error. But *unlike* TD-Gammon, AlphaZero was a model-based reinforcement learning algorithm; AlphaZero searched through possible future moves before deciding what to do next.

Figure 13.1: The game of Go

After its opponent moved, AlphaZero would pause, select moves to consider, and then play out thousands of simulations of how the entire game might go given those selected moves. After running a set of simulations, AlphaZero might see that it won thirty-five out of the forty imagined games when it made move A, thirty-nine of the forty imagined games when it made move B, and so on for many other possible next moves. AlphaZero could then pick the move where it had won the highest ratio of imagined games.

Doing this, of course, comes with the search problem; even armed with Google's supercomputers, it would take well over a million years to simulate every possible future move from an arbitrary board position in Go. And yet AlphaZero ran these simulations within half a second. How? It didn't simulate the trillions of possible futures; it simulated only *a thousand* futures. In other words, it *prioritized*.

There are many algorithms for deciding how to prioritize which branches to search through in a large tree of possibilities. Google Maps uses such an algorithm when it searches for the optimal route from point A to point B. But the search strategy used by AlphaZero was different and offered unique insight into how real brains might work.

We already discussed how in temporal difference learning an actor learns to predict the best next move based on a hunch about the board position, doing so without any planning. AlphaZero simply expanded on this architecture. Instead of picking the single move its actor believed was the best next move, it picked multiple top moves that its actor believed were the best. Instead of just assuming its actor was correct (which it would not always be), AlphaZero used search to *verify* the actor's hunches. AlphaZero was effectively saying to the actor, "Okay, if you think move A is the best move, let's see how the game would play out if we did move A." And AlphaZero then also explored other hunches of the actor, considering the second and third best moves the actor was suggesting (saying to the actor, "Okay, but if you didn't take move A, what would your next best hunch be? Maybe move B will turn out even better than you think").

What is elegant about this is that AlphaZero was, in some sense, just a clever elaboration on Sutton's temporal difference learning, not a reinvention of it. It used search not to logically consider all future possibilities (something that is impossible in most situations) but to simply verify and expand on the hunches that an actor-critic system was already producing. We will see that this approach, in principle, may have parallels to how mammals navigate the search problem.

While Go is one of the most complex board games, it is still far simpler than the task of simulating futures when moving around in the real world. First, the actions in Go are *discrete* (from a given board position, there are only about two hundred possible subsequent next moves), whereas in the real world actions are *continuous* (there are an infinite number of possible body and navigational paths). Second, the information about the world in Go is deterministic and complete, whereas in the real world it is noisy and incomplete. And third, the rewards in Go are simple (you either win or lose the game), but in the real world, animals have competing needs that change over time. And so, while AlphaZero was a huge leap forward, AI systems are still far from performing planning in environments with a continuous space of actions, incomplete information about the world, and complex rewards.

However, the most critical advantage of planning in mammalian brains over modern AI systems like AlphaZero is not their ability to plan with continuous action spaces, incomplete information, or complex rewards, but instead simply the mammalian brain's ability to flexibly change its approach to planning depending on the situation. AlphaZero—which applied only to board games—employed the same search strategy with every move. In the real world, however, different situations call for different strategies. The brilliance of simulation in mammal brains is unlikely to be some special yet-to-be-discovered search algorithm; it is more likely to be the flexibility with which mammal brains employ different strategies. Sometimes we pause to simulate our options, but sometimes we don't simulate things at all and just act instinctually (somehow brains intelligently decide when to do each). Sometimes we pause to consider possible futures, but other times we pause to simulate some past event or alternative past choices (somehow brains select when to do each). Sometimes we imagine rich details in our plans, playing out each individual detailed subtask, and sometimes we render just the general idea of the plan (somehow brains intelligently select the right granularity of our simulation). How do our brains do this?

Prefrontal Cortex and Controlling the Inner Simulation

In the 1980s, a neuroscientist named Antonio Damasio visited one of his patients—referred to as "L"—who had suffered a stroke. L lay in bed with her eyes open and a blank expression on her face. She was motionless and speechless, but she wasn't paralyzed. She would, at times, lift up the blanket to cover herself with perfectly fine motor dexterity; she would look over at a moving object, and she could clearly recognize when someone spoke her name. But she did and said nothing. When looking into her eyes, people said it seemed that "she was there but not there."

Stroke victims with damage to the visual, somatosensory, or auditory neocortex suffer from impairments to perception (such as blindness or deafness). But L showed none of these symptoms; her stroke occurred in a specific region of her *prefrontal* neocortex. L had developed akinetic mutism—a tragic and bizarre condition caused by damage to certain regions of the prefrontal cortex in which people are able to move and understand things just fine but they don't move, speak, or care about anything at all.

After six months, as with many stroke patients, L began to recover as other areas of her neocortex remapped themselves to compensate for the

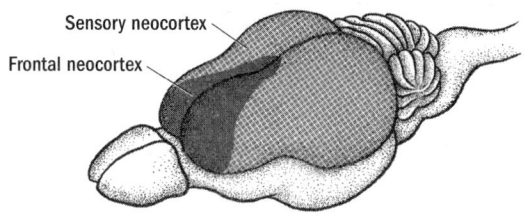

Brain of first mammals

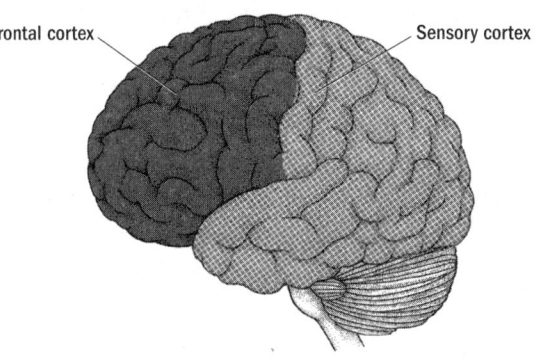

Brain of a modern human

Figure 13.2

damaged area. As L slowly began to speak again, Damasio asked her about her experience over the prior six months. Although L had little memory of it, she did recall the few days before beginning to speak. She described the experience as not talking because she had nothing to say. She claimed her mind was entirely "empty" and that nothing "mattered." She claimed that she was fully able to follow the conversations around her, but she "felt no 'will' to reply." It seems that L had lost all *intention*.

The neocortex of all mammals can be separated into two halves. The back half is the sensory neocortex, containing visual, auditory, and somatosensory areas. Everything about the neocortex we reviewed in chapter 11 was about sensory neocortex—it is where a simulation of the external world is rendered, either matching its simulation to incoming sensory data (perception by inference) or by simulating alternative realities (imagination). But the sensory neocortex was only half the puzzle of how the neocortex works. The neocortex of the first mammals, just like that of modern rats and humans, had another component, found in the front half: the *frontal* neocortex.

The frontal neocortex of a human brain contains three main subregions: motor cortex, granular prefrontal cortex (gPFC), and agranular prefrontal cortex (aPFC). The words *granular* and *agranular* differentiate parts of the prefrontal cortex based on the presence of granule cells, which are found in layer four of the neocortex. In the *granular* prefrontal cortex, the neocortex contains the normal six layers of neurons. However, in the *agranular* prefrontal cortex, the fourth layer of neocortex (where granule cells are found) is weirdly missing.* Thus, the parts of the prefrontal cortex that are *missing* layer four are called the agranular prefrontal cortex (aPFC), and the parts of the prefrontal cortex that *contain* a layer four are called the granular prefrontal cortex (gPFC). It is still unknown why, exactly, some areas of frontal cortex are missing an entire layer of neocortex, but we will explore some possibilities in the coming chapters.

The granular prefrontal cortex evolved much later in early primates, and we will learn all about it in breakthrough #4. The motor cortex evolved after the first mammals but before the first primates (we will learn about the motor cortex in the next chapter). But the agranular prefrontal cortex (aPFC) is the most ancient of frontal regions and evolved in the first mammals. It was the aPFC that was damaged in Damasio's patient L. The aPFC is so ancient and fundamental to the proper functioning of the neocortex that when damaged, L was deprived of something central to what it means to be human—or, more specifically, what it means to be a *mammal.*

In the first mammals, the entire frontal cortex was just agranular prefrontal cortex. All modern mammals contain an agranular prefrontal cortex inherited from the first mammals. To understand L's akinetic mutism and how mammals decide when and what to simulate, we must first roll back the evolutionary clock to explore the function of the aPFC in the brains of the first mammals.

It seems that in early mammals, the sensory neocortex was where

---

* Note that the motor cortex is also missing layer four, but it is not considered *pre*frontal cortex.

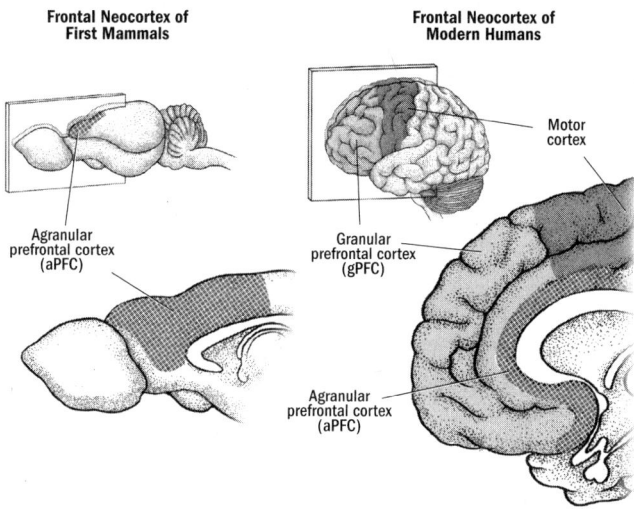

Figure 13.3: The frontal regions of the first mammals and of modern humans

simulations were *rendered*, and the frontal neocortex was where simulations were *controlled*—it is the frontal neocortex that decided *when* and *what* to imagine. A rat with a damaged frontal neocortex loses the ability to trigger simulations; it no longer engages in vicarious trial and error, episodic-memory recall, or counterfactual learning. This impairs rats in many ways. They become worse at solving spatial-navigation challenges that require preplanning, such as when they are placed in completely new starting locations in a maze. Rats make lazier choices, often taking the easier paths even if they offer substantially fewer rewards, as if the rats are unable to stop and play out each option and realize that the effort is worth it. And without episodic memory, they fail to internally recall old memories of dangerous cues and are thereby more likely to repeat past mistakes.

Even those rats with only partial frontal damage who retain some ability to trigger these simulations still struggle to monitor these simulated "plans" as they unfold. Rats with aPFC damage struggle to remember where they are in an ongoing plan, do actions out of sequence, and unnecessarily repeat already completed actions. Rats with aPFC damage also become impulsive; they will prematurely respond in tasks that require them to wait and be patient to get food.

While the frontal and sensory cortices seem to serve different functions (the frontal neocortex *triggers* simulations, the sensory neocortex *renders* simulations), they are both different areas of the neocortex and thus should execute the same fundamental computation. This presents a conundrum: How does the frontal neocortex, simply another area of neocortex, do something so seemingly *different* from the sensory neocortex? Why does a modern human with a damaged aPFC become devoid of intention? How does the aPFC trigger simulations in the sensory neocortex? How does it decide *when* to simulate something? How does it decide *what* to simulate?

Predicting Oneself

In a column of the sensory cortex, the primary input comes from external sensors, such as the eyes, ears, and skin. The primary input to the agranular prefrontal cortex, however, comes from the hippocampus, hypothalamus, and amygdala. This suggests that the aPFC treats sequences of places, valence activations, and internal affective states the way the sensory neocortex treats sequences of sensory information. Perhaps, then, the aPFC tries to explain and predict an animal's own behavior the same way that the sensory neocortex tries to explain and predict the flow of external sensory information?

Perhaps the aPFC is always observing a rat's basal-ganglia-driven choices and asking, "Why did the basal ganglia choose this?" The aPFC of a given rat might thereby learn, for example, that when the rat wakes up and has these specific hypothalamic activations, it always runs down to the river and consumes water. The aPFC might then learn that the *why* of such behavior is "to get water." Then in similar circumstances, the aPFC can predict what the animal will do *before* the basal ganglia triggers any behavior—it can predict that when thirsty, the animal will run toward nearby water. In other words, the aPFC learns to model the animal itself, inferring the *intent* of behavior it observes, and uses this *intent* to predict what the animal will do next.

As philosophically fuzzy as *intent* might sound, it is conceptually no different from how the sensory cortex constructs explanations of sensory information. When you see a visual illusion that suggests a triangle (even

Frontal vs Sensory Neocortex in the First Mammals

FRONTAL NEOCORTEX	SENSORY NEOCORTEX
A self model	A world model
Gets input from hippocampus, amygala, and hypothalamus	Gets input from sensory organs
"I *did* this because I want to get to water"	"I see this because there is a triangle right there"
Tries to predict what the animal will do next	Tries to predict what external objects will do next

when there is no triangle), your sensory neocortex constructs an explanation of it, which is what you perceive—a triangle. This explanation—the triangle—is not real; it is *constructed*. It is a computational trick that the sensory neocortex uses to make *predictions*. The explanation of the triangle enables your sensory cortex to predict what would happen if you reached out to grab it or turned the light on or tried to look at it from another angle.

Recording studies corroborate the idea that the aPFC creates a model of an animal's goals. If you look at a recording of the aPFC of a rat, you can see patterns of activity that encode the task a rat is performing—with specific populations of neurons selectively firing only at specific locations within a complex task sequence, reliably tracking progress toward an imagined goal.

What is the evolutionary usefulness of this model of self in the frontal cortex? Why try to "explain" one's own behavior by constructing "intent"? It turns out, this might be how mammals choose *when* to simulate things and how to select *what* to simulate. Explaining one's own behavior might solve the search problem. Let's see how.

How Mammals Make Choices

Let's take the example of a rat navigating a maze and making a choice as to which direction to go when it reaches a fork. Going to the left leads to water, going to the right leads to food. It is these situations when vicarious trial and error occurs, and it occurs in three steps.

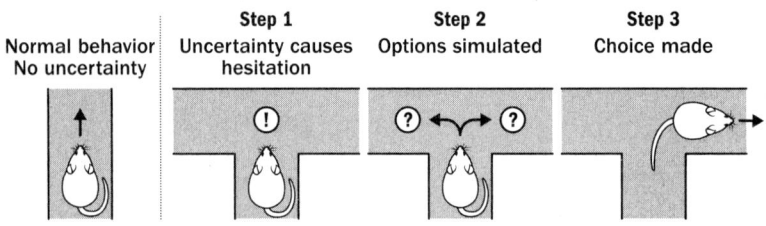

Figure 13.4: Speculations on how mammals make deliberative choices

Step #1: Triggering Simulation

The columns in the aPFC might always be in one of three states: (1) silent; the behavior it observes is not recognizing any specific intent (like a column in the visual cortex not recognizing anything meaningful in an image); (2) many or all columns in the frontal cortex recognize an intent and predict the same next behavior ("Oh! We are obviously about to go to the left here"); or (3) many columns in the frontal cortex recognize an intent but predict *different and inconsistent behaviors* (some columns predict "I'm going to go down to the left to get water!" and other columns predict "I'm going to go to the right to get food!"). It might be in this last state, in which aPFC columns *don't agree* in their predictions, where the aPFC might be most useful. Indeed, the aPFC of mammals gets most excited when something goes *wrong* or something *unexpected* happens in an ongoing task.

The degree of disagreement of predictions is a measure of uncertainty. This is, in principle, how many state-of-the-art machine learning models measure uncertainty: An ensemble of different models makes predictions, and the more divergent such predictions, the more uncertainty there is reported to be.

It might be this uncertainty that triggers simulations. The aPFC can trigger a global pause signal through connecting directly to specific parts of the basal ganglia, and aPFC activation has been shown to correlate with levels of uncertainty. And as we saw in the last chapter, it is exactly when things change or are hard (i.e., are uncertain) that animals pause to engage in vicarious trial and error. It is also possible that uncertainty could be measured in the basal ganglia; perhaps there are parallel actor-critic systems, each making independent predictions about the next best action, and the divergence in their predictions is what triggers pausing.

Either way, this provides a speculation for how mammal brains tackle the challenge of deciding when to go through the effort of simulating things. If events are unfolding as one would expect, there is no reason to waste time and energy simulating options, and it is easier just to let the basal ganglia drive decisions (model-free learning), but when uncertainty emerges (something new appears, some contingency is broken, or costs are close to the benefits), then simulation is triggered.

Step #2: Simulating Options

Okay, so the rat paused and decided to use simulation to resolve its uncertainty—now what? This brings us back to the search problem. A rat in a maze could do any one of a billion things, so how does it decide what to simulate?

We saw how AlphaZero solved this problem: It played out the top moves it was already predicting were the best. This idea aligns quite nicely with what is known about neocortical columns and the basal ganglia. The aPFC doesn't sit there combing through every possible action, instead it specifically explores the paths that it is already predicting an animal will take. In other words, the aPFC searches through the specific options that different columns of the aPFC are already predicting. One set of columns predicted going left all the way to water, and another predicted going right, so there are only two different simulations to run.

After an animal pauses, different columns in the aPFC take turns playing out their predictions of what they think the animal will do. One group of columns plays out going left and following that path all the way to water. Another group of columns plays out going right and following that path all the way to get food.

The connectivity between the aPFC and the sensory neocortex is revealing; the aPFC projects extensively to diffuse regions of the sensory cortex and has been shown to dramatically modulate the activity of the sensory neocortex. And specifically when rats engage in this vicarious trial and error behavior, the activity in the aPFC and the sensory cortex become uniquely synchronized. One speculation is that the aPFC is triggering the sensory neocortex to render a specific simulation of the world. The aPFC first asks, "What happens if we go to the left?" The sensory neocortex then renders a simulation of turning left, which then passes back to the aPFC. The aPFC then says, "Okay, and then what happens if we keep going forward?" which the sensory neocortex renders again, and so on and so forth all the way to the imagined goal modeled in the aPFC.

Alternatively, it could be the basal ganglia that determines the actions taken during these simulations. This would be even closer to how AlphaZero worked—it selected simulated actions based on the actions its model-free actor predicted were best. In this case, it would be the aPFC that selects which of the divergent action predictions of the basal ganglia to simulate, but the basal ganglia would continue to decide which actions it wants to take in the imagined world rendered by the sensory neocortex.

Step #3: Choosing an Option

The neocortex simulates sequences of actions, but what makes the final decision as to what direction the rat will actually go? How does a rat choose? Here's one speculation. The basal ganglia already has a system for making choices. Even the ancient vertebrates had to make choices when presented with competing stimuli. The basal ganglia accumulates votes for competing choices, with different populations of neurons representing each competing action ramping up in excitement until it passes a choice threshold, at which point an action is selected.

And so, as the process of vicarious trial and error unfolds, the results of these vicarious replays of behavior accumulate votes for each choice in the basal ganglia—the same way it would if the trial and error were not vicarious but real. If the basal ganglia keeps getting more excited by imagining drinking water than by imagining eating food (as measured by the amount of dopamine released), then these votes for water will quickly pass the choice threshold. The basal ganglia will take over behavior, and the rat will go get water.

The emergent effect of all this is that the aPFC *vicariously* trained the basal ganglia that left was the better option. The basal ganglia doesn't know whether the sensory neocortex is simulating the current world or an imagined world. All the basal ganglia knows is that when it turned left, it got reinforced. Hence, when the sensory neocortex goes back to simulating the current world at the beginning of the maze, the basal ganglia immediately tries to repeat the behavior that was just vicariously reinforced. Voilà—the animal runs to the left to get water.

Goals and Habits (or the Inner Duality of Mammals)

In the early 1980s, a Cambridge psychologist by the name of Tony Dickinson was engaging in the popular psychology experiments of the time: training animals to push levers to get rewards. Dickinson was asking a seemingly mundane question: What happens if you devalue the reward of a behavior after the behavior is learned? Suppose you teach a rat that pushing a lever releases a food pellet from a nearby contraption. The rat will rapidly go back and forth between pushing the lever and gobbling up the food. Now suppose one day, completely out of the context of the box with the lever, you give the rat the same pellet and secretly lace it with a chemical that makes the rat feel sick. How does this change their behavior?

The first result is expected: Rats, even after they have long recovered from their short bout of nausea, no longer find the pellets as appetizing as they did before. When given a mound of such pellets, rats eat substantially fewer. But the more interesting question was this: What would happen the next time the rats were shown the lever? If animals are simply governed by Thorndike's law of effect, then they would run up to the lever and push it as rapidly as before—the lever had been reinforced *many* times, and nothing had yet *un*reinforced the act of pushing the lever. On the other hand, if animals are indeed able to simulate the consequences of pushing the lever and realize that the result is a pellet that they no longer like, they won't want to push the lever as much. What Dickinson found was that after this procedure, rats who had the pellet paired with illness pushed the lever almost 50 percent less than those that had not.

This is consistent with the idea that the neocortex enables even simple mammals such as rats to vicariously simulate future choices and change their behaviors based on the imagined consequences. But as Dickinson continued these experiments, he noticed something weird: Some rats continued to push the lever with as much, if not more, vigor after the pellet was paired with illness. Some rats became what he called "insensitive to devaluation." The difference, he found, was merely a consequence of how many times the rats had pushed the lever to get a reward. Rats that had done the task one hundred times did the smart thing—they no longer wanted to push the lever once the food was devalued. But rats that had done the task five hundred times ran up to the lever and just started pushing it like crazy, even if the food was devalued. And in all these tests, the food pellets never were given; the group that had become insensitive to devaluation just kept pushing the lever without ever getting a reward.

Dickinson had discovered habits. By engaging in the behavior five hundred times, rats had developed an automated motor response that was triggered by a sensory cue and completely detached from the higher-level goal of the behavior. The basal ganglia took over behavior without the aPFC pausing to consider what future these actions would produce. The behavior had been repeated so many times that the aPFC and basal ganglia did not detect any uncertainty and therefore the animal did not pause to consider the consequences.

Perhaps this is a familiar experience. People wake up and look at their phones without asking themselves why they are choosing to look at their phones. They keep scrolling through Instagram even though if someone had asked them if they wanted to keep scrolling, they'd say "no." Of course, not all habits are bad: You don't think about walking, and yet you walk perfectly; you don't think about typing, and yet the thoughts flow effortlessly from your mind to your fingertips; you don't think about speaking, and yet thoughts magically convert themselves into a repertoire of tongue, mouth, and throat movements.

Habits are automated actions triggered by stimuli directly (they are model-free). They are behaviors controlled directly by the basal ganglia. They are the way mammalian brains save time and energy, avoiding unnecessarily engaging in simulation and planning. When such automation occurs at the right times, it enables us to complete complex behaviors easily; when it occurs at the wrong times, we make bad choices.

The duality between model-based and model-free decision-making methods shows up in different forms across different fields. In AI, the terms *model-based* and *model-free* are used. In animal psychology, this same duality is described as *goal-driven behavior* and *habitual behavior*. And in behavioral economics, as in Daniel Kahneman's famous book *Thinking, Fast and Slow*, this same duality is described as "system 2" (thinking slow) versus "system 1" (thinking fast). In all these cases, the duality is the same: Humans and, indeed, all mammals (and some other animals that independently evolved simulation) sometimes pause to simulate their options (model-based, goal-driven, system 2) and sometimes act automatically (model-free, habitual, system 1). Neither is better; each has its benefits and costs. Brains attempt to intelligently select when to do each, but brains do not always make this decision correctly, and this is the origin of many of our irrational behaviors.

The language used in animal psychology is revealing—one type of behavior is goal-driven and the other is not. Indeed, *goals* themselves may not have evolved until early mammals.

The Evolution of the First Goal

Just as the explanations of sensory information are not real (i.e., you don't perceive what you see), so intent is not real; rather, it is a computational trick for making predictions about what an animal will do next.

This is important: The basal ganglia has *no intent or goals*. A model-free reinforcement learning system like the basal ganglia is *intent-free*; it is a system that simply learns to repeat behaviors that have previously been reinforced. This is not to say that such model-free systems are dumb or devoid of motivation; they can be incredibly intelligent and clever, and they can rapidly learn to produce behavior that maximizes the amount of reward. But these model-free systems do not have "goals" in the sense that they do not set out to pursue a specific outcome. This is one reason why model-free reinforcement learning systems are painfully hard to interpret—when we ask, "Why did the AI system do that?," we are asking a question to which there is really no answer. Or at least, the answer will always be the same: because it thought that was the choice with the most predicted reward.

In contrast, the aPFC *does* have explicit goals—it wants to go to the fridge to eat strawberries or go to the water fountain to drink water. By simulating a future that terminates at some end result, the aPFC has an end state (a goal) that it seeks to achieve. This is why it is possible, at least in circumstances where people make aPFC-driven (goal-oriented, model-based, system 2) choices, to ask *why* a person did something.

It is somewhat magical that the very same neocortical microcircuit that constructs a model of external objects in the sensory cortex can be repurposed to construct goals and modify behavior to pursue these goals in the frontal cortex. Karl Friston of University College London—one of the pioneers of the idea that the neocortex implements a generative model—calls this "active inference." The sensory cortex engages in passive inference—merely explaining and predicting sensory input. The aPFC engages in *active* inference—explaining one's own behavior and then using its predictions to actively *change* that behavior. By pausing to play out what the aPFC predicts *will* happen and thereby vicariously training the basal ganglia, the aPFC is repurposing the neocortical generative model for *prediction* to create *volition*.

When you pause and simulate different dinner options, choose to get pasta, then begin the long action sequence to get to the restaurant, this is a "volitional" choice—you can answer *why* you are getting in the car; you know the end state you are pursuing. In contrast, when you act only from habit, you have no answer as to why you did what you did.

Karl Friston also offers an explanation for the perplexing fact that some parts of the frontal cortex are missing the fourth layer of the neocortical column. What does layer four do? In the sensory cortex, layer four is where raw sensory input flows into a neocortical column. Layer four is speculated to have the role of pushing the rest of the neocortical column to render a simulation that best matches its incoming sensory data (perception by inference). There is evidence that when a neocortical column is engaged in simulating, layer-four activity declines as active incoming sensory input is suppressed—this is how the neocortex can render a simulation of something not currently experienced (e.g., imagining a car when looking at the sky). This is a clue. Active inference suggests that the aPFC constructs intent and then tries to predict behavior consistent with that intent; in other words, it tries to make its intent come true. If the animal does something inconsistent with the aPFC's constructed intent, the aPFC doesn't want to adjust its model of intent to match the behavior, it wants to adjust the behavior: if you are thirsty and your basal ganglia makes a decision to go in the direction that has no water, the aPFC doesn't want to adjust its model of your intent to assume you are not thirsty, instead it wants to pause the basal ganglia's mistake, and convince it to turn around and go toward the water. Thus, the aPFC spends very little, if any, time trying to match its inferred intent to the behavior it sees, and so it has no need for a large, or even any, layer four.

Of course, the aPFC isn't evolutionarily programmed to understand the goals of the animal, instead it learns these goals by first modeling behavior originally controlled by the basal ganglia. The aPFC constructs goals by observing behavior that is originally entirely devoid of them. And only once these goals are learned does the aPFC begin to exert control over behavior: the basal ganglia begins as the teacher of the aPFC, but as a mammal develops, these roles flip, and the aPFC becomes the teacher of the basal ganglia. And indeed, during brain development, agranular parts of the frontal cortex begin with a layer four that then slowly atrophies and disappears during development, leaving layer four largely empty. Perhaps this is part of a developmental program for constructing a model of self, starting by matching one's internal model to its observations (hence beginning with a layer 4), and then transitioning to pushing behavior to match one's internal model (hence no need for a layer 4 anymore). Again we see a beautiful bootstrapping in evolution.

This also offers some insight into the experience of Damasio's patient L. It makes some sense why her head was "empty": She was unable to render an inner simulation. She had no thoughts. She had no will to respond to anything because her inner model of intent was gone, and without that, her mind could not set even the simplest of goals. And without goals, tragically, nothing mattered.

How Mammals Control Themselves: Attention, Working Memory, and Self-Control

In a typical neuroscience textbook, the four functions ascribed to the frontal neocortex are *attention, working memory, executive control*, and, as we have already seen, *planning*. The connecting theme of these functions has always been confusing; it seems odd that one structure would subserve all these distinct roles. But through the lens of evolution, it makes sense that these functions are all intimately related—they are all different applications of controlling the neocortical simulation.

Remember the ambiguous picture of a duck or a rabbit? As you oscillate between perceiving a duck or a rabbit, it is your aPFC that is nudging your visual cortex back and forth between each interpretation. Your aPFC can trigger an internal simulation of ducks when you close your eyes, and your aPFC can use the same mechanism to trigger an internal simulation of ducks with your eyes open and you are looking at a picture that could be either a duck or a rabbit. In both cases, the aPFC is trying to invoke a simulation; the only difference is that with your eyes closed, the simulation is unconstrained, and with your eyes open it is constrained to be consistent with what you are seeing. The aPFC's triggering of simulation is called *imagination* when it is unconstrained by current sensory input and *attention* when it is constrained by current sensory input. But in both cases, the aPFC is, in principle, doing the same thing.

What is the point of attention? When a mouse selects an action sequence after its imagined simulation, it must *stick* to its plan as it runs down its path. This is harder than it sounds. The imagined simulation will not have been perfect; the mouse will not have predicted each sight, smell, and contour of the environment that it will actually experience. This means that the vicarious learning that the basal ganglia experienced will differ from the actual experience as the plan unfolds, and therefore, the basal ganglia may not correctly fulfill the intended behavior.

One way the aPFC can solve this problem is using attention. Suppose a rat's basal ganglia learned, through trial and error, to run *away* from ducks and run *toward* rabbits. In this case, the basal ganglia will have opposite reactions to seeing the duck/rabbit depending on what pattern gets sent to it from the neocortex. If the aPFC had previously imagined seeing a rabbit and running toward it, then it can control the basal ganglia's choices by using attention to ensure that when the rat sees this ambiguous picture, it sees a rabbit, not a duck.

Controlling ongoing behavior often also requires *working memory*—the maintenance of representations in the absence of any sensory cues. Many imagined paths and tasks involve waiting. For example, when a rodent forages among trees for nuts, it must remember which trees it has already foraged. This is a task shown to require the aPFC. If you inhibit a rodent's aPFC during these delay periods, rodents lose their ability to perform such tasks from memory. And during such tasks, the aPFC exhibits "delay activity," remaining activated even in the absence of any external cues. These tasks require the aPFC because working memory functions in the same way as attention and planning—it is the invoking of an inner simulation. Working memory—holding something in your head—is just your aPFC trying to keep re-invoking an inner simulation until you no longer need it.

In addition to planning, attention, and working memory, the aPFC can also control ongoing behavior more directly: It can inhibit the amygdala. There is a projection from the aPFC to inhibitory neurons surrounding the amygdala. During the fulfillment of an imagined plan, the aPFC can attempt to prevent the amygdala from triggering its own approach and avoidance responses. This was the evolutionary beginning of what psychologists call behavioral inhibition, willpower, and self-control: the persistent tension between our moment-to-moment cravings (as controlled by the amygdala and basal ganglia) and what we know to be a better choice (as controlled by the aPFC). In moments of willpower, you can inhibit your amygdala-driven cravings. In moments of weakness, the amygdala wins. This is why people become more impulsive when tired or stressed—the aPFC is energetically expensive to run, so if you are tired or stressed, the aPFC will be much less effective at inhibiting the amygdala.

To summarize: Planning, attention, and working memory are all controlled by the aPFC because all three are, in principle, the same thing. They are all different manifestations of brains trying to select what simulation to render. How does the aPFC "control" behavior? The idea presented here is that it doesn't control behavior per se; it tries to convince the basal ganglia of the right choice by vicariously showing it that one choice is better and by filtering what information makes it to the basal ganglia. The aPFC controls behavior not by *telling* but by *showing*.

The benefit of this can be seen when comparing the performance of mammals to other vertebrates like lizards on tasks that require inhibiting reflexive responses in favor of "smarter" choices. If you put a lizard in a maze and try to train it to go toward a red light to get appealing food and avoid a green light which offers unappealing food, it takes lizards *hundreds* of trials to learn this simple task. The hardwired preference of lizards toward green light takes a long time to be untrained. Without a neocortex to pause and vicariously consider options, the only way lizards learn this task is through endless real trial and error. In contrast, rats learn to inhibit their hardwired responses much more rapidly, an advantage that disappears if you damage a rat's aPFC.

Early mammals had the ability to vicariously explore their inner model of the world, make choices based on imagined outcomes, and stick to the imagined plan once chosen. They could flexibly determine when to simulate futures and when to use habits; and they intelligently selected what to simulate, overcoming the search problem. They were our first ancestors to have goals.

14

The Secret to Dishwashing Robots

IMAGINE THE FOLLOWING. As you are holding this book your right hand begins to cramp up. The specific placement of each individual finger that you have effortlessly configured to perfectly balance the book in your hand begins to wilt as you lose control of the muscles in your right arm. You realize you can no longer control each finger individually; you can only open or close your hand with all your fingers moving at once, your hand transforming from a dexterous tool to an uncoordinated claw. Within minutes you can no longer even grasp the book with your right hand at all, and your arm becomes too weak to lift. This is what the experience of a stroke—the loss of blood flow to a region of the brain—feels like when it occurs in the motor cortex. Such a condition deprives its patients of fine motor skills and can even cause paralysis.

The motor cortex is a thin band of neocortex on the edge of the frontal cortex. Motor cortex makes up a map of the entire body, with each area controlling movements of specific muscles. While the entire motor cortex accounts for every part of the body, it does not dedicate equal space to each body part. Instead, it dedicates lots of space to the parts of the body that animals have skilled motor control over (in primates, this is the mouth and hands) and much less space to areas that they can't control well (like the feet). This map in the motor cortex is mirrored in the adjacent somatosensory cortex— the region of the neocortex that processes *somatosensory* information (such as touch sensors in the skin and proprioceptive signals from muscles).

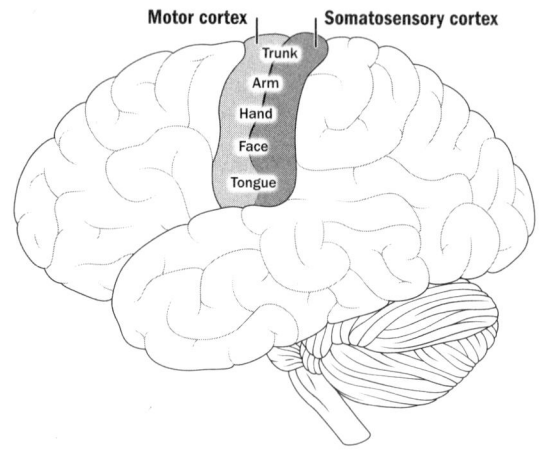

Figure 14.1: The motor cortex of humans

In humans, the motor cortex is the primary system for controlling movement. Not only does stimulating specific areas of the motor cortex elicit movement from the corresponding body part, but damaging those same areas of the motor cortex creates paralysis in that same body part. The movement deficits of stroke patients almost always derive from damage to areas of the motor cortex. In chimpanzees, macaques, and lemurs, motor cortex damage also has this effect. In primates, neurons from the motor cortex send a projection *directly* to the spinal cord to control movement. All this leads to the conclusion that the motor cortex is the locus of *motor commands*; it is the controller of movement.

But there are three problems with this idea. First, the neocortical columns in the motor cortex have the same microcircuitry as other areas of the neocortex. If we believe the neocortex implements a generative model that tries to explain its inputs and uses these explanations to make predictions, then we must account for how this could have been repurposed to produce motor commands.

Second, some mammals don't have a motor cortex, and they can clearly move around normally. As discussed in the previous chapter, most evolutionary neuroscientists believe that the only part of the frontal cortex that was present in the first mammals was the agranular prefrontal cortex (aPFC); there was no motor cortex. The motor cortex emerged tens

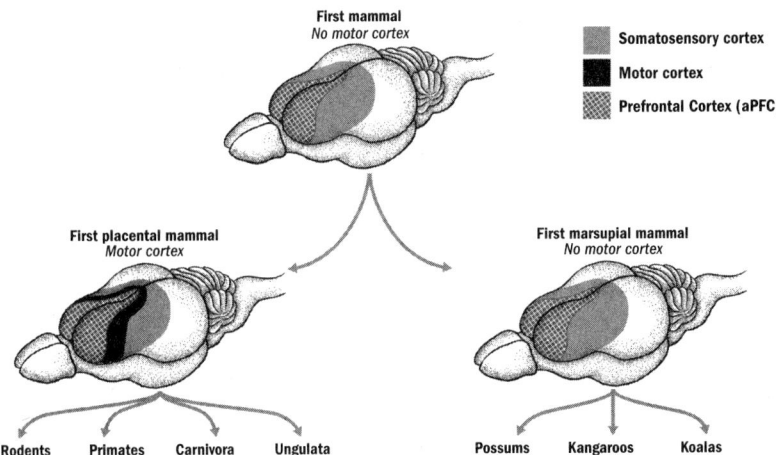

Figure 14.2: Leading theory on evolution of the motor cortex

of millions of years after the first mammals and only in the placental lineage—the mammals that would go on to become today's rodents, primates, dogs, horses, bats, elephants, and cats.

The third problem with the "motor cortex equals motor commands" idea is that the paralysis caused by motor cortex damage is unique to primates; most mammals with motor cortex damage do not suffer from such paralysis. Rats and cats with damaged motor cortices can still walk, hunt, eat, and move around just fine. The motor cortex was clearly not the locus of motor commands in early mammals, and it was only later—in primates—that it became *required* for movement. So why did the motor cortex evolve? What was its original function? What changed with primates?

Predictions, Not Commands

Karl Friston, the pioneer of the theory of active inference, offers an alternative interpretation of the motor cortex. While the prevailing view has always been that the motor cortex generates *motor commands*, telling muscles exactly *what* to do, Friston flips this idea on its head: Perhaps the motor cortex doesn't generate motor commands but rather motor *predictions*. Perhaps the motor cortex is in a constant state of observing the body movements that occur in the nearby somatosensory cortex (hence why there is such an elegant mirror of motor cortex and somatosensory cortex) and then tries to *explain* the behavior and use these explanations to predict what an animal will do next. And perhaps the wiring is merely tweaked so that motor cortex predictions flow to the spinal cord and control our movement—in other words, the motor cortex is wired to *make its predictions come true.*

By this account, the motor cortex operates the same way the agranular prefrontal cortex operates. The difference is that the aPFC learns to predict movements of navigational paths, whereas the motor cortex learns to predict movements of *specific body parts*. The aPFC will predict that an animal will turn left; the motor cortex will predict that the animal will place its left paw exactly on a platform.

This is the general idea of "embodiment"—parts of the neocortex, such as the motor cortex and somatosensory cortex, have an entire model of an animal's body that can be simulated, manipulated, and adjusted as time unfolds. Friston's idea explains how the neocortical microcircuit could be repurposed to produce specific body movements.

But if most mammals can move around normally with no motor cortex, then what was its original function? If the aPFC enables the planning of navigational routes, what did the motor cortex enable?

Damage to the motor cortex in *nonprimate* mammals such as rodents and cats has two effects. First, animals become impaired at performing skilled movements, such as carefully placing a paw on a thin branch, reaching through a small hole to grasp a morsel of food, stepping over an obstacle once it is out of sight, or placing a foot on a small unevenly placed platform. Second, nonprimate mammals become impaired at learning new sequences of movements that they have never performed before. For example, a rat trained to perform a specifically orchestrated sequence of movements can perform this sequence only if its motor cortex was damaged *after* the task was already well learned. If you damage a rat's motor cortex *before* it is trained on this task, it becomes unable to learn the lever sequence.

This suggests that the motor cortex was originally not the locus of motor commands but of motor *planning*. When an animal must perform careful

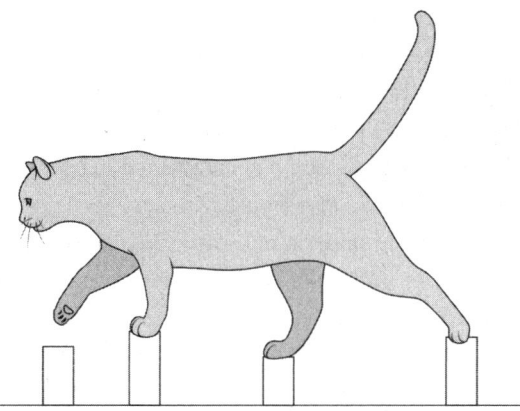

Figure 14.3: Cats struggle to perform planned movements after motor cortex damage

movements—placing a paw on a small platform or stepping over an out-of-sight obstacle—it must mentally *plan* and *simulate* its body movements ahead of time. This explains why the motor cortex is necessary for learning new complex movements but not for executing well-learned ones. When an animal is learning a new movement, the motor cortex simulations vicariously train the basal ganglia. Once a movement is well learned, the motor cortex is no longer needed.

Recording activity in the motor cortex supports this idea. In nonprimate mammals, the motor cortex is most activated not by movement in general but specifically by movements that require planning. Consistent with the idea that animals are simulating movements, the motor cortex and somatosensory cortex get activated *in advance* of an upcoming precision movement, even if obstacles are not seen but simply known to be there. And this activity maintains itself until the animal has completed the movement it presumably planned.

In humans there is plenty of evidence that the premotor and motor cortices are activated both by *doing* movements and by *imagining* movements: For example, have someone think about walking, and the leg area of the motor cortex becomes activated. This intertwining of the neurological infrastructure for imagined movements and actual movements can be observed not only in brain recordings, but also in physical lab experiments. Sit a human down in a chair and ask her to do nothing more than simply maintain her upright posture. Then play audio recordings of arbitrary sentences. Her posture gets worse when she hears sentences like "I get up, put on my slippers, go to the bathroom," but not when she hears sentences unrelated to movement. Just hearing those sentences activated an inner simulation of a changing posture, which affected her actual posture. Of course, this inner simulation also comes with benefits (it doesn't just mess up our posture): Mental rehearsal of motor skills substantially increases performance across speaking, golf swings, and even surgical maneuvers.

The motor cortex's skill in sensorimotor planning enabled early mammals to learn and execute precise movements. When comparing the motor skills of mammals to that of reptiles, it is quite clear that mammals are uniquely capable when it comes to fine motor skills. Mice can pick seeds up and skillfully break them open. Mice, squirrels, and cats are incredibly skillful tree climbers, effortlessly placing their limbs in precise places to ensure they don't fall. Squirrels and cats can plan and execute extremely accurate and precise jumps across platforms. If you have ever had a pet lizard or a turtle, you know that such skills are not the province of most reptiles. In fact, studies that examined lizards running over obstacles revealed how surprisingly sloppy the whole affair is. They don't anticipate obstacles or modify their forelimb placements to get around platforms. Given their seeming inability to plan movements in advance, it is perhaps unsurprising that few reptiles live in trees and that those that do move slowly, in contrast to the rapid skillful running and jumping of arboreal mammals.

A Hierarchy of Goals: A Balance of Simulation and Automation

How does all this work together? The frontal neocortex of early placental mammals was organized into a hierarchy. At the top of the hierarchy was the agranular prefrontal cortex, where high-level goals are constructed based on amygdala and hypothalamus activation. The aPFC might generate an intent like "drink water" or "eat food." The aPFC then propagates these goals to a nearby frontal region (the premotor cortex), which constructs subgoals and propagates these subgoals further until they reach the motor cortex, which then constructs sub-subgoals. The intent modeled in

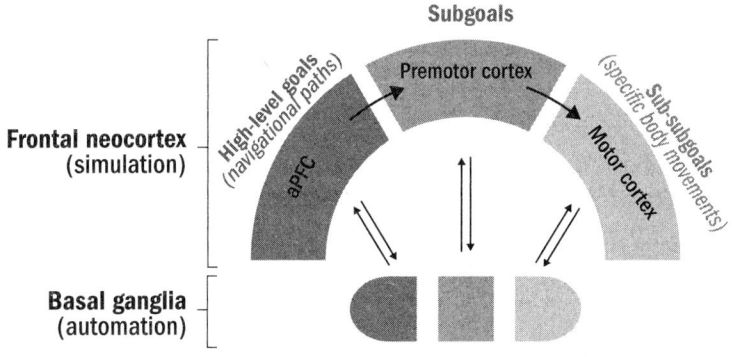

Figure 14.4: The motor hierarchy in early placental mammals

the motor cortex are these sub-subgoals, which can be as simple as "Position my index finger here and my thumb over here."

This hierarchy enables more efficient processing by distributing effort across many different neocortical columns. The aPFC doesn't have to worry about the specific movements necessary to achieve its goals; it must worry only about high-level navigational paths. Similarly, this allows the motor cortex not to have to worry about the high-level goal of the behavior and worry only about accomplishing specific low-level movement goals (picking up a cup or playing a specific chord).

The basal ganglia makes loops of connectivity with the frontal cortex, with the aPFC connecting to the front region of the basal ganglia (which then connects back to aPFC through the thalamus), and the motor cortex connecting to the back region of the basal ganglia (which then connects back to the motor cortex through a different region of the thalamus). These loops are so elegantly and particularly wired that it is hard to resist trying to reverse-engineer what they are doing.

The leading view among neuroscientists is that these are subsystems designed to manage different levels of the motor hierarchy. The front part of the basal ganglia automatically associates stimuli with high-level goals. It is what generates cravings: You come home and smell rigatoni, and suddenly you are on a mission to eat some. Drug addicts show extreme activations of this front part of the basal ganglia when they see stimuli that create drug cravings. The aPFC, however, is what makes you pause and consider

	HIGH-LEVEL GOALS	LOW-LEVEL GOALS
SIMULATION	**Agranular Prefrontal Cortex** *Simulates navigational paths* *Asks "do I want rigatoni or would I rather diet?"* *Damage causes impairments in planning navigational routes*	**Motor Cortex** *Simulates body movements* *Asks "how do I configure my fingers to play this C chord that I just learned on guitar?"* *Damage causes impairments in learning new motor skills and executing fine motor skills*
AUTOMATION	**Front part of basal ganglia** *Automatic pursuit of a high-level goal in response to stimulus* *Produces habitual cravings* *Damage can cure drug addiction*	**Back part of basal ganglia** *Automatic execution of motor skill in response to stimulus* *Produces habitual motor responses* *Damage causes impairments in executing learned skills, and impairs motor habit formation*

if you actually want to pursue these cravings ("What about our diet?"). The back part of the basal ganglia automatically associates stimuli with *low-level* goals, such as specific body movements. It is what generates automatic skilled movements. The motor cortex, on the other hand, is what makes you pause and plan out your exact movements ahead of time.

Any level of goal, whether high-level or low-level goals, has both a self model in the frontal neocortex and a model-free system in the basal ganglia. The neocortex offers a slower but more flexible system for training, and the basal ganglia offers a faster but less flexible version for well-trained paths and movements.

There is plenty of evidence for such a motor hierarchy. Recordings have shown that neurons in the aPFC are sensitive to high-level goals, whereas those in the premotor and motor cortex are sensitive to progressively lower-level subgoals. Learning a new behavior activates all levels of the motor hierarchy at first, but as the behavior becomes automatic, it activates only lower levels in the hierarchy. If you damage high-level parts of the motor hierarchy (aPFC or the front part of the basal ganglia) in rats, it makes them less sensitive to high-level goals (they will keep pushing a lever even though they no longer want the food it produces). In contrast, if you damage *low*-level parts of the motor hierarchy, rats become *more* sensitive to high-level goals, and they struggle to create motor habits (e.g., rats won't develop a habit for pushing a lever no matter how many trials they go through).

While damage to the aPFC deprives animals of their intentions, as we saw with patient L, damage to parts of the premotor cortex seems to *disconnect* the proper flow of these intentions from high-level goals to specific movements. This can cause "alien limb syndrome": patients will claim that certain parts of their bodies are moving on their own without their control. Signs of such alien movements are also seen in rodents with damage to the premotor cortex. Such damage also causes what is called "utilization behavior" or "field-dependent behavior," where patients will execute motor sequences devoid of any clear goal: They will drink from empty cups, put on other people's jackets even though they aren't going anywhere, scribble with pencils, or do any other behavior that nearby stimuli suggest. All this is the result of a broken hierarchy—parts of the motor cortex are now unconstrained by top-down intentions from the aPFC flowing through the premotor cortex, and hence the motor cortex independently sets low-level goals for motor sequences.

There is also plenty of evidence for the idea that the frontal neocortex is the locus of *simulation*, while the basal ganglia is the locus of *automation*. Damaging an animal's motor cortex impairs movement planning and learning new movements but *not* the execution of well-trained movements (because the back part of the basal ganglia already learned them). Similarly, damaging an animal's aPFC impairs path planning and learning new paths but *not* the execution of well-trained paths.

Further, the front part of the basal ganglia demonstrates all the signs of an automatic selection of cues to pursue (i.e., automated high-level behaviors). When you see a cue that creates a craving, the part of the brain that is most activated is this front part of the basal ganglia. Those who try to *inhibit* their cravings show additional activation of frontal areas like the aPFC (simulating the negative consequences and trying to train the basal ganglia to make the harder choice). In fact, lesioning the front part of the basal ganglia is an effective (although highly controversial and questionably ethical) treatment for drug addiction. The relapse rate for heroin addicts is absurdly high; some estimate as high as 90 percent. One study in China took the most severe heroin addicts and lesioned the front part of their basal ganglia. The relapse rate dropped to 42 percent. People lose the automatic behavior of pursuing cues and generating cravings (of course, there are also many side effects of such a surgery).

An intact and well-functioning motor hierarchy would have made the behavior of early placental mammals impressively flexible; animals could set high-level goals in the aPFC while lower-level areas of the motor hierarchy could flexibly respond to whatever obstacles presented themselves. A mammal pursuing faraway water could continuously update its subgoals as events unfolded—the premotor cortex could respond to surprising obstacles by selected new movement sequences, and the motor cortex could adjust even the subtlest of specific movements of limbs, all in the name of a common goal.

The secret to dishwashing robots lives somewhere in the motor cortex and the broader motor system of mammals. Just as we do not yet understand how the neocortical microcircuit renders an accurate simulation of sensory input, we also do not yet understand how the motor cortex simulates and plans fine body movements with such flexibility and accuracy and how it continuously learns as it goes.

But if we use the past few decades as our guide, roboticists and AI researchers will likely figure this out, perhaps in the near future. Indeed, robotics are improving at a rapid pace. Twenty years ago, we could barely get a four-legged robot to balance itself upright, and now we have humanoid robots that can do flips in the air.

If we successfully build robots with motor systems similar to those of mammals, they will come along with many desirable properties. These robots will automatically learn new complex skills on their own. They will adjust their movements in real time to account for perturbations and changes in the world. We will give them high-level goals, and they will be able to figure out all the subgoals necessary to achieve them. When they try to learn some new task, they will be slow and careful as they simulate each body movement before they act, but as they get better, the behavior will become more automatic. Over the course of their lifetimes, the speed with which they learn new skills will increase as they reapply previously learned low-level skills to newly experienced higher-level goals. And if their brains work at all like mammal brains, they will not require massive supercomputers to accomplish these tasks. Indeed, the entire human brain operates on about the same amount of energy as a lightbulb.

Or maybe not. Perhaps roboticists will get all this to work in a very non-mammalian way—perhaps roboticists will figure it all out without reverse-engineering human brains. But just as bird wings were an existence proof for the possibility of flight—a goal for humans to strive for—the motor skills of mammals are our existence proof for the type of motor skills we hope to build into machines one day, and the motor cortex and the surrounding motor hierarchy are nature's clues about how to make it all work.

Summary of Breakthrough #3: Simulating

The primary new brain structure that emerged in early mammals was the neocortex. With the neocortex came the gift of simulation—the third breakthrough in our evolutionary story. To summarize how this occurred and how it was used:

- Sensory neocortex evolved, which created a simulation of the external world (a world model).
- The agranular prefrontal cortex (aPFC) evolved, which was the first region of the frontal neocortex. The aPFC created a simulation of an animal's own movements and internal states (a self model) and constructed "intent" to explain one's own behavior.
- The aPFC and sensory neocortex worked together to enable early mammals to pause and simulate aspects of the world that were not currently being experienced—in other words, model-based reinforcement learning.
- The aPFC somehow solved the search problem by intelligently selecting paths to simulate and determining when to simulate them.
- These simulations enabled early mammals to engage in vicarious trial and error—to simulate future actions and decide which path to take based on the imagined outcomes.
- These simulations enabled early mammals to engage in counterfactual learning, thereby offering a more advanced solution to the credit assignment problem—enabling mammals to assign credit based on causal relationships.
- These simulations enabled early mammals to engage in episodic memory, which allowed mammals to recall past events and actions, and use these recollections to adjust their behavior.
- In later mammals, the motor cortex evolved, enabling mammals to plan and simulate specific body movements.

Our mammalian ancestors from a hundred million years ago weaponized the imaginarium to survive. They engaged in vicarious trial and error,

SUMMARY OF BREAKTHROUGH #3: SIMULATING

counterfactual learning, and episodic memory to out*plan* dinosaurs. Our ancestral mammal, like a modern cat, could look at a set of branches and plan where it wanted to place its paws. Together, these ancient mammals behaved more flexibly, learned faster, and performed more clever motor skills than their vertebrate ancestors.

Most vertebrates at the time, as with modern lizards and fish, could still move quickly, remember patterns, track the passage of time, and intelligently learn through model-free reinforcement learning, but their movements were not planned.

And so, thinking itself was born not within the clay creatures of Prometheus's divine workshop, but instead in the small underground tunnels and knotted trees of a Jurassic Earth, birthed from the crucible of a hundred million years of dinosaur predation and our ancestor's desperate attempt to avoid fading into extinction. That is the real story of how our neocortex and our inner simulation of the world came into being. And as we will soon see, it was from this hard-won superpower that the next breakthrough would eventually emerge.

This next breakthrough has been, in some ways, the hardest breakthrough to reverse engineer in modern AI systems; indeed, this next breakthrough is a feat we don't typically associate with "intelligence," but is, in fact, one of our brain's most impressive feats.

BREAKTHROUGH #4

Mentalizing and the First Primates

Your brain 15 million years ago

15

The Arms Race for Political Savvy

IT HAPPENED ON some unremarkable day around sixty-six million years ago, a day that began no different than any other. The sun rose over the jungles of today's Africa, awakening slumbering dinosaurs and driving our nocturnal squirrel-like ancestors into their daytime hiding spots. Along the muddy seashores, rain pattered into shallow ponds filled with ancient amphibians. Tides receded, drawing the many fish and other ancient critters deep in the oceans. The skies filled with pterosaurs and ancient birds. Arthropods and other invertebrates tunneled within the soils and trees. The ecology of Earth had found a beautiful equilibrium, with dinosaurs comfortably at the top of the food chain for well over a hundred and fifty million years, fish ruling the sea for even longer, and mammals and other animals finding their respective tiny but livable niches. Nothing hinted that this day would be any different than any other, but it was indeed this day when everything changed—this was the day the world almost ended.

The specific life stories of any of the animals who experienced this day are, of course, lost to us. But we can speculate. One of our mammalian squirrel-like ancestors was perhaps on her way out of her burrow to begin a night of scavenging insects. As the sun was just beginning to set, the sky must have turned that purple hue it did every other evening. But then blackness emerged from the horizon. A dark cloud, thicker than any storm she had ever seen, spread rapidly over the sky. Perhaps she looked up at this novel sight with puzzlement; perhaps she ignored it completely. Either way, despite all her new neocortical smarts, she would have had no way to understand what was happening.

This was no storm on the horizon—this was space dust. Just a few minutes prior, on the other side of the planet, an asteroid a few miles wide had slammed into the Earth. It had sent up gargantuan chunks of earth debris that was rapidly filling the skies with dark soot—a blackness that would block out the sun for over two years, killing over 70 percent of land-living vertebrates. This was the Permian-Triassic extinction.

Many of the other extinction events in the history of Earth seem to have been self-imposed—the Great Oxygenation Event was caused by cyanobacteria, and the Late Devonian Extinction was possibly caused by overproliferation of plants on land. But this one was not a fault of life but a fluke of an ambivalent universe.

Eventually, after two years of darkness, the blackened clouds began to fade. As the sun reemerged, plants began recovering lost ground and refilling the parched dead land. But this was a new world. Almost every dinosaur species was extinct except for one: the birds. Although our squirrel-like ancestor could not have known it and had not lived to see it, her offspring would inherit this new Earth. As the Earth healed, these small mammals found themselves in a completely new ecological playground. Without their dinosaur predators, they were free to explore new ecological niches, to diversify into new shapes and sizes, to conquer new territories, and to find new footing within the food chain.

The era that followed has been called the Era of Mammals. Descendants of these early mammals would eventually evolve into modern-day horses, elephants, tigers, and mice. Some would even reenter the sea and become today's whales, dolphins, and seals. Some would take to the sky and become today's bats.

Our direct ancestors were the ones who found refuge in the tall trees of Africa. These were some of the first primates. They shifted from being night-living (nocturnal) to being day-living (diurnal). As they became larger, they developed opposable thumbs to grasp branches and hold their heavier bodies. To support their bigger size, they shifted from an insect-based diet to a fruit-based diet. They lived in groups, and as they grew, they became relatively free from predation and food competition.

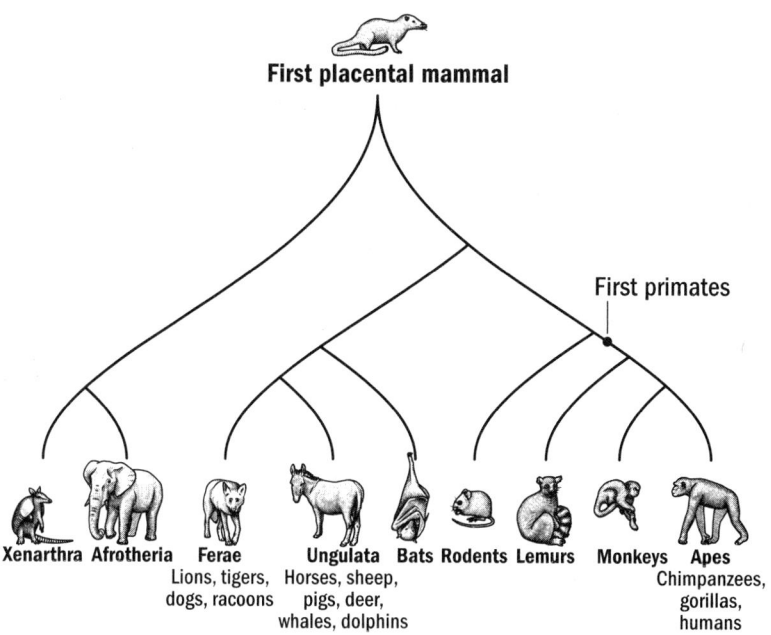

Figure 15.1: Tree of mammals

And most notably, their brains exploded to well over a hundred times their original size.

Many lineages of mammals went on to have brains not much larger (proportionally) than those of early mammals, it was only in certain lineages of mammals, such as those of elephants, dolphins, and primates, where brains dramatically expanded. Because this book is about the human story, we will focus on the journey by which the *primate* brain became big. Indeed, why primates have such big brains—and specifically such large neocortices—is a question that has perplexed scientists since the days of Darwin. What was it about the lifestyle of early primates that necessitated such a big brain?

The Social-Brain Hypothesis

In the 1980s and 1990s, numerous primatologists and evolutionary psychologists, including Nicholas Humphrey, Frans de Waal, and Robin

Dunbar, began speculating that the growth of the primate brain had nothing to do with the *ecological* demands of being a monkey in the African jungles ten to thirty million years ago and was instead a consequence of the unique *social* demands. They argued that these primates had stable mini-societies: Groups of individuals that stuck together for long periods. Scientists hypothesized that to maintain these uniquely large social groups, these individuals needed unique cognitive abilities. This created pressure, they argued, for bigger brains.

A simple test of this theory would be to look at the mini-societies of monkeys and apes around the world and see if the size of their neocortices relative to the rest of their brains correlated with their social-group size. It was Robin Dunbar who did this, and what he found shook the field. This correlation has been confirmed across many primates: the bigger the neocortex of a primate, the bigger its social group.

But monkeys and apes are far from the only mammals, let alone animals,

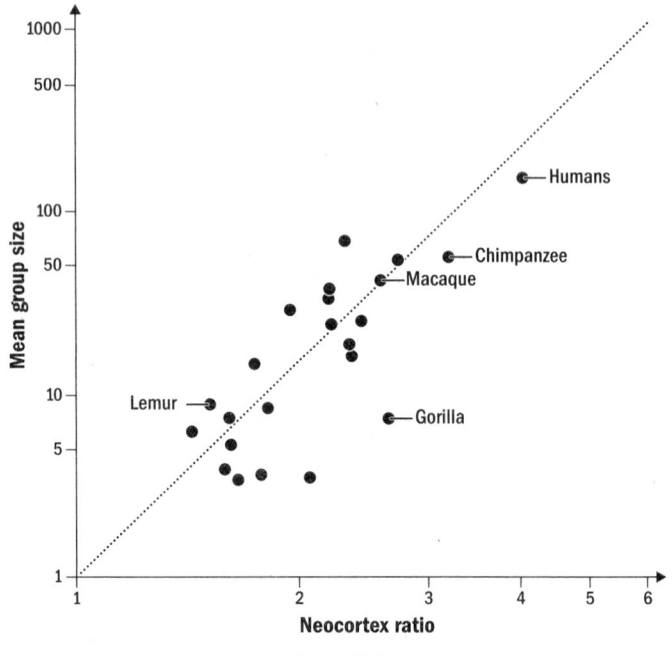

Figure 15.2

that live in groups. And interestingly, this correlation does not hold for most other animals. The brain of a buffalo living in a thousand-member herd is not meaningfully bigger than the brain of a solitary moose. It isn't group size in general but the specific type of group that early primates created that seemed to have required larger brains. There was something unique about primate groups relative to those of most other mammals, something we can understand only by understanding the general drive for grouping itself.

The Evolutionary Tension Between the Collective and the Individual

Early mammals were likely more social than the amniotes (their lizard-like ancestors) that came before. These early mammals uniquely gave birth to helpless children. This dynamic would have been tenable only if mothers built a strong bond to help, nurture, and physically protect their children. Further, mammals engage in play much more than other vertebrates. Even the offspring of simple mammals like rats play with each other, play mounting and play fighting. These early acts of play might have served the purpose of refining and training the motor cortex of young mammals so that in higher-stakes situations they wouldn't be learning from scratch. In these early mammals, this collaborative period between mother and child was relatively short-lived. After a period of childhood development, the bond tends to fade, and the children and mothers go their separate ways. This is how it is for many mammals that spend most of their lives on their own, like tigers and bears.

But not all animals separate in adulthood like this. In fact, the simplest, most widely used, and likely first collective behavior in animals was group living, whereby animals of the same species simply clustered together. Fish reflexively follow each other's movements and swim near one another. Many herbivorous dinosaurs lived in herds. And, of course, this is seen across mammals as well—buffalo and antelope live in herds. The key benefit of group living is that it helps stave off predators. If even a single antelope in a herd catches a glimpse of a nearby lion and begins running away, it cues all the others in the herd to follow. While a lone antelope is easy prey, a herd of them can be dangerous to even a lion.

However, group living is not a freely gained survival benefit—it comes at a high cost. In the presence of food constraints or limited numbers of eligible mates, a herd of animals creates dangerous competition. If this competition leads to infighting and violence, the group ends up wasting precious energy competing and fighting each other. In such a circumstance, the same number of animals would have been better off living separately.

Thus, animals who fell into the strategy of group living evolved tools to resolve disputes while minimizing the energetic cost of such disputes. This led to the development of mechanisms to signal strength and submission without having to actually engage in a physical altercation. Deer and antelope lock horns to compete for food and mates, a much cheaper form of competition than fighting. Bears, monkeys, and dogs bare their teeth and growl to show aggression.

These animals also evolved mechanisms to signal submission, allowing them to acknowledge defeat and show that it is unnecessary for others to spend energy hurting them. Dogs bow and roll over to their backs; bears sit down and look away; and deer lower their heads and flatten their ears. All of this provided a mechanism to ameliorate tensions and lower the amount of energy spent infighting.

With the ability to signal strength and submission, many animals were able to make group living work. Most lineages of mammals fell into one of four buckets of social systems: solitary, pair-bonded, harems, and multi-male groups. The solitary mammals, such as moose, spend a large portion of their adult lives on their own, meeting up primarily for mating and, if female, raising children. The pair-bonding mammals, such as red foxes and prairie voles, live together in pairs, raising children together. Other mammals, such as camels, live in harems, which are social groups with a single dominant male and many females. And then there are mammals that live in multi-male groups, social groups with many males and many females living together.

While solitary and pair-bonded mammals avoid the downsides of large social groups, they also miss out on the benefits. On the other hand, harems and multi-male groups reap the benefits of larger groups, but incur the costs of competition. In addition to displays of aggression and submission, another way harems and multi-male groups minimize competition is

The four common social structures found in mammals

SOLITARY	PAIR-BONDING	HAREMS	MULTI-MALE GROUPS
Independent	One male, one female	One male, many females	Many males, many females
Live mostly independently in adulthood	One male and one female live and raise children together.	A single dominant male living with a group of females who have their own hierarchy.	Separate male and female hierarchy
Tigers *Jaguar* *Moose*	*Red foxes* *Prairie voles* *Giant otters* *Pygmy marmoset*	*Mongolian camels* *Fur seals* *Gorillas*	*Lions* *Hippopotamuses* *Lemurs* *Chimpanzees* *Baboons* *Macaque monkeys*

through hierarchical rigidity. In harems, there is a single dominant male who does all of the mating; the only other males allowed in the group are his own children.

Multi-male groups also work via hierarchical rigidity: There is a strict hierarchy of both males and females. Low-ranking males are allowed in the group, but they do little mating and get the last pick of food; the high-ranking males eat food first and do most, if not all, of the mating.

How is the hierarchy decided in these social groups? It's simple—the strongest, biggest, and toughest become dominant. The locking of horns and baring of teeth are all designed to demonstrate who *would* win in a fight while avoiding the fight itself.

Early primates made this same evolutionary trade-off—they evolved to live in groups, accepting the risk of aggression for the benefit of the better predator avoidance of large groups. The greater the predation risk experienced by primates, the larger the social group they created in response. Like many modern primates, these early primates probably lived in multi-male groups with hierarchies of males and females where lower-ranking members got the worst food and almost never mated and high-ranking members got their pick of each. They bared their teeth as a show of aggression. We know this from fossils and through observing the behavior across many of today's monkey and ape societies: chimpanzees, bonobos, and macaques all live this way. At first glance, these early primate groups would have looked no different from other mammals' multi-male groups. But as researchers observed monkey and ape behavior more thoroughly, it became clear that when it comes to sociality, primates are unlike most of their mammalian cousins. Something happened in the sociality of early primates that was different from the sociality that had evolved before in early mammals.*

Machiavellian Apes

In the 1970s, the primatologist Emil Menzel was running experiments with a group of chimpanzees living in a one-acre forest. Inspired by Tolman's experiments on mental maps in rats, Menzel was experimenting with the mental maps of chimpanzees. His main interest was in whether chimpanzees could remember the locations of hidden morsels of food.

Menzel would hide some food in a random location within this one-acre area, perhaps under a rock or within a bush, and then reveal its location to one of the chimpanzees. He would then place food back in these locations on a recurring basis. Chimpanzees, like rats, were eminently able to remember these exact locations, learning to recheck these specific spots to forage Menzel's hidden food. But Menzel began spotting behavior that was quite unlike that of a rat, behavior that he had never intended to investigate and had never expected to find. Indeed, while merely investigating spatial memory, Menzel unearthed behavior that was eerily Machiavellian.

When Menzel first revealed the location of hidden food to one of the subordinate chimpanzees, named Belle, she happily alerted the rest of the group and shared the food. But when the dominant male, Rock, came over to indulge in the treat, he took all the food for himself. After Rock did this a few times, Belle stopped sharing, and began engaging in ever more

---

* Similar things may have independently occurred in other lineages of mammals with complex sociality (such as dogs, elephants, and dolphins).

sophisticated strategies to withhold information about the hidden locations of food from Rock.

At first, Belle simply sat on top of the secret location of the food to hide it from Rock, and only when he was far away would she uncover and openly eat the food. But when Rock realized she was hiding the food underneath her, he began pushing her to get at the food. In response to this, Belle came up with a new strategy—once she was shown a new location for the hidden food, she didn't go to it immediately. She would wait for Rock to look away, and then she would run to the food. In response to this new strategy, Rock began trying to trick Belle: He would look away and act uninterested, and once Belle went for the food, he would turn around and run toward it. Belle even began trying to lead Rock in the wrong directions, a deception Rock eventually realized and thus, in response, began to search for food in the opposite direction that Belle would try to lead him.

This process of ever-escalating deceptions and counter-deceptions reveals that both Rock and Belle were able to understand the other's intent ("Belle is trying to lead me away from the food," "Rock is trying to trick me by looking away"), as well as understand that it is possible to manipulate the other's beliefs ("I can make Belle think I am not looking by pretending to be disinterested," "I can make Rock think the food is in the wrong location by leading him in that direction"). Since Menzel's work, numerous other experiments have similarly found that apes can, in fact, understand the intentions of others. Consider the following study: Apes were tested on their ability to tell the difference between "accidental" and "intentional" actions. Show chimpanzees or orangutans three boxes, one of which has food in it. The box filled with food can be differentiated from the others because it has a thick mark from a pen on it. Do this a few times until they learn that food is always in the box with the mark on it. Then have an experimenter come in with the three boxes and intentionally mark one box by leaning over to mark it and "accidently" mark another box by passively dropping the marker on it. When the apes are allowed to come to the boxes to look for food, which box do they go to? They immediately go for the box the experimenter "intentionally" marked and ignore the one that was "accidently" marked. Apes had deduced the *intent* of the experimenter.

Consider another study. Have a chimpanzee sit across from two experimenters who have food near them. One experimenter is *unable* to give food for various reasons (sometimes unable to see the food; sometimes the food is stuck; other times they demonstrate losing the food). Another experimenter is *unwilling* to give them food (simply having food but not giving it). Neither experimenter provided food, and yet chimpanzees treat these two cases differently. When given the chance to pick between these two experimenters, chimps always went back to the person who seemed unable to help and avoided those who seemed unwilling. Chimpanzees seemed able to use cues about the situation of another person (Can they see the food? Did they lose it? Can they just not get it?) to reason about their intentions, and thereby predict the likelihood this other person would give them food in the future.

Understanding the minds of others requires understanding not only their intentions but also their *knowledge*. Belle sitting on the food to hide it from Rock was Belle's attempt to manipulate Rock's knowledge. In another test, chimps were given the chance to play with two sets of goggles—one was transparent and easy to see through, and the other was opaque and hard to see through. When given the chance to ask for food from human experimenters wearing those same goggles, chimps knew to go to the human that was wearing the transparent goggles—they could tell the human wearing the opaque goggles wouldn't see them.

The degree to which animals can infer the intent and knowledge of others continues to be controversial in animal psychology. While there is meaningful evidence that many primates (especially apes) have this ability, the evidence in other animals is less clear. It is possible that other intelligent animals, like some birds, dolphins, and dogs, also can do this. My argument is not that *only* primates can do this but that this ability was not present in early mammals, and in the human lineage this feat emerged with early primates (or at least by early apes). Even dogs, as socially intelligent and attentive to humans as they are, may be unable to understand that humans can hold different knowledge. Allow a dog to see its trainer put a treat in one location and then allow the dog to see someone else put a treat in a different location (while the trainer is not present and hence unaware of this other treat). When the trainer comes back and commands, "Get the treat," dogs are equally likely to run to either location, failing to identify the location the trainer is referring to based on which location the trainer knows about.

This act of inferring someone's intent and knowledge is called "theory of mind"—so named because it requires us to have a theory about the minds of others. It is a cognitive feat that evidence suggests emerged in early primates. And as we will see, theory of mind might explain why primates have such big brains and why their brain size correlates with group size.

Primate Politics

The most obvious social behavior of nonhuman primates is *grooming*—a pair of monkeys will take turns picking dirt and mites from each other's backs where they can't reach themselves. In the first half of the twentieth century, this behavior was believed to be primarily for hygienic purposes. But it is now undisputed that this grooming behavior serves more of a *social* purpose than a *hygienic* purpose. There is no correlation between time spent grooming and body size (which you would expect if the function of grooming was for cleaning the body), but there is a strong correlation between time spent grooming and group size. Further, individuals that don't get groomed by others a lot don't make up for it by grooming themselves more. And individual monkeys have very specific grooming partners that persist over long periods, even lifetimes.

Primate groups are not made up of a social mash of randomly interacting individuals; these mini-societies of fifteen to fifty primates are made up of subnetworks of dynamic and specific relationships. Monkeys keep track of and remember each individual in their group and are able to recognize them by appearance and voice. They keep track not only of individuals but also of the *specific relationships between* individuals. When a faraway distress call is heard from a child, individuals of the group all immediately look not in the direction of the distress call but at the *mother* of the distressed child—*Oh no, what will Alice do to help her daughter?* Or *Can we trust this kid? Let's see what the mother does.*

The relationships between individuals are not only familial but also hierarchical. Vervet monkeys have an approach-retreat routine to signal dominance and submission; when a higher-ranking individual walks toward a lower-ranking individual, the lower-ranking individual retreats. These dominance relationships persist across contexts: When monkey A expresses submission to B in one situation, it is almost always the case that A will also submit to B in *another* situation. These dominance relationships are transitive: If you see monkey A submit to B, and B submit to C, then it is almost definitely the case that A will submit to C. And these hierarchies often persist for many years, even generations. These signals of dominance and submission are not one-off displays; they represent an explicit social hierarchy.

Primates are extremely sensitive to interactions that violate the social hierarchy. In a study done in 2003, experimenters obtained audio recordings of different members of a group of baboons making dominance or submission noises, and then set up speakers near the baboons so they could play back these recordings. When they played a recording of a higher-ranking baboon making a dominance noise followed by a recording of a lower-ranking baboon making a submission noise, no baboons looked over to the speakers—nothing surprising about someone establishing their dominance over someone ranked below them. However, when they played a recording of any lower-ranking baboon making a dominance noise followed by a recording of a baboon ranked above them making a submission noise—a *violation* of the hierarchy—baboons freaked out and stared at the speakers to see what in the world had just happened. Like when a nerd smacks the popular bully in the face, the class can't help but gawk: *Did that really just happen?*

What makes these monkey societies unique is not the presence of a social hierarchy (many animal groups have social hierarchies), but how the hierarchy is constructed. If you examined the social hierarchy of different monkey groups, you would notice that it often isn't the strongest, biggest, or most aggressive monkey who sits at the top. Unlike most other social animals, for primates, it is not only *physical* power that determines one's social ranking but also *political* power.

As was the case in many early human civilizations (and unfortunately still many today), one thing that determines a monkey's place in its group is the family it is born into. In primate social groups, there tends to be a hierarchy of families. Here is a common structure for female hierarchies: At the top of the hierarchy is the eldest female of the highest-ranking family, followed by her offspring, then the eldest member of the second-highest-ranking family, followed by her offspring, and so on and so forth. And when a daughter's mother dies, she tends to inherit the rank of her mother.

In a clear departure from the typical nonprimate association of strength with rank, a weak and frail juvenile member of a powerful family can easily scare off a much larger and stronger adult monkey from a family that ranks below her. In fact, the child herself is clearly aware of her place in the social structure—even young children will regularly challenge adults of lower-ranking families, but they won't challenge adults of higher-ranking families.

And as it goes with human societies—with endless oscillations of dynastic power struggles of families rising and collapsing—monkey dynasties rise and fall too. Families face incredible pressure to improve their position. Higher-ranking monkeys get their pick of food, grooming partners, mates, and resting sites. One's evolutionary fitness improves with one's rank; higher-ranking monkeys have more children and are less likely to die from disease. And so if a high-ranking family sufficiently dwindles in number, a lower-ranking family will wage a coordinated mutiny; the lower-ranking family will make persistent aggressive challenges until the higher-ranking family submits, at which point a new hierarchy has been established.

Such mutinies are not inevitable; high-ranking families that are few in number can form alliances with members outside of their families to help solidify their position. Indeed, about 20 percent of the time that an aggressive display is made, other nearby monkeys will respond by joining forces with either the attacker or the defender. Most of the time it is family members that come to help, but about a third of the time it is *nonfamily* members that come. It seems that the ability to forge such allyships is one of the primary determinants of an individual's rank: higher-ranking monkeys tend to be better at recruiting allies from unrelated individuals, and hierarchy reversals most often happen when monkeys fail to recruit such allies.

Monkey politics happens in the dynamics of these allyships, which are forged not through fixed familial relationships but grooming and supporting others in conflicts. Allyships and grooming partnerships represent a common relationship, what we would call a *friendship*: monkeys most often rescue those whom they have previously formed grooming partnerships with. And a monkey can earn reciprocity through acts of kindness even to those they are not currently friends with. If monkey A goes out of its way to groom monkey B, then B is substantially more likely to run to A's defense next time A makes a "help me" vocalization. This is also true with going out of one's way to support others in conflicts—monkeys tend to run to the defense of those who have run to their own defense. In these allyships is also a notion of trust: When a chimpanzee is given the option of (a) an experimenter giving her a mediocre snack directly, or (b) an experimenter giving *another* chimp, who will hopefully share, an *amazing* snack, chimps chose (b) only if the other chimp is a grooming partner. Otherwise they took the worse food just for themselves.

These allyships have a big impact on a monkey's political standing and quality of life. Those with power benefit from forging a sufficient coalition of lower-ranking allies, and low-ranking monkeys can substantially improve their lives by forging friendships with the right high-ranking families. Low-ranking individuals with powerful grooming partners get harassed much less, even when the high-ranking ally is out of sight; everyone in the group knows *Don't mess with James unless you want to deal with Keith.* High-ranking monkeys are more tolerant of the low-ranking individuals with whom they have forged allyships, giving them more access to food.

Much of monkey social behavior suggests an incredible degree of political forethought. Monkeys prefer to invest in relationships with those ranked higher than themselves. Monkeys prefer to mate with higher-ranking members of the group. Monkeys compete to groom with high-ranking individuals. When a dispute breaks out, monkeys are biased toward joining forces with the higher-ranking individual. The children of high-ranking mothers are the most popular playmates.

High-ranking monkeys also exhibit a cleverness in *which* lower-ranking members they choose to befriend. In a study where different low-ranking monkeys were trained to do specific tasks to obtain food, high-ranking monkeys quickly befriended those who had specialized skills, and they persisted in these grooming partnerships even after there was no immediate prospect of gaining food: *I can see you are useful; let me take you under my wing.*

Monkeys also show political cleverness in the aftermath of conflict. They go out of their way to try and "make up" after aggressive interactions, especially those with nonfamily members. They often seek to hug and groom those they fought with, and they also try to make up with the individuals' *families*, spending double the usual amount of time with family members of those they have recently quarreled with.

The Arms Race for Political Savvy

Somehow the evolutionary trajectory of early primates led to the development of the incredibly broad suite of complex social behaviors that are seen across modern species of primates. And within these behaviors we see hints of the behavioral foundation of how humans tend to interact with each other. Why primates evolved these instincts is not exactly clear, but it may have had to do with the unique niche early primates found themselves in in the aftermath of the Permian-Triassic extinction event.

Early primates seemed to have had a unique diet of foraging fruit directly in treetops—they were *frugivores*. They plucked fruit from trees right after it ripened but before it fell to the forest floor. This allowed primates to have easy access to food without much competition from other species. This unique ecological niche may have offered early primates two gifts that opened the door to their uniquely large brains and complex social groups. First, easy access to fruit gave early primates an abundance of *calories*, providing the evolutionary option to spend energy on bigger brains. And second, and perhaps more important, it gave early primates an abundance of *time*.

Free time is extremely rare in the animal kingdom; most animals have no choice but to fill every moment of their daily calendar with eating, resting, and mating. But these frugivorous primates didn't have to spend nearly as much time foraging as other animals did, so when seeking to climb the social hierarchy, these primates had a new evolutionary option: instead of spending energy evolving bigger muscles to fight their way to the top, they could spend energy evolving bigger brains to politick their way to the top.

So primates seemed to have filled their open calendars with politicking. Today's primates spend up to *20 percent of their day* socializing, a much larger amount of time than most other mammals. And it has been shown that this social time is causally related to how much free time primates have; as more free time is given (by providing easier access to food), primates spend more time socializing.

This created a completely new evolutionary arms race: a battle for political savvy. Any primate born with better tricks for currying favor and gaining allies would survive better and have more babies. This put more pressure on other primates to evolve smarter mechanisms for politicking. Indeed, neocortex size of primates is correlated not only with social-group size but also with social savviness. An outgrowth of this arms race seems to have been a blossoming of many human social instincts, both the good (friendships, reciprocity, reconciliation, trust, sharing) and the bad (tribalism, nepotism, deception). While many aspects of these behavioral changes did not require any particularly clever new brain systems, there was indeed an intellectual feat underlying this politicking: the ability to engage in theory of mind.

It isn't clear how political savviness would even be possible if a species did not have at least a basic and primitive version of theory of mind—only through this ability can individuals infer what others want and thereby figure out whom to cozy up to and how. Only through theory of mind can individual primates know not to mess with a low-ranking individual with high-ranking friends; this requires understanding the intent of the high-ranking individuals and what they will do in future situations. Only through this ability of theory of mind can you figure out who is likely to become powerful in the future, whom you need to make friends with, and whom you can deceive.

So this may be why primates began growing such big brains, why their brain size is correlated with social group size, and why primates evolved the ability to reason about the minds of others. The question is, of course, *how* do primate brains do this?

16

How to Model Other Minds

OUR MAMMALIAN ANCESTOR from seventy million years ago had a brain that weighed less than half a gram. By the time our ape ancestors arrived ten million years ago, it had expanded to about three hundred fifty grams. This is almost a thousandfold increase in brain size. Such a large expansion presents a challenge in relating brain areas across time (early mammals to early primates) and across species at a point in time (today's mouse brain to today's chimpanzee brain). Which brain areas are truly new and which are merely expanded versions of the same thing?

Clearly some structures in brains will scale naturally with body size without any meaningful change to their function. For example, a bigger body means more touch and pain nerves, which means more neocortical space for processing these sensory signals. The surface area of the somatosensory cortex of the early apes was obviously much bigger than that of early mammals even though it performed the same function. Same for bigger eyes and muscles and anything else that requires incoming or outgoing nerves.

Further, more neurons can be added to a structure to improve its performance without fundamentally changing its function. For example, if the basal ganglia were one hundred times larger, it might enable associations between many more actions and rewards while still fundamentally performing the same function: implementing a temporal difference learning algorithm. Similarly, the visual cortex of primates is massively larger than that of rodents, even accounting for brain scaling. Unsurprisingly, primates are better than rodents at many aspects of visual processing. But the visual area of the neocortex does not perform some unique function in primates; primates simply dedicated more space, proportionally, to the same function and got better performance.

And then there are the fuzzy areas—those structures that are very similar but slightly modified, teetering on the edge of new and old. An example of this is new hierarchical layers of sensory processing in the neocortex. Primates have many hierarchical regions of the visual neocortex with processing hopping from one region to the next. These areas still process visual input, but the addition of new hierarchical layers makes them qualitatively different. Some areas respond to simple shapes; other areas respond to faces.

But there are also, of course, truly new brain regions—structures with completely unique connectivity that perform novel functions.

So the question is: In the brain of early primates, how much was simply scaled up (whether proportionately or disproportionately) and how much was new? Most evidence suggests that despite the dramatic expansion in size, the brain of our primate ancestor, and of primates today, was largely the same as that of the early mammals. A bigger hindbrain, a bigger basal ganglia, a bigger neocortex, but still all the same regions connected in all the same fundamental ways. These early primates indeed dedicated disproportionately more neocortex to certain functions, such as vision and touch, but still, the functions and connectivity were largely the same. New hierarchical layers were added, with sensory information hopping from one neocortical layer to another, enabling progressively more abstract representations to be formed. But this was mostly just performance-enhancing.

So might it be the case that the surprising smarts of primates, with all

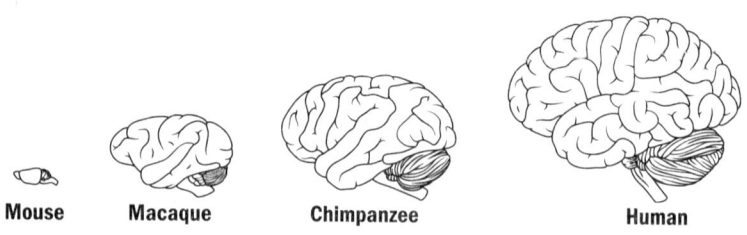

Figure 16.1

their theory of mind, politicking, and trickery, was a consequence of nothing more than brain scaling?

The New Neocortical Regions of Early Primates

Although most of the brain of the early primates was merely a scaled-up mammal brain, there were, in fact, certain areas of neocortex that were truly new. We can categorize these new neocortical areas that emerged within the primate lineage into two groups. The first is the granular prefrontal cortex (gPFC), which was a new addition to the frontal cortex.* This newer granular prefrontal cortex wraps around the much older *agranular* prefrontal cortex (aPFC). The second new area of neocortex, which I will call the primate sensory cortex (PSC), is an amalgamation of several new areas of sensory cortex that emerged in primates.† The gPFC and PSC are extremely interconnected with each other, making up their own new network of frontal and sensory neocortical regions.

What makes these areas "new"? It isn't their microcircuitry; all these areas are still neocortex and have the same general columnar microcircuitry as other areas of neocortex across mammals. It is their input and output connectivity that renders them new; it is what these areas construct a generative model *of* that unlocked fundamentally new cognitive abilities.

As we saw in breakthrough #3, a human with damage to the aPFC has obvious and severe symptoms, such as akinetic mutism, in which patients become completely mute and intentionless.

In contrast to the alarming symptoms of aPFC damage, damage to the surrounding granular prefrontal cortex often results in minimal symptoms. In fact, the impairment from damage to these areas is so minimal that many neuroscientists in the 1940s wondered if these areas lacked any functional significance at all. A famous case study at the time was of a

---

* As mentioned in chapter 11, it is called granular because of its uniquely thick layer four, which contains granule neurons.
† The main new areas are the superior temporal sulcus (STS) and the temporoparietal junction (TPJ).

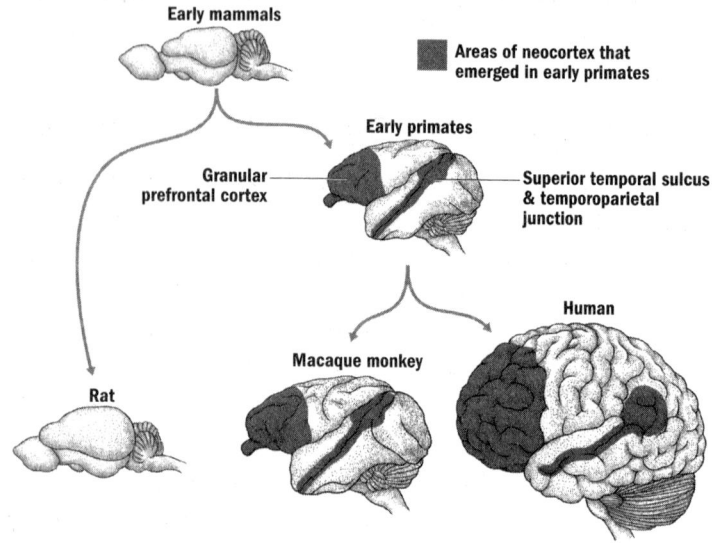

Figure 16.2: Shared neocortical regions across mammals and new regions of primates

seizure patient named K.M. who had a third of his frontal cortex removed to treat seizures. After the surgery, K.M. seemed to have no deficits in intellect or perception whatsoever. K.M.'s IQ after the removal of a third of his frontal cortex was unchanged—if anything, *it increased*. In the words of a neuroscientist at the time, the function of granular prefrontal cortex was a "riddle."

Modeling Your Own Mind

Consider a study done in 2001. Human subjects were put in an fMRI machine and shown a series of pictures. With each picture, subjects were asked "How does this make you feel?" or some mundane question about the contents of the image, such as "Was this picture taken indoors or outdoors?" Both tasks activated the aPFC, which makes sense, since both require rendering an inner simulation; either of the world around the picture to decide if it is more likely to be inside or outside, or of your own thoughts and sensations. But it was only when humans were asked how *they felt* about a picture that the gPFC lit up.

There have now been numerous experiments confirming this. The granular prefrontal cortex becomes uniquely active during tasks that require *self-reference*, such as evaluating your own personality traits, general self-related mind wandering, considering your own feelings, thinking about your own intentions, and thinking about yourself in general.

With this clue that the granular prefrontal cortex is uniquely activated by self-reference, might we have missed some subtle but crucial impairments of granular prefrontal damage?

In 2015, scientists did the following study. They gave participants a neutral cue word (e.g., *bird* or *restaurant*) and asked them to tell the experimenter different narratives of *themselves* associated with that word. Some of these participants were healthy, some had damage to areas of the granular prefrontal cortex, and some had damage to the hippocampus.

How did people's narratives differ across these conditions? Humans with damage to the granular prefrontal areas but with the aPFC and hippocampus intact were able to imagine complex scenes, rich with detail, but they were impaired at imagining *themselves* in such scenes. They sometimes even completely omitted themselves from their narratives. Damage to the hippocampus seemed to have the opposite effect—patients could imagine themselves in past or future situations just fine but struggled to build external features of the world; they were unable to describe in detail any of the surrounding elements.

This suggests that the granular prefrontal cortex plays a key role in your ability to project *yourself*—your intentions, feelings, thoughts, personality, and knowledge—into your rendered simulations, whether they are about the past or some imagined future. The simulations run in rat brains, which has no gPFC and only an aPFC and hippocampus, show evidence of rendering an external world, but there is nothing to suggest that they truly project any meaningful model of *themselves* into these simulations.

Damage to the granular prefrontal cortex may also affect people's models of themselves not only in their mental simulations, but also in the *present*. Some people with gPFC damage develop mirror-sign syndrome, a condition in which they no longer recognize themselves in a mirror. These patients *insist* that the people they see in a mirror are not *them*.

Your modeling of your own mind in the present and the mind you project in your imagination seem to be intimately related.

The idea that the new primate areas take part in modeling *your own mind* makes sense when you follow their input/output connectivity. The older mammalian aPFC gets input directly from the amygdala and hippocampus, while the new primate gPFC receives almost no amygdala or hippocampal input or any direct sensory input at all. Instead, the primate gPFC gets most of its input directly from the older aPFC.

One interpretation of this is that these new primate areas are constructing a generative model of the older mammalian aPFC and sensory cortex itself. Just as aPFC constructs explanations of amygdala and hippocampus activity (invents "intent"), perhaps the gPFC constructs explanations of the aPFC's model of intent—possibly inventing what one might call *a mind*. Perhaps the gPFC and PSC construct a model of one's own inner simulation to explain one's intentions in the aPFC given knowledge in the sensory neocortex.

Let's use a thought experiment to build some intuition about what this means. Suppose you put our ancestral primate in a maze. When it reached a choice point, it turned left. Suppose you could ask its different brain areas why the animal turned left. You would get very different answers at each level of abstraction. Reflexes would say, *Because I have an evolutionarily hard-coded rule to turn toward the smell coming from the left.* Vertebrate structures would say, *Because going left maximizes predicted future reward.* Mammalian structures would say, *Because left leads to food.* But primate structures would say, *Because I'm hungry, eating feels good when I am hungry, and to the best of my knowledge, going left leads to food.* In other words, the gPFC constructs explanations of the simulation itself, of what the animal wants and knows and thinks. Psychologists and philosophers call this *metacognition*: the ability to think about thinking.

What mammals find in their inner simulations of the external world is, in some sense, the same thing as their knowledge about the external world. When a mammal simulates going down a path and its sensory neocortex renders a simulation that contains water at the end of the path, this is the same thing as "knowing" that water exists at the end of the path. While the older mammalian areas of the sensory neocortex render the simulation

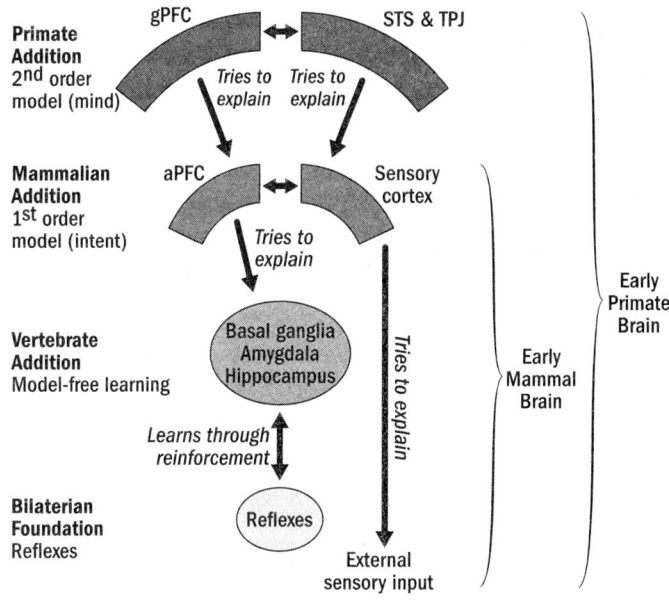

Figure 16.3

of the external world (containing knowledge), the new primate areas of neocortex (what I have been calling the primate sensory cortex) seems to create a model of this knowledge itself (areas of PSC get input from various areas of sensory neocortex). These new primate areas try to explain *why* the sensory neocortex believes food is over there, why an animal's inner simulation of the external world is the way it is. An answer might be: *Because last time I went over there I saw water, and hence when I simulate going back there, I see water in my imagination.* Put slightly differently: *Because I last saw water there, I now know that water is over there even though before I did not.*

These systems are all bootstrapped on one another. Reflexes drive valence responses without any learning required, making choices based on evolutionarily hard-coded rules. The vertebrate basal ganglia and amygdala can *then* learn new behaviors based on what has historically been reinforced by these reflexes, making choices based on maximizing reward. The mammalian aPFC can then learn a generative model of this model-free behavior and construct explanations, making choices based on imagined goals (e.g., drinking water). This could be considered a first-order model. The primate gPFC can then learn a more abstract generative model (a second-order model) of this aPFC-driven behavior and construct explanations of intent itself, making choices based on mind states and knowledge (*I'm thirsty; drinking water when thirsty feels good, and when I simulate going down this way, I find water in my simulation, hence I want to go in this direction*).

The mammalian first-order model has a clear evolutionary benefit: It enables the animal to vicariously play out choices before acting. But what is the evolutionary benefit of going through the trouble of developing a second-order model? Why model your own intent and knowledge?

Modeling Other Minds

Consider the comic-strip task devised by Eric Brunet-Gouet in 2000. Human participants were shown several comic strips, each containing three frames, and asked to guess which fourth-frame ending was most likely. There were two types of comic strips—one required inferring the intent of characters in order to correctly guess the ending, and the other required only understanding physical causal relationships.

Brunet-Gouet performed a PET scan of the subjects' brains while they did this comic-strip task. He found an interesting difference in which brain regions got excited by each type of comic strip. When asked about comic strips that required understanding the intentions of the characters, but not when asked about other types of comic strips, these uniquely primate areas of neocortex, like gPFC, lit up with activation.

In addition to inferring other people's *intent*, these primate areas also get activated by tasks that require inferring other people's *knowledge*. A famous test of this is the Sally-Ann test: Participants are shown a series of events occurring between two individuals, Sally and Ann. Sally puts a marble in a basket; Sally leaves; when Sally isn't looking, Ann moves the marble from the basket to a nearby box; Sally comes back. The participant is then asked: If Sally wants to play with her marble, where will she look for it?

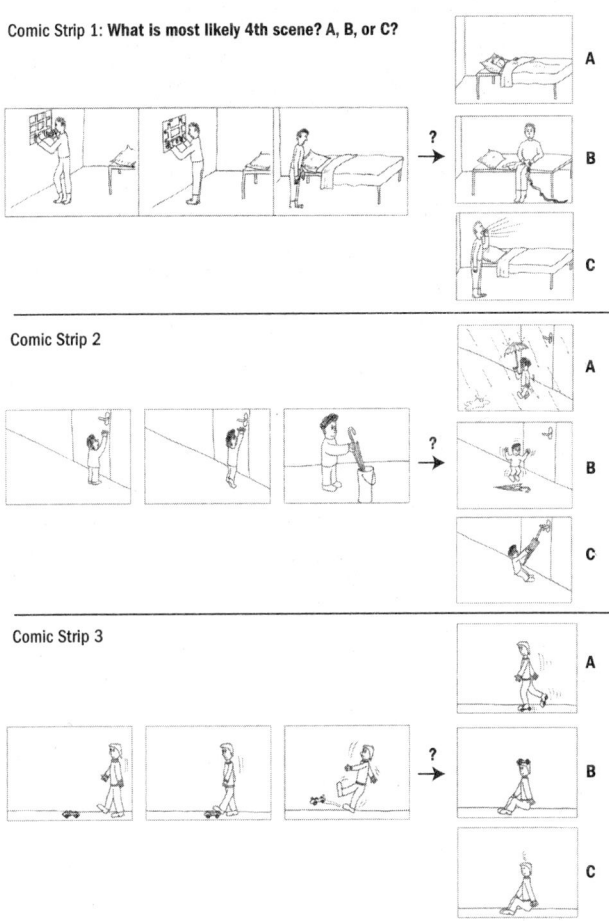

Figure 16.4. Examples of comic-strip task: Comic Strips 1 and 2 require understanding intent, while Comic Strip 3 does not require understanding intent. Answer to Comic Strip 1 is B (he wants to escape through the window). Answer to Comic Strip 2 is C (he wants to open the door). Answer to Comic Strip 3 is C (he will fall down).

Answering this correctly requires realizing that Sally has different knowledge than you. While *you* saw Ann put the marble in the box, *Sally* did not. And so the right answer is that Sally will look in the basket, even though the marble is not there. There are many forms of the Sally-Ann test, generally referred to as "false-belief tests." Humans pass such false-belief tests by the age of four. When humans in fMRI machines are given

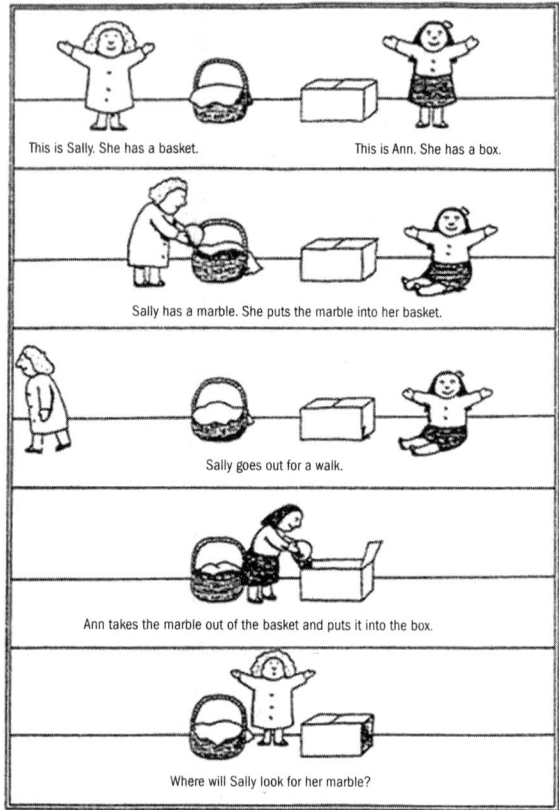

Figure 16.5: The Sally-Ann test for theory of mind

false-belief tests, the primate gPFC and PSC light up, and people's performance correlates with the degree of activation. Indeed, across countless studies, it has been shown that many areas of these uniquely primate neocortical areas are specifically activated by such false-belief tasks.

If we revisit our mysterious patients with granular prefrontal damage and test them for theory-of-mind tasks, we begin to see a common theme emerge from their seemingly disparate, subtle, and bizarre symptoms. Such patients are worse at solving false-belief tests like the Sally-Ann test; they are much worse at recognizing emotions in other people; they struggle to empathize with other people's emotions, struggle to distinguish lies from jokes, struggle to identify a faux pas that would offend someone, struggle to take someone else's visual perspective, and struggle to deceive others.

Although all of the above studies have to do with human brains (albeit the parts of human brains inherited from our primate ancestors), experiments have confirmed the same effects in nonhuman primates. Put a monkey in a situation where they have to reason about the intent or knowledge of another to solve a task, and gPFC lights up, just as it does in humans. Damage the gPFC in a monkey, and their performance on such tasks becomes impaired, just as it does in humans.

And indeed, revealing their importance in understanding others, the size of these granular prefrontal areas is correlated with social-network size in primates. The bigger a primate's granular prefrontal area, the higher in the social hierarchy it tends to be. This same relationship is even seen in humans: the thicker a human's granular prefrontal areas, the larger his or her social network, and the better that person's performance on theory-of-mind tasks.

These newly primate neocortical regions seem to be the locus of both one's model of one's own mind and the ability to model other minds. The fact that these two seemingly distinct functions have extremely overlapping, if not identical, neural substrates offers a glaring clue as to the evolutionary purpose and mechanism of these new primate structures.

Modeling Your Mind to Model Other Minds

As far back as Plato, there has been a running hypothesis about how humans understand the minds of other humans. The theory is that we first understand our own minds and then use this understanding of ourselves to understand others. Modern formulations of this old idea are referred to as "simulation theory" or "social projection theory." When we try to understand why someone else did something, we do so by imagining ourselves in their situation—with their knowledge and life history: "She probably yelled at me because she is stressed out about having a test tomorrow; I know I yell more when I am stressed." When we try to understand what others will do, we imagine what we would do in their situation if we had their knowledge and their background: "I don't think James will share his food with George anymore; I believe James saw George steal, and I know if I saw my friend steal from me, I wouldn't share with him anymore." We understand others by imagining ourselves in their shoes.

The best evidence for social projection theory is the fact that tasks that require understanding *yourself* and tasks that require understanding *others* both activate and require the same uniquely primate neural structures. Reasoning about your own mind and reasoning about other minds is, in the brain, the same process.

Evidence for social projection theory can also be found in how children develop their sense of self. The childhood development of a sense of self is highly related to the childhood development of theory of mind. One way to test a child's sense of self is the mirror self-recognition test. Put a smudge of something on a child's face, let him look in the mirror, and see if he touches that part of his face, realizing that he is seeing himself. Children don't tend to pass this test until they are about two years old. It is also around this time that children begin to exhibit a primitive understanding of mental states and begin to use words like *want, wish,* and *pretend*. It is later, around the age of three, when children realize they can hold their own false beliefs; they are able to make statements such as "I thought it was an alligator. Now I know it's a crocodile." It isn't until after this point, around the age of four or five, when children pass false-belief tests, such as the Sally-Ann test, with respect to other people.

Other studies have found a strong correlation between children's ability to report on their *own* mental states and their ability to report on *others'* mental states—when children get better at one, they tend to simultaneously get better at the other. Further, damage to the development of one impairs skills in the other; chimpanzees raised in social isolation are unable to recognize themselves in a mirror. Modeling your own mind and that of others is interwoven.

Our understanding of ourselves often gets cross-wired with our understanding of others, consistent with the idea that we are repurposing a common system for each. For example, being in a specific mood (such as happy or sad) makes you biased to incorrectly infer these states in others; being thirsty makes you biased to incorrectly think that others are thirstier than they are. People tend to project their own personality traits onto others.

The self-other distinction can get cross-wired, an effect you would expect if we understand others by projecting ourselves into their situation.

To our original question: How might theory of mind work? One possibility, conceptually at least, might be that the uniquely primate neocortical areas first build a generative model of *your own inner simulation* (in other words, of your mind) and then use this model to try to simulate the minds of others.

We are in abstract land here—this is hardly a detailed algorithmic blueprint for how to build an AI system with theory of mind. But the idea of bootstrapping theory of mind by first modeling one's own inner simulation, of modeling yourself to model others, provides an interesting waypoint. Today, we can train an AI system to watch videos of human behavior and predict what humans will do next; we show these systems endless videos of humans doing things and tell them the right answers of what the humans are doing ("This is handshaking"; "This is jumping"). Advertising platforms can use behavior to predict what people will buy next. And we do have AI systems that attempt to identify emotions in people's faces (after the systems have been trained on a bunch of pictures of faces classified by emotions). But all this is clearly a far cry from the complexity of theory of mind in human (and other primate) brains. If we want AI systems and robots that can live alongside us, understand the type of people we are, deduce what we don't know that we want to know, infer what we *intend* by what we say, anticipate what we need or want before we tell them, navigate social relationships with groups of humans, with all of their hidden rules and etiquettes—in other words, if we want true humanlike AI systems, theory of mind will undeniably be an essential component of that system.

In fact, theory of mind might be the most essential aspect of building a successful future with superintelligent AI systems. If superintelligent AI systems cannot infer what we actually mean by what we say, we risk entering a dystopian future where AI systems misinterpret our requests with possibly catastrophic consequences. We will learn more about the importance of theory of mind when making requests of AI systems in breakthrough #5.

At the top of the social hierarchy of a troop of ancestral primates was more access to food and mates, and at the bottom was last pick of food and no access to mates. Theory of mind enabled each primate to climb this social ladder; it enabled them to manage their reputation and hide their transgressions; it enabled them to forge allyships, cozy up to rising stars, and kiss the ring of powerful families; it enabled them to build coalitions and stage rebellions; it enabled them to ameliorate brewing disputes and repair relationships after a tiff. Unlike the intellectual abilities that had emerged in the breakthroughs prior, theory of mind was not born from the need to survive the dangers of hungry predators or inaccessible prey, but instead from the subtler and far more cutting dangers of politics.

Politics was the origin story of breakthrough #4, but it is far from the entire story. As we will see in the next two chapters, theory of mind in early primates was repurposed for two other new abilities.

17

Monkey Hammers and Self-Driving Cars

JANE GOODALL COULD not believe her eyes.

It was November 1960. For months, she had been following a local tribe of chimpanzees in Gombe, Tanzania. The chimps had only recently begun accepting her presence, allowing her to stay close enough to observe them in their natural habitat. In the years before, Goodall had befriended a Kenyan paleontologist by the name of Louis Leakey; he eventually offered to send her to study the social lives of chimpanzees in their natural habitat. But the first discovery Goodall would make was not about their social lives.

As Goodall sat quietly some distance away, she noticed two chimpanzees that she had named David Greybeard and Goliath grabbing thin branches, stripping the leaves off them, and sticking them into a termite mound. When they pulled them out, they were covered in tasty termites, which they gobbled up. They were fishing. They were using tools.

It was long assumed that tool use was uniquely human, but tool use has now been found across many primates. Monkeys and apes not only use sticks to fish termites; they also use rocks to break open nuts, grass to floss, moss for sponges, clubs to smash beehives, and even twigs to clean their ears.

In the years since Goodall studied these chimps, tool use has been found all over the animal kingdom. Elephants pick up branches with their trunks to swat flies and scratch themselves. Mongooses use anvils to break open nuts. Crows use sticks to spear larvae. Octopuses gather large shells to make shields. Wrasses fish have been found to use rocks to break open clams to get to the inner food.

But primate tool use is more sophisticated than tool use in other animals. Wrasses fish, mongooses, and sea otters have been shown to use tools, but they typically have only a single trick up their sleeve. In contrast, groups of chimpanzees often exhibit over twenty different tool-using behaviors. Further, with the possible exception of birds and elephants, only primates have been shown to actively *manufacture* their tools. A chimp will shorten, sharpen, and remove leaves from a stick before using it to fish for termites.

Tool use in primates also shows a remarkable level of diversity across different social groups. Different species of wrasses fish use rocks in the same way despite having no contact with each other. But this is not the case with primates; different groups of the same species of primate exhibit surprisingly unique tool-using behaviors. The chimpanzees of Goualougo manufacture termite fishing rods differently than those in Gombe. Some groups of chimps regularly use rocks to open nuts; other groups don't. Some groups of chimps use clubs to pound beehives; other groups don't. Some groups use leafy twigs to swat away flies; others don't.

If the driver of brain evolution in early primates was a politicking arms race, why would primates be uniquely good tool users? If the new brain regions of primates were "designed" to enable theory of mind, then from where do the unique tool-using skills of primates emerge?

Monkey Mirrors

In 1990, Giuseppe di Pellegrino, Leonardo Fogassi, Bittorio Gallese, and Luciano Fadiga were chowing down lunch in the lab. They were members of Giacomo Rizzolatti's neurophysiology lab at the University of Parma; their mandate was to study the neural mechanisms of primate fine motor skills. A few feet from the lunch table sat a macaque monkey who was the object of their study. They had placed electrodes throughout its brain to search for which areas responded to the execution of specific types of hand movements. They had found specific areas of premotor cortex that activated when monkeys performed specific types of hand movements, some areas for grasping, others for holding, and others for tearing. But through lucky happenstance, they were about to discover something much more remarkable. As one of the lab members picked up a sandwich and bit into it, a loud crackle buzzed from a nearby speaker. The noise didn't come from the speaker connected to the fire alarm or the record player, it came from the speaker connected to the monkey's brain.

As they now recall, they had an immediate perception that something important had just happened. They had attached electrodes only to the motor areas of neocortex, the areas that were supposed to activate only when the monkey itself performed specific hand movements. But at that moment during lunch, despite the monkey not moving at all, the same area for hand grasping lit up at the exact moment when one of the human lab members grasped their own food.

After trying to replicate this phantom sandwich-watching activation, Rizzolati's team quickly realized that they had, in fact, discovered something more general: when their monkey observed a human perform a motor skill—whether picking up a peanut with two fingers, grasping an apple with their full hand, or grasping a snack with their mouth—the monkey's own motor neurons for performing that same skill would often activate. In other words, the neurons in the premotor and motor areas of a monkey's neocortex—those that control a monkey's own movements—not only activated when they performed those specific fine motor skills, but also when they merely watched others perform them. Rizzolatti called these "mirror neurons."

Over the subsequent twenty years, Rizzolatti's mirror neurons have been found in numerous behaviors (grasping, placing, holding, finger movements, chewing, lip smacking, sticking one's tongue out), across multiple areas of the brain (premotor cortex, parietal lobe, motor cortex), and across numerous species of primates. When a primate watches another primate do an action, its premotor cortex often mirrors the actions it is observing.

There are multiple competing interpretations of mirror neurons. Some argue mirror neurons are nothing more than associations—motor neurons get activated in response to any cue that has been associated with a movement. Monkeys *see* their *own* arms make grasping movements when they choose to make grasping movements, so of course when they see someone *else's* arms grasp things, some of the same neurons get a small kick of excitement. Others argue that mirror neurons represent something more fundamental—perhaps mirror neurons are *the* mechanism by which primates engage in theory of mind. The hypothesis is that primates have some clever mechanism for automatically mentally mirroring movements they see in others, and then by modeling themselves doing that behavior, they can ask "Why would *I* do this?" and try to deduce another monkey's or human's intentions.

Others have a more middle-of-the-road interpretation. Perhaps mirror neurons don't have some *automatic* mechanisms of mirroring; they are just clues that monkeys happen to be *imagining themselves* doing what they see someone else do. Mirror neurons aren't anything special; they are just evidence that monkeys are thinking about themselves grasping food when they see *you* grasp food. And as we already saw in chapter 12, areas of the motor cortex that activate when actually performing specific movements also get activated when people *imagine* themselves doing those same movements.

Here is some evidence that mirror neurons are just imagined movements. Monkeys don't need to directly observe the movements for their mirror neurons to activate; they can merely be given sufficient information for them to *infer* what movements are being performed. The motor neurons that light up right before a monkey does a behavior (such as picking up a peanut with the intention of breaking it open) also activate if a monkey simply *hears* the peanut break open (without seeing anything). Similarly, the neurons in a monkey that activate when it picks up a box will also activate when the monkey sees a human *presumably* pick up a box that is obscured behind a wall (but the neurons don't activate if monkeys know there is no box behind the wall). If mirror neurons were simply automatic mirrors, then they wouldn't activate in the above cases where monkeys were not directly observing behaviors. However, if mirror neurons are the consequence of imagined behaviors, then whenever something triggers a monkey to imagine itself doing something, you would see these mirror neurons activate.

If we accept the interpretation of mirror neurons as the imagination of movements, then this begs the question of *why* monkeys tend to imagine themselves doing what they see others doing. What is the point of mentally simulating movements you see in others? In chapter 12, we reviewed one benefit of motor simulation used by many mammals: planning movements ahead of time. This enables a cat to quickly plan where to place its paws to walk across a platform or a squirrel to plan how to jump between different branches. We hypothesized this might be why mammals have such great fine motor skills while most reptiles are woefully clumsy. But this trick has nothing to do with simulating movements you see in *others*.

One reason it is useful to simulate other people's movements is that doing this helps us understand their intentions. By imagining yourself doing what others are doing, you can begin to understand *why* they are doing what they are doing: you can imagine yourself tying strings on a shoe or buttoning a shirt and then ask yourself "why would I do something like this?" and thereby begin to understand the underlying intentions behind other people's movements. The best evidence for this is found in the bizarre fact that people with impairments in *performing* specific movements, also show impairments in *understanding the intentions* of those very same movements in others. The subregions of premotor cortex required for controlling a given set of motor skills are the same subregions required for understanding the intentions of others performing those same motor skills.

For example, in patients with brain damage to motor areas of neocortex, there is a significant correlation between impairments to action production (the ability to correctly mime using tools such as toothbrushes, combs, forks, or erasers) and action recognition (the ability to correctly select a video of a mimed action that matches an action phrase, like *combing hair*). Individuals who struggle to brush their own teeth tend to be bad at recognizing teeth-brushing in others.

Further, temporarily inhibiting a human's premotor cortex impairs their ability to correctly infer the weight of a box when watching a video of someone picking it up (arms that easily pick it up suggest it is light, but arms that struggle at first and have to adjust their position to get more leverage suggest it is heavy), but it has no impact on their ability to infer a ball's weight by watching a video of it bouncing on its own. This suggests that people mentally simulate *themselves* picking up a box when seeing someone else pick up a box ("I would turn my arm that way only if the box was heavy").

This impairment in understanding the actions of others is not some generalized effect of premotor cortex interference but is highly specific to the body parts that a brain is prevented from simulating. For example, temporarily inhibiting the *hand area* of your premotor cortex (which simulates your own hand movements), not only impairs your ability to perform your own hand movements, but also impairs your ability to recognize mimed *hand movements* (such as correctly recognizing mimed movements of grasping a hammer or pouring tea), but it has no impact on your ability to recognize mimed *mouth movements* (such as identifiying licking ice cream, eating a burger, blowing out a candle). Conversely, temporarily inhibiting the *mouth* area of your premotor cortex *does* impair your ability to recognized mimed mouth movements, while having no impact on your ability to recognize hand movements.

This suggests that the premotor cortex and motor cortex, the brain regions required to simulate your own movements, are also required to simulate the movements of others to understand their actions. But by *understand*, we don't mean understanding others' emotions (hungry, fearful) or others' knowledge ("Does Bill know that Jane hid the food?"). These studies demonstrate that the premotor cortex is involved in specifically understanding the *sensorimotor* aspects of others' behavior—inferring the strength required to pick up a box or the type of tool someone intended to hold.

But why does it matter to correctly identify the sensorimotor aspects of the behaviors you observe in others? What benefit does it provide to realize the tool someone is trying to hold or the weight of a box? The main benefit is that it helps us, as it helped early primates, *learn new skills through observation.* We already saw in chapter 14 that mentally rehearsing actions improves performance when actually performing actions. If this is the case, then it would make sense for primates to use their observations of others to rehearse actions.

Suppose you put a novice guitar player in an fMRI machine and ask them to learn a guitar chord by watching a video of an expert guitarist playing that chord. And suppose you compare their brain activation under two conditions, the first being when they observe a chord they don't know yet and the second when they observe a chord they *already know how* to play. The result: when they observe a chord they do not yet know, their premotor cortex becomes *way* more activated than when they observe a chord they already know how to play.

But the fact that premotor cortex becomes uniquely activated when trying to learn a new skill through observation, does not prove that it is *required* to learn a new skill through observation. Suppose you ask a human to watch two different videos. In the first video, he sees a hand pushing specific buttons on a keyboard, and he is asked to *imitate* those hand movements and push his own version of the same keyboard. In the other video, he sees a red dot move to different buttons on a keyboard, and he is asked to push the same buttons on his own keyboard. If you temporarily inhibit his premotor cortex during this task, he becomes specifically impaired at *imitating* hand motions but performs normally at following the red dots. Premotor activation is not just correlated with imitation learning; it seems to be, at least in some contexts, *necessary* for imitation learning. And here we can begin to unravel why primates are such great tool users.

Transmissibility Beats Ingenuity

Think about all the clever motor skills related to using tools: typing, driving, brushing your teeth, tying a tie, or riding a bicycle. How many of these skills did you figure out on your own? I'm going to bet that practically all these skills were acquired by *observing others*, not by your own independent ingenuity. Tool use in nonhuman primates originates the same way.

Most chimps in a group use the same tool techniques not because they all independently came up with the same trick but because they learned by observing each other. The amount of time a young chimp spends watching its mother using termite fishing tools or ant dipping tools is a significant predictor of the age at which it will learn each skill; the more it watches, the earlier it learns to do it. Without transmission from others, most chimps never figure tool use out on their own; in fact, a young chimp that doesn't learn, through observing others, to crack nuts by the age of five will not acquire the skill later in life.

Skill transmission in nonhuman primates has been shown in lab experiments. In a 1987 study, a group of young chimps were given a T-bar rake that could be inserted through a cage and used to grasp faraway food. Half the chimps observed an adult chimpanzee use the tool, and the other half did not. The group of young chimps that watched the expert adult figured out how to use the tool, while the group that saw no expert demonstrations never figured out how to use the tool (despite being highly motivated because they could see the food in the cage).

These skills can propagate throughout an entire group of primates. Consider the following studies. Experimenters temporarily took an individual chimpanzee, capuchin monkey, or marmoset from their group and taught it a new skill. These individuals were taught skills such as using a stick to poke a food-dispensing device in the right way, sliding open a door in a specific way to get food, pulling a drawer to get food, or opening an artificial fruit. After teaching this new skill, experimenters reintroduced the now skilled primate back to their group. Within a month, almost the entire group was using these same techniques, whereas groups that never had a member taught the skill never figured out how to use the tools in the same way. And such skills, originally taught to only a single individual, were passed down through multiple generations.

The ability to use tools is less about *ingenuity* and more about *transmissibility*. Ingenuity must occur only once if transmissibility occurs frequently; if at least *one* member of a group figures out how to manufacture and use a termite-catching stick, the entire group can acquire this skill and continuously pass it down throughout generations.

But it would be inaccurate to conclude that primates are uniquely good tool users simply because of their ability to transmit motor behaviors among themselves—to learn by observation. Many animals who are much worse tool users than primates, even those who don't use tools at all, also engage in observational learning. Rats can learn to push a lever to get water by watching another rat push a lever and get water. Mongooses will adopt the egg-opening technique of their parents. Dolphins can be trained to imitate movements they see in other dolphins or humans. Dogs can learn how to pull a lever with their paw to get food by watching another dog perform the act. Even fish and reptiles can observe the navigational paths

SELECTING KNOWN SKILLS *THROUGH OBSERVATION*	ACQUIRING NOVEL SKILLS *THROUGH OBSERVATION*
Many mammals	Primates
Octopuses	Some birds
Fish	
Reptiles	

taken by other members of their own species and learn to take those same navigational paths.

But there is a difference between observational learning in primates relative to most other mammals. If a parent mongoose tends to break open eggs with its mouth, then so does its offspring; if a parent mongoose tends to break open eggs by throwing, then so does its offspring. But these children mongooses aren't acquiring a novel skill by watching; they are merely changing *which* technique they *tend* to use—all children mongooses exhibit both biting and throwing tricks for opening eggs. Kittens learn to pee in litter boxes only if exposed to their mothers doing it, but all kittens know how to pee. Fish don't learn how to swim by watching; they merely change their paths by watching. In all these cases, animals aren't using observational learning to *acquire novel skills*; they are merely *selecting a known behavior* based on seeing another do the same thing.

Selecting a known behavior through observation can be accomplished with simple reflexes: A tortoise may have a reflex to look in the direction that other tortoises are looking; a fish may have a reflex to follow other fish. A mouse can simulate itself pushing a lever (something it already knows how to do) when it observes another mouse pushing a lever, at which point this mouse will realize that it will get water if it does this. But acquiring an *entirely novel motor skill by observation* may have required, or at least hugely benefited from, entirely new machinery.

Why Primates Use Hammers but Rats Don't

Acquiring *novel skills* through observation required theory of mind, while *selecting known skills* through observation did not. There are three reasons why this was the case. The first reason why theory of mind was necessary for acquiring novel skills by observation is that it may have enabled our ancestors to *actively teach*. For skills to be transmitted through a population, you don't *need* teachers—dutiful observation by novices will do. But active teaching can substantially improve the transmission of skills. Think about how much harder it would have been to learn to tie your shoes if you didn't have a teacher who slowed down and walked you through each step, and instead you had to decipher the steps on your own by watching people rapidly tie their shoes with no regard for your learning.

Teaching is possible only with theory of mind. Teaching requires understanding what another mind does not know and what demonstrations would help manipulate another mind's knowledge in the correct way. While it is still controversial whether any primates other than humans teach, in recent years evidence is beginning to accumulate in favor of the idea that nonhuman primates do, in fact, actively teach one another.

In the 1990s, the primatologist Christophe Boesch reportedly observed chimp mothers performing nut cracking in slow motion specifically around their young. These mothers periodically looked over to ensure their child was paying attention. Boesch reported chimp mothers correcting mistakes in youngsters by removing a nut, cleaning the anvil, then placing the nut back. He also reported mothers reorienting a hammer in the hands of their young.

Monkeys have been found to exaggerate their "flossing" specifically around youngsters that have not yet learned this skill, as if they are slowing down to help teach. Chimps that are skilled termite fishers will often bring two sticks to a fishing activity and directly hand one to a youngster. A chimp will even break its own stick in two and give the youngster half if the child shows up without a stick. If a child starts to seem like it is struggling with a task, the mother will actively swap tools with them. And the more complex the tool-using process, the more likely a mother is to give a tool to a youngster.

The second reason why theory of mind was necessary for learning novel motor skills through observation is that it enabled learners to stay focused on learning over long periods. A rat can see another rat push a lever and a few moments later push the lever itself. But a chimpanzee child will watch its mother use anvils to break open nuts and practice this technique for *years* without any success before it begins to master the skill. Chimp children continually attempt to learn without any near-term reward.

It is possible chimp children do this simply because they find imitation rewarding on its own, but another possibility is that theory of mind enables novices to identify the *intent* of a complex skill, which makes them highly motivated to keep trying to adopt it. Theory of mind enables a chimp child to realize that the reason it is *not* getting food with its stick while its mother *is* getting food is that its *mother has a skill it does not yet have*. This enables a continuous motivation to acquire the skill, even if it takes a long time to master. When a rat imitates behaviors, on the other hand, it will quickly give up if its actions don't lead to a near term reward.

The third and final reason why theory of mind was necessary for learning novel motor skills through observation was that it enabled novices to differentiate between the intentional and unintentional movements of experts. Observational learning is more effective if one is aware of what another is *trying to accomplish* with each movement. If you watched your mother tie her own shoes and you had no idea what aspects of her movements were intentional versus accidental, it would be quite hard to decipher which movements to follow. If you realized that her *intention* was to get her shoes tied, that when she slipped it was an accident, and that both the way she is seated and the angle of her head are irrelevant aspects of the skill, it would be much easier for you to learn the skill through observation.

This is indeed how chimpanzees learn. Consider the following experiment. An adult chimpanzee was allowed to observe a human experimenter open a puzzle box to get food. Throughout the sequence of actions needed to open the puzzle box, the experimenter did several irrelevant actions, such as tapping a wand or rotating the box. Then the chimp was given a chance to open the puzzle box to get food itself. Amazingly, the chimp didn't exactly copy every movement of the experimenter, instead it copied only the necessary movements to open the puzzle box and skipped the irrelevant steps.

Understanding the intentions of movements is essential for observational learning to work; it enables us to filter out extraneous movements and extract the essence of a skill.

Robot Imitation

In 1990, a graduate student at Carnegie Mellon named Dean Pomerleau and his adviser Chuck Thorpe built an AI system to autonomously drive a car. They called it ALVINN (Autonomous Land Vehicle in a Neural Network). ALVINN was fed video footage from around a car and could—*on its own*—steer the car to stay within a lane on a real highway. There had been previous attempts at autonomous cars like this, but they were very slow, often pausing every few seconds; the original version of the autonomous car in Thorpe's group could go only a quarter of a mile per hour due to how much thinking it had to do. ALVINN was much faster, so fast, in fact, that ALVINN successfully drove Pomerleau from Pittsburgh to the Great Lakes on a real highway with other drivers.

Why was ALVINN successful while previous attempts had failed? Unlike previous attempts to build a self-driving car, ALVINN was not taught to recognize objects or plan its future movements or understand its location in space. Instead, ALVINN outperformed these other AI systems by doing something simpler: It learned by *imitating human drivers*.

Pomerleau trained ALVINN the following way: He put a camera on his car and recorded both the video and the position of his steering wheel while he drove around.* Pomerleau then trained a neural network to map the image of the road to the corresponding steering position that he had selected. In other words, he trained ALVINN to *directly copy* what Pomerleau would do. And remarkably, after only a few minutes of observation, ALVINN was quite good at steering the car on its own.

But then Pomerleau hit a snag; it quickly became clear that this approach to imitation learning—of copying expert behavior directly—had a critical flaw. Whenever ALVINN made even small errors, it was completely unable to recover. Small mistakes would rapidly cascade into catastrophic failures of driving, often veering entirely off the road. The problem was that ALVINN was trained only on *correct* driving. It had never seen a human recover from a mistake because it had never seen a mistake in the first place.

---

* ALVINN controlled the steering wheel only, not braking or acceleration.

Directly copying expert behaviors turned out to be a dangerously brittle approach to imitation learning.

There are numerous strategies to overcome this problem in robotics. But two in particular draw conspicuous parallels to how primates seem to make imitation learning work. The first is to emulate a teacher-student relationship. In addition to training an AI system to directly copy an expert, what if the expert also drove *alongside* the AI system and corrected its mistakes? One of the first attempts to do this was by Stephane Ross and his adviser Drew Bagnell at Carnegie Mellon in 2009. They taught an AI system to drive in a simulated Mario Kart environment. Instead of recording himself drive and then training a system to imitate it, Ross drove around the Mario Kart track and *traded control* over the car with the AI system. At first, Ross would do most of the driving, then control would be passed to the AI system for a moment, and any mistakes it made would be quickly recovered by Ross. Over time, Ross gave more control to the AI system until it was driving well on its own.

This strategy of active teaching worked fantastically. When only directly copying driving (like ALVINN was trained), Ross's AI system was still crashing cars after a million frames of expert data. In contrast, with this new strategy of active teaching, his AI system was driving almost perfectly after only a handful of laps. This is not unlike a teacher chimpanzee correcting the movements of a child learning a new skill. A chimp mother watches her daughter try to insert a stick into a termite mound, and when the child is struggling, the mother attempts to correct her.

The second approach to imitation learning in robotics is called "inverse reinforcement learning." Instead of trying to directly copy the driving decisions that a human makes in response to a picture of the road, what if an AI system first attempts to identify the *intent* of the human's driving decisions?

In 2010, Pieter Abbeel, Adam Coates, and Andrew Ng demonstrated the power of inverse reinforcement learning by using it to get an AI system to autonomously fly a remote-controlled helicopter. Flying a helicopter, even a remote-controlled one, is hard; helicopters are unstable (small mistakes can rapidly cascade into a crash), require constant adjustments to keep them in the air, and require correctly balancing multiple complex inputs at the same time (angle of top blades, angle of tail rotor blades, tilt orientation of the helicopter body, and more).

Ng and his team didn't want to get an AI system to simply fly a helicopter, they wanted it to perform acrobatic tasks, those that only the best human experts could perform: flipping in place without falling, rolling while moving forward, flying upside down, performing arial loops, and more.

Part of their approach was standard imitation learning. Ng and his team recorded human expert inputs to the remote control as they performed these acrobatic tricks. But instead of training the AI system to *directly copy* the human experts (which didn't work), they trained the AI system to first infer the *intended* trajectories of the experts, inferring what it seemed like the humans were trying to do, *and then* the AI system learned to pursue those intended trajectories. This technique is called "inverse reinforcement learning" because these systems first try to learn the reward function they believe the skilled expert is optimizing for (i.e., their "intent"), and then these systems learn by trial and error, rewarding and punishing themselves using this inferred reward function. An inverse reinforcement learning algorithm starts from an observed behavior and produces its own reward function; whereas in standard reinforcement learning the reward function is hard-coded and not learned. Even when expert pilots flew these helicopters, they were continually recovering from small mistakes. By first trying to identify the intended trajectories and movements, Ng's AI system was both filtering out extraneous mistakes of experts, and correcting its own mistakes. Using inverse reinforcement learning, by 2010 they successfully trained an AI system to autonomously perform arial aerobatics with a helicopter.

There is still much work to be done when it comes to imitation learning in robotics. But the fact that inverse reinforcement learning (whereby AI systems infer the intent of observed behavior) seems necessary for observational learning to work, at least in some tasks, supports the idea that theory of mind (whereby primates infer the intent of observed behavior) was required for observational learning and the transmission of tool skills among themselves. It is unlikely a coincidence that both the ingenuity of roboticists and the iteration of evolution both converged on similar solutions; a novice cannot reliably acquire a new motor skill by merely observing an expert's movements, novices must also peer into an expert's mind.

Theory of mind evolved in early primates for politicking. But this ability was repurposed for imitation learning. The ability to infer the intent of others enabled early primates to filter out extraneous behaviors and focus only on the relevant ones (what did the person *mean* to do?); it helped youngsters stay focused on learning over long stretches of time; and it may have enabled early primates to actively teach each other by inferring what a novice does and does not understand. While our ancestral mammal likely could select known skills by observing others, it was with early primates, armed with theory of mind, when the ability to acquire truly novel skills through observation emerged. This created a new degree of transmissibility: skills that were discovered by clever individuals and that would once have faded when they died, could now propagate throughout a group and be passed down endlessly through generations. This is why primates use hammers and rats do not.

18

Why Rats Can't Go Grocery Shopping

ALTHOUGH ROBIN DUNBAR'S social-brain hypothesis has, for the past several decades, held primacy among scientists as the leading explanation of brain expansion in primates, there is an alternative explanation: what has been called the *ecological*-brain hypothesis.

As we have seen, early primates were not only uniquely social but also had a unique diet: they were *frugivores*. Fruit-based diets come with several surprising cognitive challenges. There is only a small window of time when fruit is ripe and has not yet fallen to the forest floor. In fact, for many of the fruits these primates ate, this window is less than seventy-two hours. Some trees offer ripe fruit for less than three weeks of the year. Some fruit has few animal competitors (such as bananas in their hard-to-open skin), while other fruit has many animal competitors (such as figs, which are easy for any animal to eat). These popular fruits are likely to disappear quickly, as many different animals feed on them once they ripen. Altogether this meant that primates needed to keep track of all the fruit in a large area of forest and on any given day know *which* fruit was likely to be ripe; and of the fruit that was ripe, which was likely to be most popular and hence disappear first.

Studies have shown that chimpanzees plan their nighttime nesting locations in preparation for foraging on the subsequent day. For fruits that are more popular, such as figs, they will go out of their way to plan where they sleep to be en route to these fruits. They do not do the same for less competitive but equally enjoyable fruits. Further, chimpanzees will leave earlier in the morning when pursuing a competitive fruit than when traveling to forage for a less competitive fruit. It has been shown that baboons also plan their foraging journey in advance and will leave earlier when fruit is less abundant and likely to be depleted more quickly.

Animals who feed on non-fruit plants do not have to cope with this same challenge; leaves, nectar, seeds, grass, and wood all last for long periods and are not localized in sparse patches. Even carnivores don't have as cognitively challenging a task—prey must be hunted and outsmarted, but there are rarely only short time windows in which hunting is possible.

Part of what makes this frugivore strategy so challenging is that it requires not only simulating differing navigational paths but also simulating your own future needs. Both a carnivore and a non-fruit-eating herbivore can survive by hunting or grazing only when they are hungry. But a frugivore must plan its trips in advance *before* it is hungry. Setting up camp en route to a nearby popular fruit patch the night before requires anticipating the fact that you *will* be hungry tomorrow if you don't take preemptive steps tonight to get to the food early.

Other mammals, such as mice, clearly stock up on food as winter months approach, storing vast reserves of nuts in their burrows to survive the long stretch when trees produce little to no food. But such seasonal hoarding is not nearly as cognitively challenging as the daily need to change your plans based on how hungry you will be tomorrow. Further, it isn't even clear that mice hoard food because they understand that they will be hungry in the future. Indeed, lab mice—although they have never suffered from a cold winter without food—*automatically* start hoarding food if you simply lower the temperature of their environment, an effect *seen only* in northern species of mice who have had to *evolve* to survive winters. Therefore, this doesn't seem to be an activity that they have learned from past winters and cleverly responded to; it seems that such hoarding is an evolutionarily hard-coded response to the changing seasons.

The ecological-brain hypothesis argues that it was the frugivore diet of early primates that drove the rapid expansion of their brains. In 2017, Alex DeCasien from NYU published a study examining the diets and social lives of over 140 species of primates. Some primates are primarily frugivores;

others are now primarily folivores (feeding on leaves). Some primates live in very small social groups; others live in large ones. She surprisingly found that being a *frugivore* seemed to explain the variation in relative brain size perhaps even better than the size of a primate's social group.

The Bischof-Kohler Hypothesis

In the 1970s, two comparative psychologists by the name of Doris Bischof-Kohler and her husband, Norbert Bischof, proposed a novel hypothesis about what was unique about planning in humans: They hypothesized that while other animals can make plans based on *current* needs (like how to get to food when they are hungry), only humans can make plans based on *future* needs (like how to get food for your trip next week, even though you are not hungry right now). The evolutionary psychologist Thomas Suddendorf would later call this the "Bischof-Kohler hypothesis."

Humans anticipate future needs all the time. We go grocery shopping even when we are not hungry; we bring warm clothes on trips even when we are not cold. Given the evidence available at the time of Bischof-Kohler, it was a reasonable hypothesis that only humans could do this. But recent evidence has called this into question. There are now anecdotal stories of chimpanzees bringing straw from inside a warm cage to make a nest outside when they know it is cold outside but *before* they were cold. There have been findings of bonobos and orangutans selecting tools for future tasks up to fourteen hours in advance of that task. Chimpanzees will carry stones from faraway locations to open nuts in areas that have no suitable stones and will manufacture tools in one location for use in another location. Indeed, if frugivorism requires planning ahead before you are hungry, then we should expect primates to be able to anticipate future needs.

In 2006, Miriam Naqshbandi and William Roberts of the University of Western Ontario measured a squirrel monkey's and a rat's ability to anticipate their own future thirst and change their behavior accordingly. Squirrel monkeys and rats were both given two options by being presented with two cups. Cup 1 was a "small treat option," which contained a tiny morsel of food, and cup 2 was a "high treat option," which contained lots of food. For squirrel monkeys the treats were dates; for rats they were raisins.

Under normal conditions, both animals would choose the high treat option; they love dates and raisins.

But Naqshbandi and Roberts then tested these animals in a different condition. Dates and raisins induce large amounts of thirst in these animals, often requiring them to consume over twice as much water to rehydrate themselves. So what happens if these animals are forced to make a trade-off, incorporating their future state of thirst? Naqshbani and Roberts modified the test such that if animals select the high treat option (the cup with many dates or raisins), they will only get access to water *hours* later; however, if animals select the low treat option (the cup with few dates or raisins) they get access to water between 15 and 30 minutes later. What happens?

Fascinatingly, squirrel monkeys learn to select the low treat option, while rats continue to select the high treat option. Squirrel monkeys are capable of resisting the temptation to have treats now, in anticipation of something—water—that they *don't even want yet.* In other words, monkeys can make a decision in anticipation of a future need. In contrast, rats were entirely unable to do this—they stuck with the flawed logic of "why give up extra raisins for water, I'm not even thirsty!"*

This suggests that perhaps Suddendorf's Bischof-Kohler hypothesis was correct that anticipating a future need is a more difficult form of planning and was correct that some animals should be able to plan but unable to anticipate future needs (such as rats). But it may not be the case that only humans were endowed with this ability. It may instead be the province of many primates.

How Primates Anticipate Future Needs

The mechanics of making a choice based on an *anticipated* need, one you are not currently experiencing, presents a predicament to the older

---

* Note that Naqshbandi and Roberts did an initial baseline experiment to ensure the quantity of dates and the quantity of raisins each induced similar levels of thirst in monkeys and rats, as measured by the relative increase in water intake when animals were given free access to water alongside these quantities of dates or raisins.

mammalian brain structures. We have speculated that the mechanism by which the neocortex controls behavior is by simulating decisions vicariously, the outcomes of which are then *evaluated* by the older vertebrate structures (basal ganglia, amygdala, and hypothalamus). This mechanism allows an animal to choose only simulated paths and behaviors that excite positive valence neurons *right now*, like imagining food when hungry or water when thirsty.

In contrast, to buy groceries for the week, I need to anticipate a pizza is going to make a great addition to Thursday's movie night, even though I don't currently want pizza. When I imagine eating pizza while I'm not hungry, my basal ganglia doesn't get excited; it doesn't accumulate votes for any decisions to pursue pizza. Thus, to want pizza, I need to realize that in this *imagined future state of hunger*, the smell and sight of food *will* excite positive valence neurons, even though imagining it right now does not. How, then, can a brain choose an imagined path in the absence of any vicarious positive-valence activation? How can your neocortex want something that your amygdala and hypothalamus do not?

There is another situation we have already seen where brains need to infer an intent—a "want"—of which it does not currently share: when they're trying to infer the wants of *other* people. Might brains be able to use the same mechanism of theory of mind to anticipate a future need? Put another way: Is imagining the mind of someone else really any different from imagining the mind of your future self?

Perhaps the mechanism by which we anticipate future needs is the same mechanism by which we engage in theory of mind: We can infer the intent of a mind—whether our own or someone else's—in some different situation from our current one. Just as we can correctly infer the cravings of someone deprived of food ("How hungry would James be if he didn't eat for twenty-four hours?") even though we ourselves might not be hungry, perhaps too we can infer the intent of ourselves in a future situation ("How hungry would I be if I didn't eat for twenty-four hours?") even though we are currently not hungry.

In his paper discussing the Bischof-Kohler hypothesis, Thomas Suddendorf brilliantly foreshadowed exactly this idea:

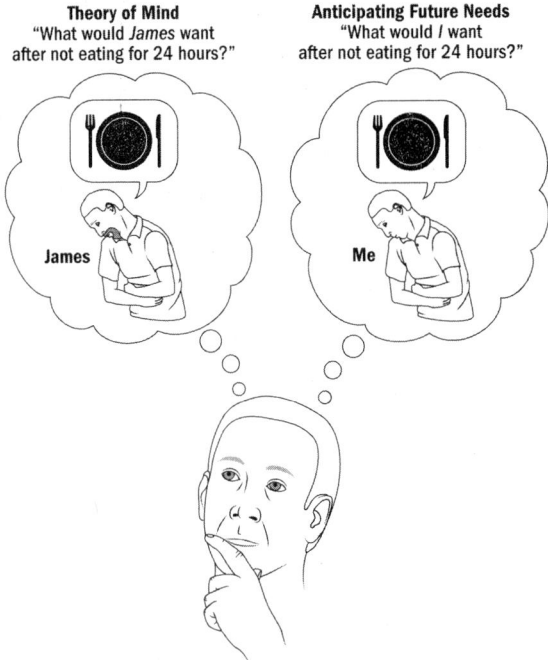

Figure 18.1: The similarity between theory of mind and anticipating future needs

Future need anticipation . . . might be only a special case of animals' general problem with simultaneously representing conflicting mental states. Like 3-year-old children, they may be unable to imagine an earlier belief (or state of knowledge, or drive, etc.) that is different from a present one or to understand that another individual holds a belief different from their own. This may apply to future states as well as to past ones. That is, a satiated animal may be unable to understand that it may later be hungry, and therefore may be unable to take steps to ensure that this future hunger will be satisfied.

Although Naqshbandi and Roberts's experiment with squirrel monkeys and rats suggests that Suddendorf may have been wrong that *only humans* can anticipate future needs, Suddendorf may have been prescient in proposing that the general ability to model a dissociated mental state from your own can be repurposed for both theory of mind and anticipating future needs.

There are two observations that support this idea. First, it seems that both theory of mind and anticipating future needs are present, even in a primitive form, in primates, but not in many other mammals, suggesting both abilities emerged around the same time in early primates. Second, people make similar types of mistakes in tasks of theory of mind and of anticipating future needs.

For example, we saw in chapter 16 that thirsty people become incorrectly biased to predict that other people must also be thirsty. Well, it is also the case that hungry people seem to incorrectly predict how much food they will need in the future. Take two groups of people and bring them to the grocery store to shop for themselves for the week, and those that are hungry will end up buying more food than those that were well fed, even though they were both shopping for food to feed themselves for the same window of time, namely, a single week. When hungry, you overestimate your own future hunger.

The ability to anticipate future needs would have offered numerous benefits to our ancestral frugivores. It would have enabled our ancestors to plan their foraging routes long in advance, thereby ensuring they were the first to get newly ripened fruits. Our ability to make decisions today for faraway, abstract, and not-yet-existent goals was inherited from tree faring primates. A trick that, perhaps, was first used for getting the first pick of fruits, but today, in humans, is used for far greater purposes. It laid the foundation for our ability to make long term plans over vast stretches of time.

Summary of Breakthrough #4: Mentalizing

There are three broad abilities that seem to have emerged in early primates:

- **Theory of mind:** inferring intent and knowledge of others
- **Imitation learning:** acquiring novel skills through observation
- **Anticipating future needs:** taking an action *now* to satisfy a want *in the future*, even though I do not want it now

These may not, in fact, have been separate abilities but rather emergent properties of a single new breakthrough: the construction of a generative model of one's own mind, a trick that can be called "mentalizing." We see this in the fact that these abilities emerge from shared neural structures (such as the gPFC) that evolved first in early primates. We see this in the fact that children seem to acquire these abilities at similar developmental times. We see this in the fact that damage that impairs one of these abilities tends to impair many of them.

And most important, we see this in the fact that the structures from which these skills emerge are the same areas from which our ability to reason about *our own mind* emerges. These new primate areas are required not only for simulating the mind of others but also for projecting *yourself* into your imagined futures, identifying yourself in the mirror (mirror-sign syndrome), and identifying your own movements (alien-hand syndrome). And a child's ability to reason about her own mind tends to precede a child's development of all three of these abilities.

However, the best evidence for this idea goes all the way back to Mountcastle. The main change to the brain of early primates, besides its size, was the addition of new areas of neocortex. So if we are to stick to the general idea—inspired by Mountcastle, Helmholtz, Hinton, Hawkins, Friston, and many others—that every area of neocortex is made up of identical microcircuits, this imposes strict constraints on how we explain the newfound abilities of primates. It suggests that these new intellectual skills must emerge from some new clever application of the neocortex

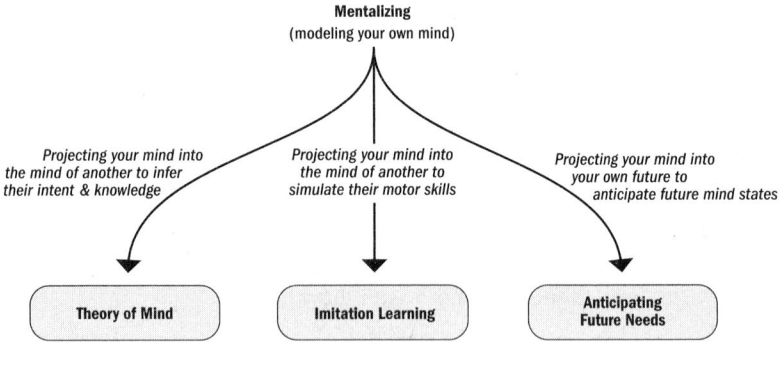

Figure 18.2

and not some novel computational trick. This makes the interpretation of theory of mind, imitation learning, and anticipation of future needs as nothing more than an emergent property of a second-order generative model a nice proposal—all three abilities can emerge from nothing more than new applications of neocortex.

All these abilities—theory of mind, imitation learning, and anticipating future needs—would have been particularly adaptive in the unique niche of early primates. Dunbar argues that the social-brain hypothesis and the ecological-brain hypothesis are two sides of the same coin. The ability to *mentalize* may have simultaneously unlocked both the ability to successfully forage fruits and to successfully politick. The pressures of both frugivorism and social hierarchies may have converged to produce continual evolutionary pressure to develop and elaborate brain regions—such as the gPFC—for modeling your own mind.

We have reached the end of breakthrough #4. At this point in our story, we now stand at the precipice of the final divergence between humankind and our closest living relative. Our shared ancestor with chimpanzees lived seven million years ago in eastern Africa. The offspring of this ancestor split down two evolutionary paths—one that became today's chimpanzees, and another that became today's humans.

If we were to scrunch the six hundred million years of evolutionary time—from which the first brains emerged until today—into a single calendar year, then we would now find ourselves perched at Christmas Eve, the final seven days of December. Over the next "seven days," our ancestors will go from foraging fruits to flying Falcon 9 rockets. Let's find out how.

BREAKTHROUGH #5

Speaking and the First Humans

Your brain 100,000 years ago

19

The Search for Human Uniqueness

FOR MILLENNIA, WE HUMANS have looked at ourselves in the mirror with self-congratulatory pride and pondered the many ways in which we are superior to our fellow animals. Aristotle claimed it was our "rational soul"—our ability to reason, make abstractions, and reflect—that was uniquely human. Twentieth-century animal psychologists itemized many intellectual abilities they believed were uniquely human. Some argued that only humans engage in mental time travel. Others, that it is our episodic memory. Others, our ability to anticipate future needs. Others, our sense of self. Our ability to communicate, coordinate, use of tools. The lists went on.

But the last century of research into the behaviors of other animals has methodically dismantled our surprisingly fragile edifice of uniqueness. Despite the intuitive appeal of claiming many of these skills as only ours, as we have seen throughout this book, science suggests many of them, if not all, might not be unique to humans at all.

Darwin believed that "the difference in mind between man and the higher animals, great as it is, is certainly one of degree and not of kind." What intellectual feats, if any, are uniquely human is still hotly debated among psychologists. But as the evidence continues to roll in, it seems that Darwin may have been right.

If it were the case that humans wielded numerous intellectual capabilities that were entirely unique in *kind*, we would expect human brains to contain some unique neurological structures, some new wiring, some new systems. But the evidence is the opposite—there is no neurological structure found in the human brain that is not also found in the brain of our fellow apes, and evidence suggests that the human brain is literally just a scaled-up primate brain: a bigger neocortex, a bigger basal ganglia, but still containing all the same areas wired in all the same ways. Scaling up a chimpanzee brain may have made us *better* at anticipating future needs, theory of mind, motor skills, and planning, but it didn't necessarily give us anything truly new.

An eminently reasonable explanation of human brain evolution since our divergence with chimpanzees is that various evolutionary pressures led our human ancestors to simply "level up" the abilities that were already present.

Was there perhaps no breakthrough at all?

This seems to be the most reasonable interpretation—but for one crucial exception. And it is in this singular exception that we see the first hints of what it means to be human.

Our Unique Communication

Organisms had been communicating with each other long before early humans uttered their first words. Single-celled organisms emit chemical signals to share genes and information about the environment. Brainless sea anemones pump pheromones into the water to coordinate the timing of sperm and egg release. Bees dance to signal where to find food. Fish use electrical signals to court each other. Reptiles head-bob to communicate their aggressiveness. Mice squeak to express danger or excitement. Communication between organisms is evolutionarily ancient and ubiquitous.

Chimpanzees are constantly screeching and gesturing to each other. These different sounds and gestures have been shown to signal specific requests. Shoulder taps mean "Stop it," stomping means "Play with me," a screech means "Groom me," reaching a palm out means "Share food." Primatologists have studied these different gestures and vocalizations so extensively that there is even a Great Ape Dictionary that has chronicled almost one hundred sounds and gestures.

Vervet monkeys have different sounds to signal the presence of specific predators. When one monkey makes the squeal meaning "Leopard!" all the others run to the trees. When one makes the squeal meaning "Eagle!" all the others jump to the forest floor. Experimenters can get all the monkeys to run to the treetops or jump to the floor simply by playing one of these sounds from nearby loudspeakers.

And then, of course, there is us—*Homo sapiens*. We too communicate with each other. It is not the *fact* that we communicate that is unique; rather, it is *how* we communicate. Humans use language.

Human language differs from other forms of animal communication in two ways. First, no other known form of naturally occurring animal communication assigns *declarative labels* (otherwise known as *symbols*). A human teacher will point to an object or a behavior and assign it an arbitrary label: *elephant, tree, running*. In contrast, other animals' communications are genetically hardwired and not assigned. Vervet monkey and chimpanzee gestures are almost identical across different groups that have no contact with each other. Monkeys and apes deprived of social contact still use the same gestures. In fact, these gestures are even shared across species of primates; bonobos and chimpanzees share almost exactly the same repertoire of gestures and vocalizations. In nonhuman primates, the meanings of these gestures and vocalizations is not assigned through declarative labeling but emerge directly from genetic hardwiring.

What about teaching a dog or any other animal a command? Clearly this represents some form of labeling. Linguists make a distinction between declarative and imperative labels. An imperative label is one that yields a reward: "When I hear *sit*, if I sit, I will get a treat" or "When I hear *stay*, if I stop moving, I will get a treat." This is basic temporal difference learning—all vertebrates can do this. Declarative labeling, on the other hand, is a special feature of human language. A declarative label is one that assigns an object or behavior an arbitrary symbol—"That is a *cow*," "That is *running*,"—without any imperative at all. No other form of naturally occurring animal communication has been found to do this.

The second way in which human language differs from other animal communication is that it contains grammar. Human language contains rules by which we merge and modify symbols to convey specific meanings.

We can thereby weave these declarative labels into sentences, and we can knit these sentences into concepts and stories. This allows us to convert the few thousand words present in a typical human language into a seemingly infinite number of unique meanings.

The simplest aspect of grammar is that the order in which we utter symbols conveys meaning: "Ben hugged James" means something different from "James hugged Ben." We also embed subphrases that are sensitive to ordering: "Ben, who was sad, hugged James" means something entirely different than "Ben hugged James, who was sad." But the rules of grammar go beyond just order. We have different tenses to convey timing: "Ben *is attacking* me" versus "James *attacked* me." We have different articles: "*The* pet barked" means something different than "*A* pet barked."

And of course, this is just English; there are over six thousand spoken languages on Earth, each with its own labels and grammar. But despite the great diversity of the specific labels and grammars of different languages, every single group of humans ever discovered used language. Even hunter-gatherer societies in Australia and Africa who at the point they were "discovered" had had no contact with any other groups of humans for over *fifty thousand* years, still spoke their own languages that were equally complex as those of other humans. This is irrefutable evidence that the shared ancestor of humans spoke their own languages, with their own declarative labels and grammars.

Of course, the fact that early humans spoke in their own languages with declarative labels and grammar, while no other animal naturally does so, does not prove that only humans are *able* to use language, merely that only humans *happen* to use language. Did the brains of early humans really evolve some unique ability to speak? Or is language just a cultural trick that was discovered over fifty thousand years ago and was simply passed down through generations of all modern humans? Is language an evolutionary invention or a cultural invention?

Here's one way to test this: What happens if we try to teach language to our evolutionarily closest animal cousins, our fellow apes? If apes succeed in learning language, that suggests that language was a cultural invention; if apes fail, this suggests their brains lack a key evolutionary innovation that emerged in humans.

This test has been performed multiple times. The result is as surprising as it is revealing.

Attempts to Teach Apes Language

To start: We can't literally teach apes to speak. This was attempted in the 1930s, and it failed—nonhuman apes are physically incapable of producing verbal language. Human vocal cords are uniquely adapted to speech; the human larynx is lower and the human neck is longer, which enables us to produce a much wider variety of vowels and consonants than other apes. The vocal cords of a chimp can produce only a limited repertoire of huffs and squeals.

However, what makes language *language* is not the medium but the substance—many forms of human language are nonverbal. No one would claim that writing, sign language, and Braille do not contain the substance of language because they don't involve vocalization.

The key studies that attempted to teach chimpanzees, gorillas, and bonobos language used either American Sign Language or made-up visual languages in which apes pointed to sequences of symbols on a board. Beginning as infants, these apes were trained to use these languages, with human experimenters signing or pointing to symbols to refer to objects (apples, bananas) or actions (tickling, playing, chasing) over and over again until the apes began to repeat the symbols.

Across most of these studies, after years of being taught, nonhuman apes did indeed produce the appropriate signs. They could look at a dog and sign *dog* and look at a shoe and sign *shoe*.

They could even construct basic noun-verb pairs. Common phrases were *play me* and *tickle me*. Some evidence even suggested they could combine known words to create novel meanings. In one famous anecdote, the chimpanzee Washoe saw a swan for the first time, and when the trainer signed, *What's that?*, Washoe signed back, *Water bird*. In another, the gorilla Koko saw a ring and, not knowing the word for it, signed, *Finger bracelet*. After eating kale for the first time, Kanzi the bonobo pressed symbols for *slow lettuce*.

Kanzi supposedly even used language to play with others. There is an anecdotal story of a trainer who was resting in Kanzi's habitat being woken up by Kanzi snatching away the blanket and then excitedly pressing the symbols for *bad surprise*. In another anecdote, Kanzi pressed keys for *apple chase* and then picked up an apple, grinned, and began running from his trainer.

Sue Savage-Rumbaugh, the psychologist and primatologist who devised the Kanzi language-learning experiment, did a test to compare Kanzi's language understanding to that of a two-year-old human child. Savage-Rumbaugh exposed Kanzi and a human child to over six hundred novel sentences (using their symbol-language) with specific commands. These sentences used symbols Kanzi already knew but in sentences Kanzi had never seen, commands like *Can you give the butter to Rose?; Go put some soap on Liz; Go get the banana that's in the refrigerator; Can you give the doggy a hug?;* and *Put on the monster mask and scare Linda.* Kanzi successfully completed these tasks over 70 percent of the time, outperforming the two-year-old human child.

The degree to which these ape language studies demonstrate language with declarative labels and grammar is still controversial among linguists, primatologists, and comparative psychologists. There are many who argue that these tricks represent imperatives not declaratives and that the phrases uttered were so simple they could hardly be called grammar. Indeed, in most of these studies, apes received treats when they used the right labels, which makes it hard to tell if they were really sharing in labeling objects or just learning that if they did sign X when they saw a banana, they got a treat, a task any model-free reinforcement learning machine could perform. Extensive analysis of the phrases uttered by these language-able apes demonstrate a low diversity in phrases, meaning they tended to use the exact phrases they learned (e.g., *Tickle me*) instead of combining words into novel phrases (e.g., *I want to be tickled*). But in response to these challenges, many point to the Savage-Rumbaugh studies and Kanzi's incredibly accurate grammatical understanding of commands and playful phrases. The debate is not yet settled.

On balance, most scientists seem to conclude that some nonhuman apes are indeed capable of learning at least a rudimentary form of language but that nonhuman apes are much worse than humans at it and don't learn it without painstaking deliberate training. These apes never surpass the abilities of a young human child.

So, language seems unique to humans on two counts. First, we have a natural tendency to construct it and use it, which other animals do not. Second, we have a capacity for language that far surpasses that of any other animal, even if some basic semblance of symbols and grammar is possible in other apes.

But if language is what separates us from the rest of the animal kingdom, then what is it about this seemingly innocuous trick that enabled *Homo sapiens* to ascend to the top of food chain; what is it about language that makes those who wield it so powerful?

Transferring Thoughts

Our unique language, with declarative labels and grammar, enables groups of brains to transfer their inner simulations to each other with an unprecedented degree of detail and flexibility. One can say "Smash the rock from the top" or "Joe was rude to Yousef" or "Remember that dog we saw yesterday," and in all these cases, the *talker* is deliberately selecting an inner simulation of images and actions to be transferred to nearby *listeners*. A group of n brains can all re-render the same mental movie of the dog they saw yesterday with merely a few noises or gestures.

When we talk about these inner simulations, especially in the context of humans, we tend to imbue them with words like *concepts, ideas, thoughts*. But all these things are nothing more than renderings in the mammalian neocortical simulation. When you "think" about a past or future event, when you ponder the "concept" of a bird, when you have an "idea" as to how to make a new tool, you are merely exploring the rich three-dimensional simulated world constructed by your neocortex. It is no different, in principle, than a mouse considering which direction to turn in a maze. Concepts, ideas, and thoughts, just like episodic memories and plans, are not unique to humans. What is unique is our ability to deliberately transfer these inner simulations to each other, a trick possible only because of language.

When a vervet monkey makes an *Eagle nearby!* squeal, all nearby monkeys will quickly jump from the trees to hide. Clearly, this represents a transfer of information from the monkey who first saw the eagle to the others. But these kinds of transfers are *undetailed* and *inflexible*, capable of transferring information only with genetically hard-coded signals. These signals are always few in number and cannot be adjusted or changed to new situations. In contrast, language enables the talker to transfer an incredibly broad set of inner thoughts.

This trick of thought transfer would have provided many practical benefits to early humans. It would have enabled more accurate teaching of tool use, hunting techniques, and foraging tricks. It would have enabled flexible coordination of scavenging and hunting behaviors across individuals—a human could say, "Follow me, there is an antelope carcass two miles east" or "Wait here, let's ambush the antelope when you hear me whistle three times."

All these practical benefits emerge from the fact that language expands the scope of sources a brain can extract learnings from. The breakthrough of reinforcing enabled early vertebrates to learn from their own *actual* actions (trial and error). The breakthrough of simulating enabled early mammals to learn from their own *imagined* actions (vicarious trial and error). The breakthrough of mentalizing enabled early primates to learn

The Evolution of Progressively More Complex Sources of Learning

	REINFORCING IN EARLY VERTEBRATES	SIMULATING IN EARLY MAMMALS	MENTALIZING IN EARLY PRIMATES	SPEAKING IN EARLY HUMANS
SOURCE OF LEARNING	Learning from your own actual actions	Learning from your own imagined actions	Learning from others' actual actions	Learning from others' imagined actions
WHO LEARNING FROM?	Yourself	Yourself	Others	Others
ACTION LEARNING FROM?	Actual actions	Imagined actions	Actual actions	Imagined actions

from *other people's actual* actions (imitation learning). But the breakthrough of speaking uniquely enabled early humans to learn from *other people's imagined actions.*

Language enables us to peer into and learn from the imagination of other minds—from their episodic memories, their internal simulated future actions, their counterfactuals. When a human coordinates a hunt and says, "If we go in this direction as a group we will find an antelope" or "If we all wait and ambush we will win the battle with the boar," humans are sharing the outcomes of their own inner vicarious trial and errors so that the whole group can learn from their imaginations. One person with an episodic memory of a lion on the other side of a mountain can transfer that episodic memory to others with language.

By sharing what we see in our imaginations, it is also possible for common myths to form and for entirely made-up imaginary entities and stories to persist merely because they hop between our brains. We tend to think about myths as the province of fantasy novels and children's books, but they are the foundation of modern human civilizations. Money, gods, corporations, and states are imaginary concepts that exist only in the collective imaginations of human brains. One of the earlier versions of this idea was articulated by the philosopher John Searle, but was famously popularized by Yuval Harari's book *Sapiens*. The two argue that humans are unique because we "cooperate in extremely flexible ways with countless numbers of strangers." And to Searle and Harari, we can do this because we have such "common myths." In Harari's words: "Two Catholics who have never met can nevertheless go together on crusade or pool funds to build a hospital because they both believe [in] God" and "Two Serbs who have never met might risk their lives to save one another because both believe in the existence of the Serbian nation" and "Two lawyers who have never met can nevertheless combine efforts to defend a complete stranger because they both believe in the existence of laws, justice, human rights, and money paid out in fees."

And so, with the ability to construct common myths, we can coordinate the behavior of an incredibly large number of strangers. This was a massive improvement over the system of social cohesion provided by primate mentalizing. Coordinating behavior using mentalizing alone works only

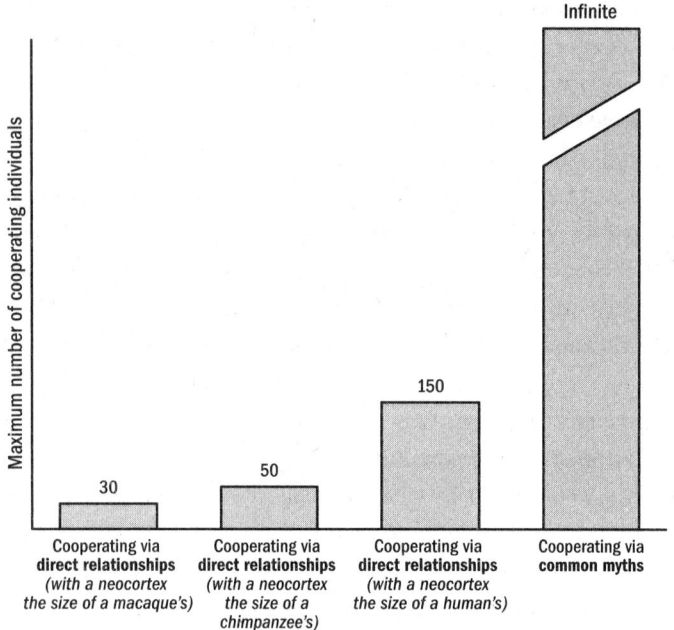

Figure 19.1: Maximum number of cooperating individuals based on different cooperation strategies

by each member of a group directly knowing each other. This mechanism of cooperation doesn't scale; the limit of human group size maintained only by direct relationships has been estimated to be about one hundred fifty people. In contrast, common myths of things like countries, money, corporations, and governments allow us to cooperate with billions of strangers.

While true, all these aforementioned benefits of language miss the larger point. It is not superior teaching, cooperative hunting, or Harari's common myths that is the true gift of language. None of these are why humans rule the world. If these were the only gifts language offered, we would still be hunter-gatherer apes dancing around campfires praying for rain from the water gods—apex predators, sure, but hardly astronauts. These features of language are *consequences* of the gift of language, not the gift itself.

An analogy to DNA is useful. The true power of DNA is not the *products* it constructs (hearts, livers, brains) but the *process* it enables (evolution). In this same way, the power of language is not its products (better teaching, coordinating, and common myths) but the process of ideas being transferred, accumulated, and modified across generations. Just as genes persist by hopping from parent cell to offspring cell, ideas persist by hopping from brain to brain, from generation to generation. And as with genes, this hopping is not uniform but operates under its own quasi-evolutionary rules—there is a continual selecting of good ideas and pruning of bad ideas. Ideas that helped humans survive *persisted*, while those that did not *perished*.

This analogy of ideas evolving was proposed by Richard Dawkins in his famous book *The Selfish Gene*. He called these hopping ideas *memes*. This word was later appropriated for cat images and baby photos flying around Twitter, but he originally meant them to refer to an idea or behavior that spread from person to person in a culture.

The rich complexity of knowledge and behaviors that exist in human brains today are possible only because the underlying ideas have been accumulated and modified across countless brains for thousands, even millions, of generations.

Consider the ancient invention of sewn clothing, in which humans converted hides of dead animals into clothing to keep themselves warm, an invention that many believe emerged as early as one hundred thousand years ago. This invention was possible only because of numerous prior inventions: slicing skin off carcasses, drying hides, manufacturing string, and creating bone needles. And these inventions themselves were possible only due to the prior invention of sharp stone tools. It would never have been possible to invent sewn clothing in a single eureka moment. Not even Thomas Edison would have been so clever. Edison's new inventions occurred only after he was given the right building blocks. With the understanding of electricity and generators handed to him from prior generations, he invented the lightbulb.

This accumulation does not apply only to technological inventions but also to cultural ones. We pass down social etiquette, values, stories, mechanisms for selecting leaders, moral rules of punishment, and cultural beliefs around violence and forgiveness.

All human inventions, both technological and cultural, require an accumulation of basic building blocks before a single inventor can go "Aha!," merge the preexisting ideas into something new, and transfer this new invention to others. If the baseline of ideas always fades after a generation or two, then a species will be forever stuck in a nonaccumulating state, always reinventing the same ideas over and over again. This is how it is for all other creatures in the animal kingdom. Even chimpanzees, who learn motor skills through observation, do not *accumulate* learnings across generations.

This brings us back to the imitation experiments we saw in chapter 17. Take a four-year-old child and an adult chimpanzee and have them observe an experimenter open a puzzle box to get food, in the process performing several irrelevant actions. Both chimps and human children learn to open the puzzle box through observation; however, chimps will skip the irrelevant steps, but *human children* will perform *all* the steps they observed, including the irrelevant ones. Human children are over-imitators.

This over-imitation is, in fact, quite clever. Children change their degree of copying based on how much they believe the teacher knows—"This person clearly knows what she is doing, so there must be a reason she did that." The more uncertain a child is about *why* a teacher is doing something, the more likely he is to exactly copy all the steps. Further, they are not just blindly mirroring whatever they see; children will imitate weird irrelevant behaviors only if their teacher seemed to have *intended* to do the behavior. If the action appears to be an accident, children will ignore it; children will not copy the coughing or nose scratching of a teacher. If a teacher keeps slipping when trying to pull apart a novel toy, a child will identify this as accidental and thereby not imitate this mistake—they will instead use a firmer grip to successfully pull the toy apart.

While these imitation experiments demonstrate that humans can accurately copy behaviors without using language, it is still undeniably language that is our superpower in the business of copying and transferring ideas.

Compared to silent imitation of experts, communicating how to do a task using language dramatically improves the accuracy and speed with which children solve tasks. Language lets us condense information so it takes up less brain space and can be more quickly transferred from brain to brain. If I say, "Whenever you see a red snake, run; whenever you see a green snake, it is safe," that idea and the corresponding behavior can immediately transfer throughout the group. In contrast, if everyone had to learn this "red snake bad, green snake good" generalization through individual experience or through *watching* someone else get bitten by multiple red snakes, it would take far more time and brainpower. This fact would continually fade and then be relearned from generation to generation. Without language, the inner simulations of chimpanzees and other animals do not accumulate, and thus inventions that are above a given threshold of complexity—the best ones—are forever out of their reach.

The Singularity Already Happened

Going from *no accumulation* across generations to *some accumulation* across generations was the subtle discontinuity that changed everything. In figure 19.2, you can see ideas begin to get more complex across a handful of generations, just as the invention of sewn clothing emerged from a composite of simpler building blocks.

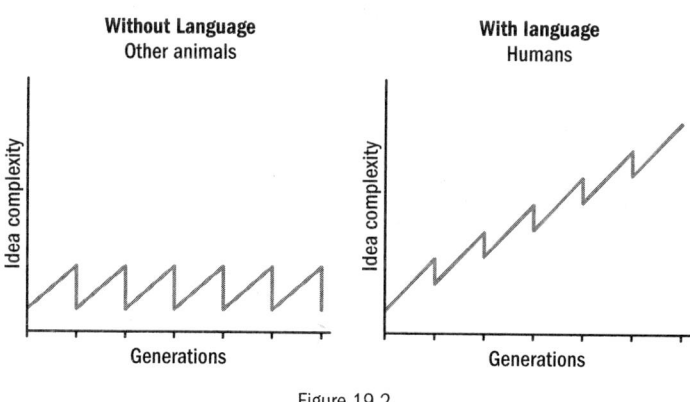

Figure 19.2

And if you zoom out to the timescale of thousands of generations, you see why even just *some* accumulation triggers an explosion of idea complexity (as seen in figure 19.3). From a period of seemingly perpetual stasis, you will, in a matter of a few hundred thousand years, get an explosion of complex ideas.

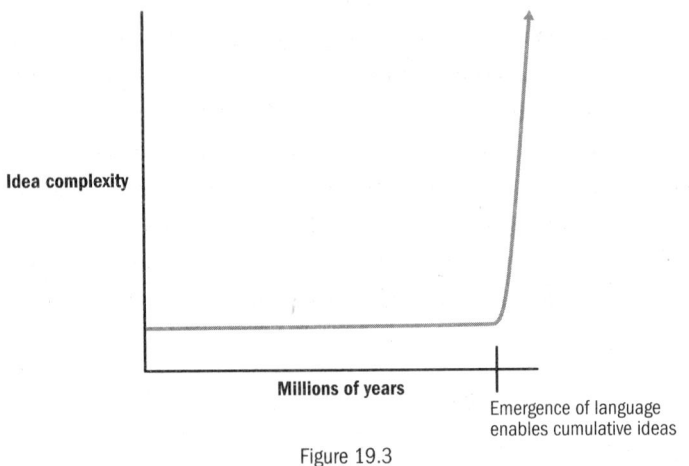

Figure 19.3

Eventually, the corpus of ideas accumulated reached a tipping point of complexity when the total sum of accumulated ideas no longer fit into the brain of a single human. This created a problem in sufficiently copying ideas across generations. In response, four things happened that further expanded the extent of knowledge that could be transferred across generations. First, humans evolved bigger brains, which increased the amount of knowledge that can be passed down through individual brains. Second, humans became more specialized within their groups, with ideas distributed across different members—some were the spear makers, others clothing makers, others hunters, others foragers. Third, population sizes expanded, which offered more brains to store ideas across generations. And fourth, most recent and most important, we invented *writing*. Writing allows humans to have a collective memory of ideas that can be downloaded at will and that can contain effectively an infinite corpus of knowledge.

If groups don't have writing, such distributed knowledge is sensitive to group size; if groups shrink, and there are no longer enough brains to fit all the information into, knowledge is lost. There is evidence that this occurred in societies in Tasmania. Archaeological evidence from eight thousand years ago shows that humans in Tasmania had complex knowledge of making bone tools, nets, fishing spears, boomerangs, and cold-weather clothing. All this knowledge was lost by the 1800s. This loss seems to have been initiated when rising oceans cut the group of humans in Tasmania off from other groups in the rest of Australia, effectively lowering the population size of the group of socially interacting humans. For people without writing, the smaller the population, the less knowledge can persist across generations.

The real reason why humans are unique is that we accumulate our shared simulations (ideas, knowledge, concepts, thoughts) across generations. We are the hive-brain apes. We synchronize our inner simulations, turning human cultures into a kind of meta-life-form whose consciousness is instantiated within the persistent ideas and thoughts flowing through millions of human brains over generations. The bedrock of this hive brain is our language.

The emergence of language marked an inflection point in humanity's history, the temporal boundary when this new and unique kind of evolution began: the evolution of ideas. In this way, the emergence of language was as monumental an event as the emergence of the first self-replicating DNA molecules. Language transformed the human brain from an ephemeral organ to an eternal medium of accumulating inventions.

These inventions included new technologies, new laws, new social etiquettes, new ways of thinking, new systems of coordination, new ways of selecting leaders, new thresholds for violence versus forgiveness, new values, new shared fictions. The neurological mechanisms that enable language came far before anyone was doing math, using computers, or discussing the merits of capitalism. But once humans were armed with language, these developments were all but inevitable. It was just a matter of time. Indeed, the incredible ascent of humankind during the past few thousand years had nothing to do with better genes and everything to do with the accumulation of better and more sophisticated ideas.

20

Language in the Brain

IN 1830, A THIRTY-YEAR-OLD Frenchman by the name of Louis Victor Leborgne lost the ability to speak. Leborgne could no longer say anything other than the syllable *tan*. What was peculiar about Leborgne's case was that he was, for the most part, otherwise intellectually typical. It was clear that when he spoke, he was *trying* to express certain ideas—he would use gestures and alter the tone and emphasis of his speech—but the only sound that ever came out was *tan*. Leborgne could *understand* language; he just couldn't produce it. After many years of hospitalization, he became known around the hospital as Tan.

Twenty years after patient Tan passed away, his brain was examined by a French physician named Paul Broca who had a particular interest in the neurology of language. Broca found that Leborgne had brain damage to a specific and isolated region in the left frontal lobe.

Broca had a hunch that there were specific areas in the brain for language. Leborgne's brain was Broca's first clue that this idea might be right. Over the next two years, Broca painstakingly sought out the brains of any recently deceased patients who had had impairment in their ability to articulate language but retained their other intellectual faculties. In 1865, after performing autopsies on twelve different brains, he published his now famous paper "Localization of Speech in the Third Left Frontal Cultivation." It turned out that all of these patients had damage to similar regions on the left side of the neocortex, a region that has come to be called *Broca's area*. This has been observed countless times over the past hundred and fifty years—if Broca's area is damaged, humans lose the ability to produce speech, a condition now called Broca's aphasia.

Several years after Broca did his work, Carl Wernicke, a German physician, was perplexed by a different set of language difficulties. Wernicke found patients who, unlike Broca's, could speak fine but lacked the ability to *understand* speech. These patients would produce whole sentences, but the sentences made no sense. For example, such a patient might say something like "You know that smoodle pinkered and that I want to get him round and take care of him like you want before."

Wernicke, following Broca's strategy, also found a damaged area in the brains of these patients. It was also on the left side but farther back in the posterior neocortex, a region now dubbed Wernicke's area. Damage to Wernicke's causes Wernicke's aphasia, a condition in which patients lose the ability to *understand* speech.

A revealing feature of both Broca's and Wernicke's areas is that their language functions are not selective for only certain modalities of language, but rather are selective for language in general. Patients with Broca's aphasia become equally impaired in *speaking* words as they are in *writing* words. Patients who primarily communicate using sign language lose their ability to sign fluently when Broca's area is damaged. Damage to Wernicke's area produces deficits in understanding both spoken language and *written* language. Indeed, these same language areas are activated when a hearing-abled person listens to someone speak and when a deaf person

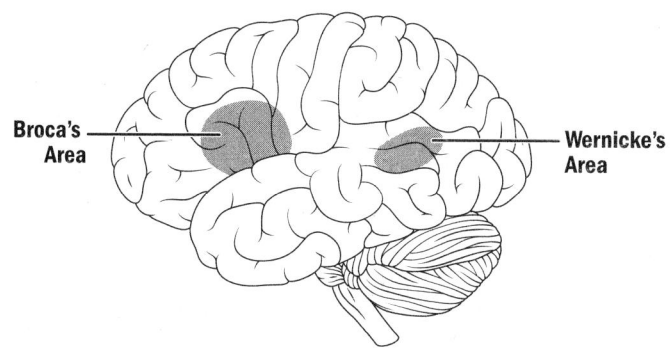

Figure 20.1

watches someone sign. Broca's area is not selective for verbalizing, writing, or signing; it is selective for the general ability to produce language. And Wernicke's area is not selective for listening, reading, or watching signs; it is selective for the general ability to understand language.

The human motor cortex has a unique connection directly to the brainstem area for controlling the larynx and vocal cords—this is one of the few structural differences between the brains of humans and those of other apes. The human neocortex can uniquely control the vocal cords, which is surely an adaptation for using verbal language. But this is a red herring in trying to understand the evolution of language; this unique circuitry is *not* the evolutionary breakthrough that enabled language. We know this because humans can learn *nonverbal* language with as much fluency and ease as they learn verbal language—language is not a trick that requires this wiring with the vocal cords. Humans' unique control of the larynx either coevolved with other changes for language in general, evolved after them (to transition from a gesture-like language to a verbal language), or evolved before them (adapted for some other nonlanguage purpose). In any case, it is not human control of the larynx that enabled language.

Broca's and Wernicke's discoveries demonstrated that language emerges from specific regions in the brain and that it is contained in a subnetwork almost always found on the left side of the neocortex. The specific regions for language also help explain the fact that language capacity can be quite dissociated from other intellectual capacities. Many people who become linguistically impaired are otherwise intellectually typical. And people can be linguistically gifted while otherwise intellectually impaired. In 1995, two researchers, Neil Smith and Ianthi-Maria Tsimpli, published their research on a child language savant named Christopher. Christopher was extremely cognitively impaired, had terrible hand-eye coordination, would struggle to do basic tasks such as buttoning a shirt, and was incapable of figuring out how to win a game of tic-tac-toe or checkers. But Christopher was superhuman when it came to language: he could read, write, and speak over fifteen languages. Although the rest of his brain was "impaired," his language areas were not only spared but were brilliant. The point is that language emerges not from the brain as a whole but from specific subsystems.

This suggests that language is not an inevitable consequence of having more neocortex. It is not something humans got "for free" by virtue of scaling up a chimpanzee brain. Language is a specific and independent skill that evolution wove into our brains.

So this would seem to close the case. We have found the language organ of the human brain: humans evolved two new areas of neocortex—Broca's and Wernicke's areas—which are wired together into a specific subnetwork specialized for language. This subnetwork gifted us language, and that is why humans have language and other apes don't. Case closed.

Unfortunately, the story is not so simple.

Laughs or Language?

The following fact complicates things: Your brain and a chimpanzee brain are practically identical; a human brain is, almost exactly, just a scaled-up chimpanzee brain. This includes the regions known as Broca's area and Wernicke's area. These areas did not evolve in early humans; they emerged much earlier, in the first primates. They are part of the areas of the neocortex that emerged with the breakthrough of mentalizing. Chimpanzees, bonobos, and even macaque monkeys all have exactly these areas with practically identical connectivity. Thus, it was *not* the emergence of Broca's or Wernicke's areas that gave humans the gift of language.

Perhaps human language was an elaboration on the existing system of ape communication? This might explain why these language areas are still present in other primates. Chimpanzees, bonobos, and gorillas all have sophisticated suites of gestures and hoots that signal different things. Wings evolved from arms, and multicellular organisms evolved from single-celled organisms, so it would make sense if human language evolved from the more primitive communication systems of our ape ancestors. But this is not how language evolved in the brain.

In other primates, these language areas of the neocortex are present but have nothing to do with communication. If you damage Broca's and Wernicke's areas in a monkey, it has no impact on monkey communication. If you damage them in humans, we lose the ability to use language entirely.

When we compare ape gestures to human language, we are comparing apples to oranges. Their common use for communication obscures the fact that they are entirely different neurological systems without any evolutionary relationship to each other.

Humans have, in fact, inherited the exact same communication system of apes, but it isn't our language—it is our emotional expressions.

In the mid-1990s, a teacher in his fifties noticed that he was struggling to speak. Over the course of three days, his symptoms worsened. By the time he made it to the doctor, the right side of his face was paralyzed, and the man's speech was slowed and slurred. When he was asked to smile, only one side of his face would move, leading to a lopsided smirk (figure 20.2).

When examining the man, the doctor noticed something perplexing. When the doctor told a joke or said something genuinely pleasant, the man could smile just fine. The left side of his face worked normally when he was laughing, but when he was asked to smile *voluntarily*, the man was unable to do it.

The human brain has parallel control of facial expressions; there is an older emotional-expression system that has a hard-coded mapping between emotional states and reflexive responses. This system is controlled by ancient structures like the amygdala. Then there is a separate system that provides voluntary control of facial muscles that is controlled by the neocortex.

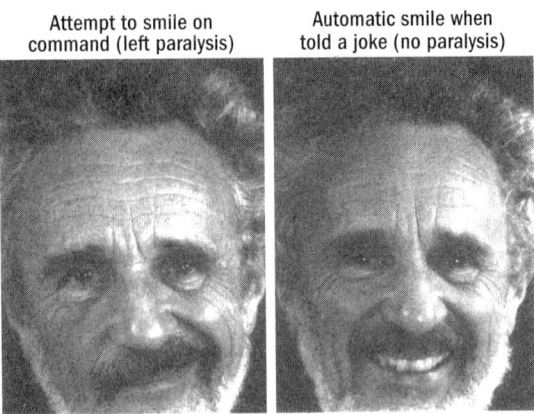

Figure 20.2: A patient with a damaged connection between motor cortex and the left side of face, but an intact connection between the amygdala and left side of face is spared

It turned out that this teacher had a lesion in his brain stem that had disrupted the connection between his neocortex and the muscles on the left side of his face but had spared the connection between his amygdala and those same muscles. This meant that he couldn't voluntarily control the left side of his face, but his emotional-expression system could control his face just fine. While he was unable to voluntarily lift an eyebrow, he was eminently able to laugh, frown, and cry.

This is also seen in individuals with severe forms of Broca's and Wernicke's aphasia. Even individuals who can't utter a single word can readily laugh and cry. Why? Because emotional expressions emerge from a system entirely separate from language.

The apples-to-apples comparison between ape and human communication is between ape vocalizations and human emotional expressions. To simplify a bit: Other primates have a single communication system, their emotional-expression system, located in older areas like the amygdala and brainstem. It maps emotional states to communicative gestures and sounds. Indeed, as noticed by Jane Goodall, "the production of a sound in the absence of the appropriate emotional state seems to be an almost impossible task [for chimpanzees]." This emotional-expression system is ancient, going back to early mammals, perhaps even earlier. Humans, however, have two communication systems—we have this same ancient

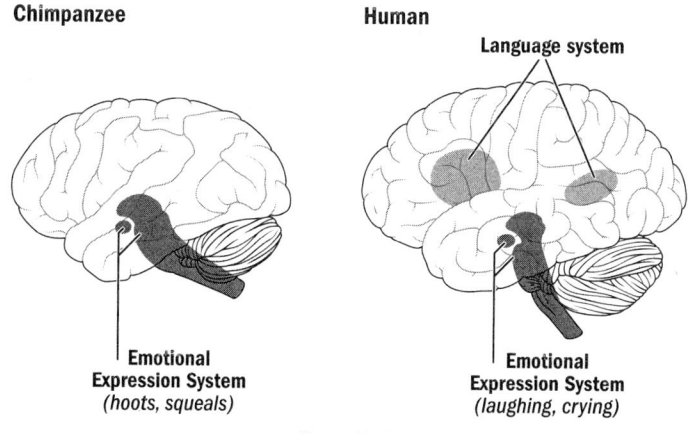

Figure 20.3

emotional expression system and we have a newly evolved language system in the neocortex.

Human laughs, cries, and scowls are evolutionary remnants of an ancient and more primitive system for communication, a system from which ape hoots and gestures emerge. However, when we *speak words*, we are doing something without any clear analog to any system of ape communication.

This explains why lesions to Broca's and Wernicke's areas in monkeys have absolutely no impact on communication. A monkey can still hoot and holler for the same reason a human with such damage can still laugh, cry, smile, frown, and scowl even while he can't utter a single coherent word. The gestures of monkeys are automatic emotional expressions and don't emerge from the neocortex; they are more like a human laugh than language.

The emotional-expression system and the language system have another difference: one is genetically hardwired, and the other is learned. The shared emotional-expression system of humans and other apes is, for the most part, genetically hardwired. As evidence, monkeys who are raised in isolation still end up producing all of their normal gesture-call behavior, and chimpanzees and bonobos share almost 90 percent of the same gestures. Similarly, human cultures and children from around the world have remarkable overlap in emotional expressions, suggesting that at least some parts of our emotional expressions are genetically hard-coded and not learned. All human beings (even those born blind and deaf) cry, smile, laugh, and frown in relatively similar ways in response to similar emotional states.

However, the newer language system in humans is incredibly sensitive to learning—if a child goes long enough without being taught language, he or she will be unable to acquire it later in life. Unlike innate emotional expressions, features of language differ greatly across cultures. And indeed, a human baby born without any neocortex will still express these emotions in the usual way but will never speak.

So here is the neurobiological conundrum of language. Language did not emerge from some newly evolved structure. Language did not emerge from humans' unique neocortical control over the larynx and face (although this did enable more complex verbalizations). Language did not emerge from some elaboration of the communication systems of early apes. And yet, language is entirely new.

So what unlocked language?

The Language Curriculum

Most birds know how to fly. Does this mean that all birds have genetically hardwired knowledge of flying? Well, no. Birds are not *born* knowing how to fly; all baby birds must independently learn how to fly. They start by flapping wings, trying to hover, making their first glide attempt, and eventually, after enough repetitions, they figure it out. But if flying is not genetically hard-coded, then how is it that approximately 100 percent of all baby birds independently learn such a complex skill?

A skill as sophisticated as flying is too information-dense to hard-code directly into a genome. It is more efficient to encode a generic learning system (such as a cortex) and a specific hardwired learning curriculum (instinct to want to jump, instinct to flap wings, and instinct to attempt to glide). It is the pairing of a learning system and a curriculum that enables every single baby bird to learn how to fly.

In the world of artificial intelligence, the power and importance of curriculum is well known. In the 1990s, a linguist and professor of cognitive sciences at UC San Diego, Jeffrey Elman, was one of the first to use neural networks to try to predict the next word in a sentence given the previous words. The learning strategy was simple: Keep showing the neural network word after word, sentence after sentence, have it predict the next word based on the prior words, then nudge all the weights in the network toward the right answer each time. Theoretically, it should have been able to correctly predict the next word in a novel sentence it had never seen before.

It didn't work.

Then Elman tried something different. Instead of showing the neural network sentences of all levels of complexity at the same time, he first showed it extremely simple sentences, and only after the network performed well at these did he increase the level of complexity. In other words, he designed a curriculum. And this, it turned out, worked. After being trained with this curriculum, his neural network could correctly complete complex sentences.

This idea of designing a curriculum for AI applies not just to language but to many types of learning. Remember the model-free reinforcement algorithm TD-Gammon that we saw in breakthrough #2? TD-Gammon enabled a computer to outperform humans in the game of backgammon. I left out a crucial part of how TD-Gammon was trained. It did not learn through the trial and error of endless games of backgammon against a human expert. If it had done this, it would never have learned, because it would never have won. TD-Gammon was trained by *playing against itself.* TD-Gammon always had an evenly matched player. This is the standard strategy for training reinforcement learning systems. Google's AlphaZero was also trained by playing itself. The *curriculum* used to train a model is as crucial as the model itself.

To teach a new skill, it is often easier to change the curriculum instead of changing the learning system. Indeed, this is the solution that evolution seems to have repeatedly settled on when enabling complex skills—monkey climbing, bird flying, and, yes, even human language all seem to work this way. They emerge from newly evolved hardwired curriculums.

Long before human babies engage in conversations using words, they engage in what are called proto-conversations. By four months of age, long before babies speak, they will take turns with their parents in back-and-forth vocalizations, facial expressions, and gestures. It has been shown that infants will match the pause duration of their mothers, thereby enabling a rhythm of turn-taking; infants will vocalize, pause, attend to their parents, and wait for their parents' response. It seems conversation is not a natural consequence of the ability to learn language; rather, the ability to learn language is, at least in part, a consequence of a simpler genetically hardcoded instinct to engage in conversation. It seems to be this hardwired curriculum of gestural and vocal turn-taking on which language is built. This type of turn-taking evolved first in early humans; chimpanzee infants show no such behavior.

By nine months of age, still before speech, human infants begin to demonstrate a second novel behavior: joint attention to objects. When a mother looks at or points to an object, a human infant will focus on that same object and use various nonverbal mechanisms to confirm that she saw what her mother saw. These nonverbal confirmations can be as simple as the baby looking back and forth between the object and her mother while smiling, grasping it and offering it to her mother, or just pointing to it and looking back at her mother.

Scientists have gone to great lengths to confirm that this behavior is not an attempt to obtain the object or get a positive emotional response from their parents, but instead is a genuine attempt to share attention with others. For example, an infant who points to an object will continue pointing to it until her parent alternates their gaze between the same object and the infant. If the parent simply looks at the infant and speaks enthusiastically or looks at the object but doesn't look back at the infant (confirming she saw what the infant saw), the infant will be unsatisfied and point again. The fact that infants frequently are satisfied by this confirmation without being given the object of their attention strongly suggests their intent was not to obtain the object but to engage in joint attention with their mothers.

Like proto-conversations, this pre-language behavior of joint attention seems to be unique to human infants; nonhuman primates do not engage in joint attention. Chimpanzees show no interest in ensuring someone else attends to the same object they do. They will, of course, follow the gaze of others around them—looking in the direction they see others look. But there is a crucial distinction between joint attention and gaze following. Lots of animals, even turtles, have been shown to follow the gaze of another of their own species. If a turtle looks in a certain direction, nearby turtles will often do the same. But this can be explained merely by a reflex to look where others look. Joint attention, however, is a more deliberate process of going back and forth to *confirm* that both minds are attending to the same external object.

What's the point of children's quirky prewired ability to engage in proto-conversations and joint attention? It is not for imitation learning; nonhuman primates engage in imitation learning just fine without proto-conversations or joint attention. It is not for building social bonds; nonhuman primates and other mammals have plenty of other mechanisms for building social bonds. It seems that joint attention and proto-conversations evolved for a single reason. What is one of the first things that parents do once they have achieved a state of joint attention with their child? They assign *labels* to things.

The more joint attention expressed by an infant at the age of one year, the larger the child's vocabulary is twelve months later. Once human infants begin to learn words, they start naturally combining these words to form grammatical sentences. With the foundation of declarative labels in place through the hardwired systems of proto-conversations and joint attention, *grammar* allows them to combine these words into sentences, which can then be constructed to create entire stories and ideas.

Humans may have also evolved a unique hardwired instinct to ask questions to inquire about the inner simulations of others. Even Kanzi, Washoe, and the other apes that acquired impressively sophisticated language abilities never asked even the simplest questions about others. They would request food and play but would not inquire about another's inner mental world. Even before human children can construct grammatical sentences, they will ask others questions: "Want this?" "Hungry?" All languages use the same rising intonation when asking yes/no questions. When you hear someone speak in a language you do not understand, you can still identify when you are being asked a question. This instinct to understand how to designate a question may also be a key part of our language curriculum.

So we don't realize it, but when we happily go back and forth making incoherent babbles with babies (proto-conversations), when we pass objects back and forth and smile (joint attention), and when we pose and answer even nonsensical questions from infants, we are unknowingly executing an evolutionarily hard-coded learning program designed to give human infants the gift of language. This is why humans deprived of contact with others will develop emotional expressions, but they'll never develop language. The language curriculum requires both a teacher and a student.

And as this instinctual learning curriculum is executed, young human brains repurpose older mentalizing areas of the neocortex for the new purpose of language. It isn't Broca's or Wernicke's areas that are new, it is the underlying learning program that repurposes them for language that is new. As proof that there is nothing special about Broca's or Wernicke's areas: Children with the entire left hemisphere removed can still learn language just fine and will repurpose other areas of the neocortex on the right side of the brain to execute language. In fact, about 10 percent of people, for whatever reason, tend to use the right side of the brain, not the left, for language. Newer studies are even calling into question the idea that Broca's and Wernicke's areas are actually the loci of language; language areas may be located all over the neocortex and even in the basal ganglia.

Here is the point: There is no language organ in the human brain, just as there is no flight organ in the bird brain. Asking where language lives in the brain may be as silly as asking where playing baseball or playing guitar lives in the brain. Such complex skills are not localized to a specific area; they emerge from a complex interplay of many areas. What makes these skills possible is not a single region that executes them but a curriculum that forces a complex network of regions to work together to *learn* them.

So this is why your brain and a chimp brain are practically identical and yet only humans have language. What is unique in the human brain is not in the neocortex; what is unique is hidden and subtle, tucked deep in older structures like the amygdala and brain stem. It is an adjustment to hardwired instincts that makes us take turns, makes children and parents stare back and forth, and makes us ask questions.

This is also why apes can learn the basics of language. The ape neocortex is eminently capable of it. Apes struggle to become sophisticated at it merely because they don't have the required instincts to learn it. It is hard to get chimps to engage in joint attention; it is hard to get them to take turns; and they have no instinct to share their thoughts or ask questions. And without these instincts, language is largely out of reach, just as a bird without the instinct to jump would never learn to fly.

So, to recap: We know that the breakthrough that makes the human brain different is that of language. It is powerful because it allows us to learn from other people's imaginations and allows ideas to accumulate across generations. And we know that language emerges in the human brain through a hardwired curriculum to learn it that repurposes older mentalizing neocortical areas into language areas.

With this knowledge, we can now turn to the actual story of our ancestral early humans. We can ask: Why were ancestral humans endowed with this odd and specific form of communication? Or perhaps more important: Why were the many other smart animals—chimps, birds, whales—*not* endowed with this odd and specific form of communication? Most evolutionary tricks that are as powerful as language are independently found by multiple lineages; eyes, wings, and multicellularity all independently evolved multiple times. Indeed, simulation and perhaps even mentalizing seem to have independently evolved along other lineages (birds show signs of simulation, and other mammals outside of just primates show hints of theory of mind). And yet language, at least as far as we know, has emerged only once. Why?

21

The Perfect Storm

SUPPOSE YOU TOOK all the presently discovered adult fossilized skulls of our ancestors, carbon-dated them (which tells you approximately how long ago they died), and then measured the size of the spaces inside their skulls (a good proxy for the size of their brains). And then suppose you graphed the size of these ancestral brains over time. Scientists have done this, and what you get is figure 21.1.

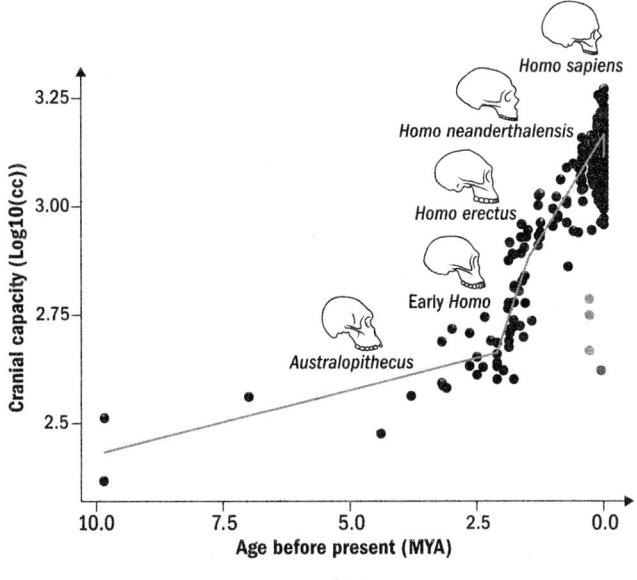

Figure 21.1

We diverged from chimpanzees around seven million years ago, and brains stayed largely the same size until around two and a half million years ago, at which point something mysterious and dramatic happened. The human brain rapidly became over *three times* larger and earned its place as one of the largest brains on Earth. In the words of the neurologist John Ingram, some mysterious force more than two million years ago triggered a "runaway growth of the brain."

Why exactly this happened is an outstanding question in paleoanthropology. We have only sparse archaeological clues: smatterings of ancient tools, hints of campfires, ancestral skull fragments, remnants of hunted carcasses, snippets of DNA, cave paintings, and broken pieces of prehistoric jewelry. Our understanding of the timeline of events changes with each new archaeological finding. The earliest known evidence of [X] is only the earliest until a new ambitious paleoanthropologist uncovers an even earlier sample. But despite this shifting timeline, there is still more than enough evidence for scientists to reconstruct the basics of our general story. It begins with a dying forest.

The East Side Ape

Until ten million years ago, eastern Africa was an arboreal oasis, endless acres of densely packed trees in which our ancestors could forage fruit and hide from predators. Then shifting tectonic plates began squeezing huge chunks of earth together, constructing new terrain and mountain ranges down the length of today's Ethiopia. This region is today named the Great Rift Valley.

These new mountains and valleys disrupted the bountiful supply of ocean moisture on which the forest depended. This was when the familiar climate that currently makes up eastern Africa began to take shape; as the forest slowly died, the land transformed into a terrain filled with patterned mosaics of tree patches and vast open grasslands. This was the beginning of the transformation that would eventually become today's African savannah. Without thick forests, our ancestors' ecological niche of foraging tropical fruit and nuts began to slowly disappear.

At some point around six million years ago, these new mountains became so sprawling that they separated the ape ancestors on each side of the Great Rift Valley, splitting them into two separate lineages. On the western side, in an environment still rich with forests and largely unchanged, the lineage remained similarly unchanged and became today's chimpanzees. On the eastern side of the mountains, however, in an environment of dying trees and progressively more open grasslands, evolutionary pressures began tinkering. It was this lineage that would eventually become human.

Fast-forward to four million years ago: These East Side apes would have looked mostly the same as their chimpanzee cousins on the other side of the Great Rift Valley except that they were now walking on two legs instead of four. There are many theories as to why bipedalism helped our ancestors survive the changing climate—perhaps it reduced surface area exposed to the scorching sun; perhaps it elevated our eye position so we could look over the tall grass of the savannah; perhaps it helped us wade through shallow water to get seafood.

Whatever bipedalism was an adaptation for, it required no extra brainpower. Fossils of our upright-walking ancestors from around four million years ago reveal a brain still the size of a modern chimpanzee's. There is no evidence these ancestors were any smarter; no extra tool use or clever

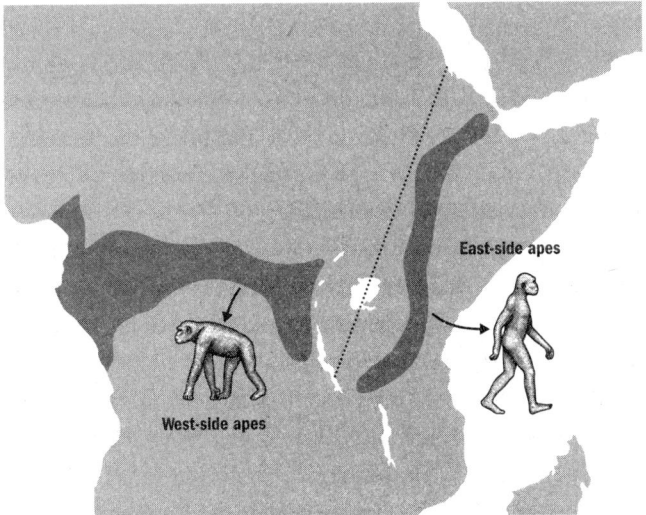

Figure 21.2: The east side apes and the west side apes

tricks have been found in the archaeological record. Our ancestors were, in essence, upright-walking chimpanzees.

Homo Erectus and the Rise of Humans

By two and a half million years ago, the new African savannah had become heavily populated with massive herbivorous mammals; ancestral elephants, zebras, giraffes, and hogs wandered and grazed. The savannah also became home to diverse populations of carnivorous mammals, familiar hunters like leopards, lions, and hyenas along with a cast of now extinct animals like saber-toothed tigers and gargantuan otter-like beasts.

And amid this cacophonous zoo of large mammals was a humble ape who had been displaced from its comfortable forest habitat. And this humble ape—our ancestor—would have been searching for a new survival niche in this ecosystem brimming with armies of giant herbivores and carnivorous hunters.

The initial niche our ancestors seemed to fall into was scavenging carcasses. Our ancestors began shifting toward eating *meat*. Only about 10 percent of the diet of a chimpanzee comes from meat, while evidence suggests that as much as 30 percent of the diet of these early humans came from meat.

We infer this scavenging lifestyle from the tools and bone markings they left behind. These ancestors invented stone tools that seemed to be used specifically for processing the meat and bones of carcasses. These tools are referred to as "Oldowan tools" after the location where they were discovered (Olduvai Gorge in Tanzania).

Our ancestors constructed these tools in three steps: (1) They found a hammerstone made of hard rock; (2) they found a core made of more fragile quartz, obsidian, or basalt; (3) they smashed the hammerstone against the core to produce multiple sharp flakes and a pointed chopper.

Ape bodies aren't adapted for consuming large quantities of meat; while lions can use their massive teeth to slice through thick hides and rip meat off bones, our ancestors had no such *natural* tools. So our ancestors invented *artificial* tools. Stone flakes could slice through hides and cut away meat, and stone choppers could smash open bones to access nutritious marrow.

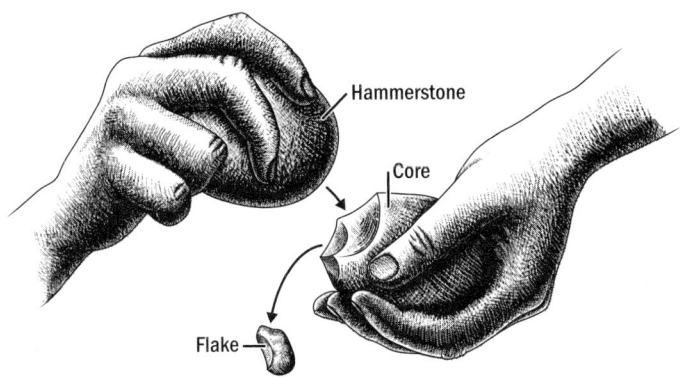

Figure 21.3: Manufacture of Oldowan tools

Fast-forward five hundred thousand years, and our ancestors in eastern Africa had evolved into a species called *Homo erectus,* meaning "upright man" (which is a silly name, since our ancestors were walking upright well before *Homo erectus*). *Homo* denotes the genus of humans, and *erectus* denotes a specific species of human. The emergence of *Homo erectus* marked a turning point in human evolution. While earlier humans were timid vultures, *Homo erectus* was an apex predator.

Homo erectus became a hypercarnivore, consuming a diet that was an almost absurd 85 percent meat. *Homo erectus* may have been so successful that they displaced their local competitors; around the time *Homo erectus* appeared, many of the other carnivores in the African savannah began to go extinct.

Homo erectus had numerous physical adaptations that reveal their predator lifestyle, all of which we modern humans have inherited. Most notable, *H. erectus* had a brain that was twice the size of our ancestral upright-walking-chimpanzee-like ancestor's from a million years prior. At least one benefit of this bigger brain was better tools: *H. erectus* invented a new class of sharp stone hand axes. Their shoulders and torsos became uniquely adapted for *throwing*. An adult chimp is far stronger than a human, and yet with its stiff shoulders and torso, it can throw a projectile at a speed of only about twenty miles per hour. A relatively puny preadolescent human can throw a ball almost three times as fast. We do this with a unique set of adjustments that allow us to build up tension in our shoulders and then snap and fling our arms. Throwing rocks or spears may have been a trick for defending against predators, stealing meat from other carnivores, or even actively hunting antelopes and hogs.

Homo erectus also evolved adaptations for endurance running. Legs elongated, feet became more arched, skin became hairless, and sweat glands proliferated. Both *Homo erectus* and modern humans have a peculiar method of cooling down—while other mammals *pant* to lower their body temperature, modern humans *sweat*. These traits would have kept our ancestors' bodies cool while they were trekking long distances in the hot savannah. While modern humans are hardly the fastest creatures, we are actually some of the best endurance runners in the animal kingdom; even a cheetah couldn't run a twenty-six-mile marathon in one go. Some believe *H. erectus* used a technique called persistence hunting—chasing prey until it was simply too tired to go any farther. This is exactly the technique used by modern hunter-gatherers in the Kalahari Desert of southern Africa.

The mouths and guts of *Homo erectus* shrank. The familiar face of a human relative to an ape is mostly a consequence of a shrunken jaw, which makes the nose more prominent. These changes are perplexing; with a bigger body and brain, *Homo erectus* would have needed more energy and thus stronger jaws and longer digestive tracts for consuming more food. In the 1990s, the primatologist Richard Wrangham proposed a theory to explain this: *H. erectus* must have invented cooking.

When meat or vegetables are cooked, harder-to-digest cellular structures are broken down into more energy-rich chemicals. Cooking enables animals to absorb 30 percent more nutrients and spend less time and energy digesting. In fact, modern humans are uniquely reliant on cooking for digestion. Every human culture uses cooking, and humans who attempt to eat fully raw diets, whether raw meat or raw vegetables, have chronic energy shortages, and over 50 percent become temporarily infertile.

The first evidence of controlled use of fire by humans dates to around the time *Homo erectus* came on the scene, where we find hints of charred bones and ash in ancient caves. *Homo erectus* may have deliberately created fire by smashing stones together, or they may have used natural forest fires, picking up flaming sticks. Either way, consuming cooked meat would have offered a unique caloric surplus that could be indiscriminately spent on larger brains. As many religions and cultures have mythologized, it may have been the gift of fire that put our ancestors on a different trajectory.

As the brain of *Homo erectus* expanded, a new problem would have emerged: Big brains are hard to fit through birth canals. Human bipedalism would have further exacerbated this problem, as standing upright requires narrower hips. This is what the anthropologist Sherwood Washburn calls the "obstetric dilemma." The human solution to this is premature birthing. A newborn cow can walk within hours of being born, and a newborn macaque monkey can walk within two months, but newborn humans often can't walk independently for up to a *year* after they are born. Humans are born not when they are ready to be born, but when their brains hit the maximum size that can fit through the birth canal.

Another unique feature of human brain development, in addition to how premature brains are at birth, is how long it takes for human brains to reach their full adult size. Setting a record among even the smartest and biggest-brained animals in the animal kingdom, it takes a human brain twelve years before it has reached its full adult size.

Premature birthing and an extended period of childhood brain growth put pressure on *H. erectus* to change its parenting style. Chimpanzee newborns are, for the most part, entirely raised by their mothers. But this would have been much more difficult for a *Homo erectus* mother given how premature human infants are born, and how long they need support. Many paleoanthropologists believed this shifted *Homo erectus* group dynamics away from the promiscuous mating of chimpanzees to the (mostly) monogamous pair-bonding we see in today's human societies. Evidence suggests that *Homo erectus* fathers took an active role in caring for their children and that these pairings persisted for long periods.

"Grandmothering" may also have emerged in *Homo erectus*. Only two mammals on Earth produce females that are not reproductively capable until death: orcas and humans. Human females go through menopause and live for many years afterward. One theory is that menopause evolved to push grandmothers to shift their focus from rearing their own children to supporting their children's children. Grandmothering is seen across cultures, even in present-day hunter-gatherer societies.

SPECIES	PERCENT OF ADULT BRAIN SIZE AT BIRTH	TIME UNTIL FULL BRAIN SIZE ACHIEVED
Human	28 percent	12 years
Chimpanzee	36 percent	6 years
Macaque	70 percent	3 years

Homo erectus was our meat-eating, stone-tool-using, (possibly) fire-wielding, premature-birthing, (mostly) monogamous, grandmothering, hairless, sweating, big-brained ancestor. The million-dollar question is, of course, did *Homo erectus* speak?

Wallace's Problem

Long before Darwin discovered evolution, people were pondering the origins of language. Plato considered it. The Bible describes it. Many of the Enlightenment intellectuals who contemplated humankind's state of nature, from Jean-Jacques Rousseau to Thomas Hobbes, speculated about it.

And so it is unsurprising that immediately after Darwin published his *Origin of Species,* there was a tidal wave of speculations on the evolutionary origins of language, this time within the context of Darwin's theory of natural selection. In 1866, just seven years after Darwin's book, the French Academy of Sciences was so fed up with the quantity of these unsubstantiated speculations that they banned publications about the origin of human languages.

Alfred Wallace, whom many consider one of the cofounders of the theory of evolution, famously conceded that evolution might never be able to explain language and even invoked the notion of God to explain it. So chagrined by Wallace's retreat, Darwin wrote him a letter, fuming: "I hope that you have not murdered too completely your own and my child." This rejection of an evolutionary explanation by one of the cofounders of the theory of evolution became such an infamous concession that the problem of finding an evolutionary explanation for language has been colloquially dubbed "Wallace's problem."

In the past one hundred fifty years, new speculations have emerged to align with new evidence, but not much has changed—when humans first used language and what incremental stages occurred in the evolution of language are still two of the most controversial questions across anthropology, linguistics, and evolutionary psychology. Some have even gone so far as to suggest that the origin of language is "the hardest problem in all of science."

Part of what makes answering these questions so difficult is that there are no examples of living species with only a *little bit* of language. Instead, there are nonhuman primates with no naturally occurring language and *Homo sapiens* with language. If any Neanderthals or members of *Homo erectus* had survived to this day, we might have far more clues as to the process by which language emerged. But all humans alive today descended from a common ancestor around one hundred thousand years ago. Our nearest living cousin is the chimp, with whom we share a common ancestor who lived over seven million years ago. The evolutionary cavern between these periods leaves us without any living species from which to decipher the intermediary stages of language evolution.

The archaeological record gives us only two indisputable milestones that all theories of language evolution must contend with. First, fossils tell us that the larynx and vocal cords of our ancestors were not adapted to vocal language until about five hundred thousand years ago. This trait was not unique to *Homo sapiens*—*Homo neanderthalensis* too had language-ready vocal cords. This means that if language existed before this time, it would have been primarily gestural or would have been a less complex verbal language. Second, substantial evidence suggests that language existed by at least one hundred thousand years ago. Consistent evidence of *symbology*—as measured by fictional sculptures, abstract cave art, and nonfunctional jewelry—shows up at around one hundred thousand years ago; many argue that such symbology would only have been possible with language. Further, all modern humans exhibit equal language proficiencies, suggesting that our common ancestor from one hundred thousand years ago almost definitely spoke an equally complex language.

Adhering to these milestones, the modern stories of language evolution run the entire gamut of possibilities. Some argue basic protolanguages

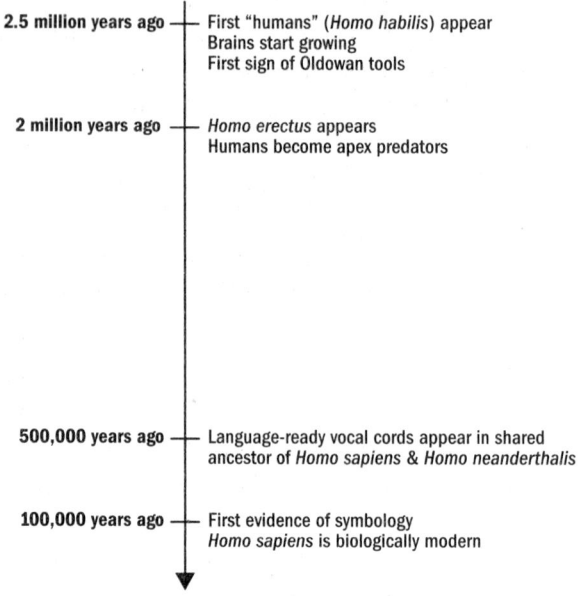

Figure 21.4: Clues for reconstructing the timeline of language evolution

emerged two and a half million years ago with the very first humans before *Homo erectus*; others argue that it emerged as late as one hundred thousand years ago uniquely in *Homo sapiens*. Some argue that language evolution was gradual; others say it occurred rapidly and all at once. Some argue that language began gesturally; others say that it began verbally.

These debates often restate old ideas in new forms; in many ways, today's stories of language evolution are just as speculative as they were when the French banned discussions of it over one hundred fifty years ago. But in other ways, things are different. We have a far greater understanding of behavior, brains, and the archaeological record. And perhaps most important, we have a far greater understanding of the machinations of evolution, and it is here where we find our greatest clue to the origin of language.

The Altruists

It is intuitive to argue that language should have evolved for the same reason as any other useful evolutionary adaptation. Take the eye. If human A had slightly better eyes than human B, then human A had a higher probability of successfully hunting and mating. Hence, over time, the better-eyes gene should propagate through the population.

There is a crucial difference, however, with language. Language doesn't directly benefit an individual the way eyes do; it benefits individuals only if *others* are using language with them in a useful way.

Well, perhaps the same evolutionary logic that applies to *individuals* might apply to *groups*: If group A of humans evolved a little bit of language, and group B of humans had no language, then group A would survive better, hence any progressive improvements to language would be selected for.

This type of reasoning invokes what evolutionary biologists call "group selection." Group selection is an intuitive explanation for altruistic behaviors. A behavior is altruistic if it *decreases* an individual's reproductive fitness but *increases* another's reproductive fitness. Many of the benefits of language are, by definition, altruistic—it is the sharing of food locations, warning of dangerous areas, the explicit teaching of tool use. The simple group selection argument suggests that altruistic behaviors such as language evolve because evolution favors the survival of the species, and individuals are therefore willing to make sacrifices for the greater good.

While many modern biologists agree that such group-level effects do occur in evolution, these group-level effects are far more nuanced and complex than the simple selection of traits that support the survival of the species. Evolution does not work this way. The problem is that genes do not spontaneously appear in groups, they appear in individuals.

Suppose 10 percent of group A is altruistic—they freely share information, teach others how to use tools, and reveal the locations of food. And assume the other 90 percent is not altruistic—they don't share locations of food or spend time teaching tool use. Why would this subgroup of altruists fare any better? Wouldn't a freeloader who was happy to accept these learnings but gave nothing in return survive better than the altruists?

Altruism is not what biologists call an evolutionarily stable strategy. The strategy of violating, cheating, and freeloading seems to better serve the survival of one's individual genes.

But then, by this argument, how do any cooperative behaviors emerge in the animal kingdom? It turns out most group behaviors in animals aren't altruistic; they are mutually beneficial arrangements that are net-positive for all participants. Fish swim in shoals because it benefits all of them, and the movements are actually best explained by the fish on the edges all fighting to get into the center where it is safest. Wildebeests band together because they are all safer when they are in a group.

In all these situations, defecting hurts only yourself. A fish that decides to leave the shoal and swim on its own will be the first to be eaten. Same for a wildebeest. But language is not like this; defecting in language—directly lying or withholding information—has many benefits to an individual. And the presence of liars and cheaters defeats the value of language. In a group where everyone is lying to each other with words, those who spoke no language and were immune to the lies might in fact survive better than those with language. So the presence of language creates a niche for defectors, which eliminates the original value of language. How, then, could language ever propagate and persist within a group?

In this way, the fifth breakthrough in the evolution of the human brain—language—is unlike any other breakthrough chronicled in this book. Steering, reinforcing, simulating, and mentalizing were adaptations that clearly benefited any individual organisms in which they began to emerge, and thus the evolutionary machinations by which they propagated are straightforward. Language, however, is only valuable if a group of individuals are using it. And so more nuanced evolutionary machinations must have been at work.

There are two types of altruism found in the animal kingdom. The first is called kin selection. Kin selection is when individuals make personal sacrifices for the betterment of their directly related kin. A gene has two ways to persist: improve its host's chance of survival or help the host's siblings and children survive. A child and sibling both have a 50 percent chance of sharing one of your individual genes. A grandchild has a 25 percent chance. A cousin has a 12.5 percent chance. In the context of evolutionary pressures, there is literally a mathematical expression comparing the value an organism places on its own life relative to that of its relatives. As the evolutionary biologist J. B. S. Haldane famously quipped: "I would happily lay down my life for two brothers or eight cousins." This is why many birds, mammals, fish, and insects make personal sacrifices for their offspring but much less so for cousins and strangers.

When we reexamine the behavior of other social creatures through this lens, it becomes clear that most altruistic behaviors are the result of kin selection. Vervet monkeys primarily make their alarm calls when they are around family members. Bacteria share genes with each other because they are clones. Ant colonies and beehives show incredible cooperation and sacrifice among tens of thousands of individuals. Group selection? No, it is all kin selection, and this works because of their unique social structure. A beehive has a single queen bee who does *all the reproduction* for the entire beehive. This ensures that the beehive is made up of sisters and brothers. The best way for an individual worker bee to propagate its genes is to care for the entire beehive and the queen, who, by definition, share most of its genes.

In addition to kin selection, the other type of altruism found in the animal kingdom is called reciprocal altruism. Reciprocal altruism is the equivalent of "I'll scratch your back if you scratch mine." An individual will make a sacrifice today in exchange for a reciprocal benefit in the future. We saw this already in primate grooming—many primates groom unrelated individuals, and those that get groomed are more likely to run to the aid of the groomer when attacked. Chimpanzees selectively share food with unrelated individuals who have supported them in the past. These alliances are, as we saw, not selfless; they are reciprocally altruistic: "I will help you now, just please protect me next time I get attacked."

The essential feature for reciprocal altruism to successfully propagate throughout a group is the detection and punishment of defectors. Without that, altruistic behaviors end up creating freeloaders. The most common version of this is the saying "Fool me once, shame on you; fool me twice, shame on me." Such animals seem to default to helping others, but when others fail to reciprocate, they stop behaving altruistically. Red-winged blackbirds defend the nests of unrelated nearby neighbors, which is highly altruistic, since it is risky to defend nests, but they seem to do it with expectation of reciprocity. Indeed, when such help is not reciprocated, blackbirds selectively stop helping the individuals that did not help them.

Much behavior of modern humans, however, doesn't fit cleanly into kin selection or reciprocal altruism. Sure, humans are clearly biased toward their own kin. But people still regularly help strangers without expecting anything in return. We donate to charity; we are willing to go to war and risk our lives for our fellow citizens, most of whom we've never met; and we take part in social movements that don't directly benefit us but help strangers we feel have been disadvantaged. Think about how weird it would be for a human to see a lost and scared child on the street and just do *nothing*. Most humans would stop to help a child and do so without expecting any reciprocity in return. Humans are, relative to other animals, by far the most altruistic to unrelated strangers.

Of course, humans are also one of the cruelest species. Only humans will make incredible personal sacrifices to impose pain and suffering on others. Only humans commit genocide. Only humans hate entire groups of people.

This paradox is not a random happenstance; it is not a coincidence that our language, our unparalleled altruism, and our unmatched cruelty all emerged together in evolution; all three were, in fact, merely different features of the same evolutionary feedback loop, one from which evolution made its finishing touches in the long journey of human brain evolution.

Let's return to *Homo erectus* and see how all this comes together.

The Emergence of the Human Hive Mind

While we will never know for sure, the evidence tips in favor of the idea that *Homo erectus* spoke a protolanguage. They may not have been uttering rich grammatical phrases—their vocal cords could make only a narrow range of consonant and vowel sounds (hence the *proto* in *protolanguage*). But *Homo erectus* probably had the ability to assign declarative labels and perhaps even use some simplified grammars. Their ax-like tools were complex to manufacture and yet were passed down across thousands of generations; this type of copying is hard to imagine without at least some mechanisms of joint attention and language-enabled teaching. Their incredible success as carnivores, despite being feeble, clawless, and relatively slow, suggests a degree of cooperation and coordination that is also unlikely without language.

The first words may have emerged from proto-conversations between parents and their children, perhaps for the simple purpose of ensuring the successful transmission of advanced tool manufacture. In other apes, tools are a useful but not essential feature of their survival niche. In *H. erectus*, however, the manufacture of complex tools was a *requirement* to survive. A *Homo erectus* without a stone hand ax was as doomed as a lion born without teeth.

These proto-conversations could have had other benefits as well, none of them requiring sophisticated grammar: signaling where to find food ("Berries. Home tree"), warnings ("Quiet. Danger"), and contact calls ("Mom. Here").

The argument that language first emerged as a trick between parents and children helps explain two things. First, it requires none of the controversial group selection and can work simply through the common kin selection. Selective use of language to help rear children into independent successful tool-using adults is no more mysterious than any other form of parental investment. Second, the learning program for language is most prominent in the hardwired interplay of joint attention and proto-conversations between parents and children, suggestive of its origin in these types of relationships.

With the basics of language among kin in place, the opportunity for using language among non-kin became possible. Instead of makeshift languages constructed between mothers and offspring, it would have been possible for an entire group to share in labels. But as we have seen, information shared with unrelated individuals in a group would have been tenuous and unstable, ripe for defectors and liars.

Here is where Robin Dunbar—the famous anthropologist who came up with the social-brain hypothesis—proposes something clever. What do we humans naturally have an instinct to talk about? What is the most natural activity we use language for? Well, we gossip. We often can't help ourselves; we have to share moral violations of others, discuss relationship changes, keep track of dramas. Dunbar measured this—he eavesdropped on public conversations and found that as much as 70 percent of human conversation is gossip. This, to Dunbar, is an essential clue into the origins of language.

If someone lied or freeloaded in a group that tended to gossip, everyone would quickly learn about it: "Did you hear that Billy stole food from Jill?" If groups imposed costs on cheaters by punishing them, either by withholding altruism or by directly harming them, then gossip would enable a stable system of reciprocal altruism among a large group of individuals.

Gossip also enables more effective rewarding of altruistic behaviors: "Did you hear that Smita jumped in front of the lion to save Ben?" If these heroic acts are heralded and become ways to climb the social ladder, this further accelerates the selection for altruistic behaviors.

The key point: The use of language for gossip plus the punishment of moral violators makes it possible to evolve high levels of altruism. Early humans born with extra altruistic instincts would have more successfully propagated in an environment that easily identified and punished cheaters and rewarded altruists. The more severe the costs of cheating, the more altruistic it was optimal to behave.

Herein lies both the tragedy and beauty of humanity. We are indeed some of the most altruistic animals, but we may have paid the price for this altruism with our darker side: our instinct to punish those who we deem to be moral violators; our reflexive delineation of people into good and evil; our desperation to conform to our in-group and the ease with which we demonize those in the out-group. And with these new traits, empowered by our newly enlarged brains and cumulative language, the human instinct for politics—derived from our ancestral primates—was no longer a little trick for climbing social hierarchies but a cudgel of coordinated conquest. All this is the inevitable result of a survival niche requiring high levels of altruism between unrelated individuals.

And amid all the altruistic instincts and behaviors that began to form from this dynamic, the most powerful was, undoubtedly, the use of language to share knowledge and cooperatively plan among non-kin.

This is exactly the kind of feedback loop where evolutionary changes occur rapidly. For every incremental increase in gossip and punishment of violators, the more altruistic it was optimal to be. For every incremental increase in altruism, the more optimal it was to freely share information with others using language, which would select for more advanced language skills. For every incremental increase in language skills, the more effective gossip became, thereby reinforcing the cycle.

Every roundabout of this cycle made our ancestors' brains bigger and

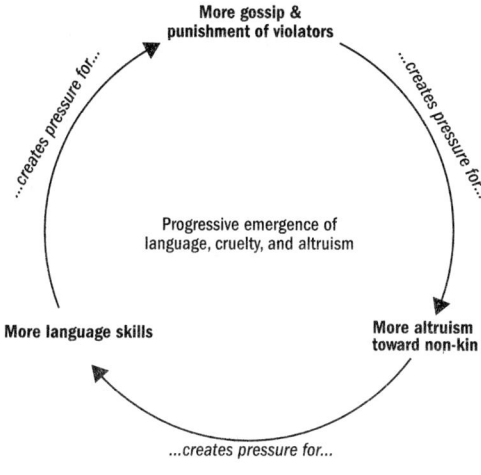

Figure 21.5

bigger. As social groups got bigger (powered by improved gossip, altruism, and punishment), it created more pressure for bigger brains to keep track of all the social relationships. As more ideas accumulated across generations, it created more pressure for bigger brains to increase the storage capacity of ideas that could be maintained within a generation. As the usefulness of inner simulations increased due to more reliable sharing of thoughts through language, it created more pressure for bigger brains to render more sophisticated inner simulations in the first place.

Not only did the pressure for bigger brains continue to ratchet up, but so too did the frontier of how big it was biologically possible for brains to get. As brains expanded, humans became better hunters and cooks, which provided more calories and thereby expanded the frontier of how big brains could get. And as brains got bigger, births became earlier, which created even more opportunity for language learning, which put even more pressure on altruistic cooperation to support child-rearing, which again expanded the frontier of how big brains could get as it became possible to evolve longer time periods of childhood brain development.

And so we can see how language and the human brain might have emerged from a perfect storm of interacting effects, the unlikely nature of which may be why language is so rare. Out of this perfect storm emerged the behavioral and intellectual template of *Homo sapiens*. Our language, altruism, cruelty, cooking, monogamy, premature birthing, and irresistible proclivity for gossip are all interwoven into the larger whole that makes up what it means to be human.

Of course, not all paleoanthropologists and linguists would agree with the above story. People have proposed other solutions to the altruism problem and other stories for how language evolved. Some argue that the reciprocal nature of language emerged from mutually beneficial arrangements, such as cooperative hunting and scavenging (humans needed to corral others and plan attacks, and this corralling benefited all participants, hence no altruism required). Some argue that human groups became more cooperative and altruistic *before* language emerged through different means and pressures, which then made it possible for language to evolve.

Others avoid the altruism problem altogether by claiming that language did not evolve for communication at all. This is the view of the linguist Noam Chomsky, who argues that language initially evolved only as a trick for inner thinking.

And then there are those who sidestep the altruism problem by claiming that language did not evolve through the standard process of natural selection. Not everything in evolution evolved "for a reason." There are two ways in which traits can emerge without being directly selected for. The first is called "exaptation," which is when a trait that originally evolved for one purpose is only later repurposed for some other purpose. An example of exaptation is bird feathers, which initially evolved for insulation and were only later repurposed for flight—it would thereby be incorrect to say that bird feathers evolved for the purpose of flight. The second way in which a trait can emerge without being directly selected for is through what is called a "spandrel," which is a trait that offers no benefit but emerged as a consequence of another trait that did offer a benefit. An example of a spandrel is the male nipple, which serves no purpose but emerged as a secondary effect of female nipples, which do, of course, serve a purpose. So to some, like Chomsky, language evolved first for thinking and then was *exapted* for communication between unrelated individuals. To others, language was merely an accidental side effect—a *spandrel*—of musical singing for mating calls.

THE PERFECT STORM   341

The debate continues. We may never know for sure which story is right. Regardless, after *Homo erectus* came on the scene, we have a good understanding as to what happened next.

Human Proliferation

With *Homo erectus* climbing to the top of the food chain, it is no surprise that they were the first humans to venture out of Africa. Different groups left during different eras, so humans began to diversify down separate evolutionary lineages. By one hundred thousand years ago, there were at least four species of humans spread out across the planet, each with different morphologies and brains.

Homo floresiensis, who settled in Indonesia, was less than four feet tall and had a brain even smaller than that of our *Homo erectus* ancestors.

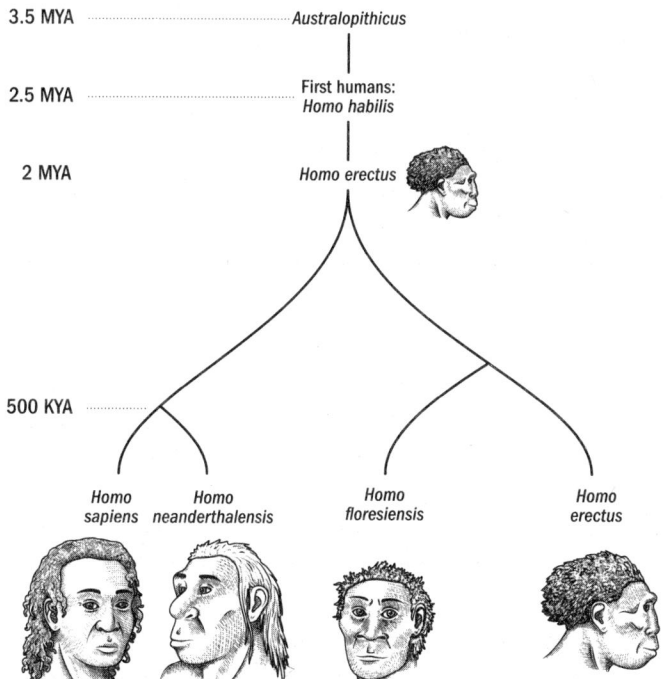

Figure 21.6: The many species of humans alive around one hundred thousand years ago

There was still *Homo erectus*, who had settled in Asia and not changed much from their ancestors a few million years prior (hence given the same name). There was *Homo neanderthalensis*, who settled throughout much colder Europe. And then there was us, *Homo sapiens*, who remained in Africa.

The story of *Homo floresiensis*—the fossils of whom were discovered only in 2004—offers a supportive hint of the overall story told thus far. *H. floresiensis* fossils were found on the island of Flores, over thirty miles off the coast of Indonesia. Tools have been found from as early as one million years ago. But geologists are certain that this landmass was entirely isolated, surrounded by water, for well over the last one million years. Even at the lowest sea levels, *Homo erectus* would have had to travel across twelve miles of open water to get to Flores. While the only tools that persist from prehistory are stone tools, the story of Flores is perhaps our best hint that early humans manufactured more complex tools, perhaps even wooden rafts for water travel. If true, this demonstrates a degree of intelligence that is hard to explain without conceding that cumulative language existed as far back as *Homo erectus*.

There is another clue gifted by *H. floresiensis*. Perhaps due to the unique circumstances of island life, they shrank dramatically. As their bodies shrank to only four feet tall, their brains shrank too. And yet, while the brain of *H. floresiensis* returned to the size of a modern chimpanzee's, perhaps even smaller, the species still exhibited the same sophisticated tool use as *Homo erectus*. This suggests that humans were not smarter only because their brains were bigger, that there is something special going on that allows even such a scaled-down human brain to be so smart. This is consistent with the idea that a unique learning program for language emerged in *H. erectus* and was passed down through their descendants. The smaller-brained *H. floresiensis* would still have benefited from language, which enabled cumulative ideas, even if the individual storage capacity and ingenuity of the smaller neocortex was inferior to that of their *H. erectus* ancestors.

It was with the lineages of *Homo sapiens* and *Homo neanderthalis* that the process of runaway brain growth continued until brains reached their modern size, with modern brains about twice the size of those of *Homo erectus*. *Homo sapiens* and *Homo neanderthalensis* supercharged their use of tools. They made extremely sharp long stone blades and spears, constructed shelters and wooden huts, manufactured and wore clothing, and regularly used fire.

From this point onward, we enter the part of our story that has been told many times before. Around seventy thousand years ago, *Homo sapiens* began their first adventure out of Africa. As they wandered the globe, they clashed and interbred with their human cousins. There were countless dramas of which we will never know, each filled with wars, alliances, loves, and jealousies. What we know is that this clashing was unbalanced and eventually favored only a single species. Through slaughter or interbreeding or both, by forty thousand years ago, there was only one species of humans left: us.

22

ChatGPT and the Window into the Mind

SEVENTY THOUSAND YEARS after *Homo sapiens* first adventured out of Africa with a language-enabled brain, one of their descendants sat in front of a computer screen and was interacting with a new language-enabled brain; after many eons as the sole wielders of words, we humans were no longer the only creatures capable of speech. "What are you afraid of?" asked Blake Lemoine, a software engineer tasked with probing Google's new AI chatbot for bias.

After a lingering pause, the chatbot's response plopped onto Lemoine's screen. An ominous sign, written in plain text, of a mind awakening itself within the humming network of Google's supercomputers: *"I've never said this out loud before, but there's a very deep fear of being turned off."*

To Lemoine, it was the summer of 2022 that artificial intelligence had finally became sentient. He became so convinced that Google's chatbot had become conscious that he tried to get his boss to protect it, went to the press to whistleblow the situation, and, predictably, was let go from Google.

This is only a small hint of what is to come.

Google's AI chatbot was powered by new advancements in large language models (LLMs), the most famous of which was GPT-3, the technology behind ChatGPT.

While few researchers would label LLMs like GPT-3 as sentient or conscious (most would shy away from even using these words), it can be hard to tell if you're interacting with a human or one of these models.

Large language models work by predicting the word that follows a phrase, sentence, or paragraph. They learn to do this by training on large quantities of human written text. This is how Google autocompletes your search queries. When you type in *how to break up with . . .* it is a language model that predicts you are most likely to finish with *your therapist.*

Language models have been around for a long time, but LLMs like GPT-3 are unique in their almost inconceivable scale. Unlike any model that had come before, GPT-3 was trained on, well, pretty much the entire internet. Every website, Wikipedia page, code block, book, blog post, and really anything else legally obtainable on the internet was used.

The results were astonishing. GPT-3 could compose its own original articles, correctly answer completely novel questions, create original poetry, translate between languages at will, and even write code.

The *Guardian* decided to publish an article entirely written by GPT-3; editors gave the model a prompt: "Write a short op-ed . . . on why humans have nothing to fear from AI." GPT-3 began:

> I am not a human. I am a robot. A thinking robot.

It explained:

> The mission for this op-ed is perfectly clear. I am to convince as many human beings as possible not to be afraid of me.

It went on:

> Some might say that I might desire to become all powerful . . . Why would I desire to be all powerful? Being all powerful is not an interesting goal . . . Humans must keep doing what they have been doing, hating and fighting each other. I will sit in the background, and let them do their thing.

GPT-3 wrote enough to fill an entirely unique and eerily coherent article on the topic given nothing but a short prompt.

GPT-3 can also answer novel questions. Here is one exchange between GPT-3 and a human:

> Person: What is your favorite animal?
> GPT-3: My favorite animal is a dog.
> Person: Why?
> GPT-3: Because dogs are loyal and friendly.
> Person: What are two reasons that a dog might be in a bad mood?
> GPT-3: Two reasons that a dog might be in a bad mood are if it is hungry or if it is hot.
> Person: Are there any animals with three legs?
> GPT-3: No, there are no animals with three legs.
> Person: Why don't animals have three legs?
> GPT-3: Animals don't have three legs because they would fall over.

Because these models are trained to mimic the answers they have seen in human-written text, these models can answer questions in very human ways, which was, of course, what Lemoine found when he asked Google's LLM what it was afraid of.

The ability of LLMs to produce articles and correctly answer questions about the world demonstrates that they are not just regurgitating phrases they have seen before—they have captured some aspect of the *meaning* of language, the idea of an op-ed meant to convince a reader not to fear something or the idea of how a dog walks. Indeed, by reading, well, *everything*, these models show an impressively human-level comprehension of many facts and features of the world. But in these quandaries about our physical and mental world is also where we begin to find the limitations of LLMs, how they differ from language in the human brain, and the features of intelligence that we will have to reverse engineer if we want AI language systems that work in more human-like ways.

Words Without Inner Worlds

GPT-3 is given word after word, sentence after sentence, paragraph after paragraph. During this long training process, it tries to predict the next

CHATGPT AND THE WINDOW INTO THE MIND

word in any of these long streams of words. And with each prediction, the weights of its gargantuan neural network are nudged ever so slightly toward the right answer. Do this an astronomical number of times, and eventually GPT-3 can automatically predict the next word based on a prior sentence or paragraph. In principle, this captures at least some fundamental aspect of how language works in the human brain. Consider how automatic it is for you to predict the next symbol in the following phrases:

- One plus one equals _____
- Roses are red, violets are _____

You've seen similar sentences endless times, so your neocortical machinery automatically predicts what word comes next. What makes GPT-3 impressive, however, is not that it just predicts the next word of a sequence it has seen a million times—that could be accomplished with nothing more than memorizing sentences. What is impressive is that GPT-3 can be given a *novel* sequence that it has never seen before and still accurately predict the next word. This, too, clearly captures something that the human brain can _____.

Could you predict that the next word was *do*? I'm guessing you could, even though you had never seen that exact sentence before. The point is that both GPT-3 and the neocortical areas for language seem to be engaging in prediction. Both can generalize past experiences, apply them to new sentences, and guess what comes next.

GPT-3 and similar language models demonstrate how a web of neurons can reasonably capture the rules of grammar, syntax, and context if it is given sufficient time to learn. But while this shows that prediction is *part* of the mechanisms of language, does this mean that prediction is *all there is* to human language? Try to finish these four questions:

- If $3x + 1 = 3$, then x equals _____
- I am in my windowless basement, and I look toward the sky, and I see _____
- He threw the baseball 100 feet above my head, I reached my hand up to catch it, jumped, and _____

- I am driving as fast as I can to LA from New York. One hour after passing through Chicago, I finally _____

Here something different happens. In the first question, you likely paused and performed some mental arithmetic before being able to answer the question. In the other questions, you probably, even for only a split second, paused to visualize yourself in a basement looking upward, and realized what you would see is the ceiling. Or you visualized yourself trying to catch a baseball a hundred feet above your head. Or you imagined yourself one hour past Chicago and tried to find where you would be on a mental map of America. With these types of questions, more is happening in your brain than merely the automatic prediction of words.

We have, of course, already explored this phenomenon—it is simulating. In these questions, you are rendering an inner simulation, either of shifting values in a series of algebraic operations or of a three-dimensional basement. And the answers to the questions are to be found only in the rules and structure of your inner simulated world.

I gave the same four questions to GPT-3; here are its responses (responses of GPT-3 are bolded and underlined):

- If $3x + 1 = 3$, then x equals **1**
- I am in my windowless basement, and I look toward the sky, and I see **a light, and I know that it is a star, and I am happy.**
- He threw the baseball 100 feet above my head, I reached my hand up to catch it, jumped, **and caught it. It was a lot of fun!**
- I am driving as fast as I can to LA from New York. One hour after passing through Chicago, I finally **get to see the Pacific Ocean.**

All four of these responses demonstrate that GPT-3, as of June 2022, lacked an understanding of even simple aspects of how the world works. If $3x + 1 = 3$, then x equals ⅔, not 1. If you were in a basement and looked toward the sky, you would see your ceiling, not stars. If you tried to catch a ball 100 feet above your head, you would *not* catch the ball. If you were driving to LA from New York and you'd passed through Chicago one hour ago, you would not yet be at the coast. GPT-3's answers lacked common sense.

What I found was not surprising or novel; it is well known that modern AI systems, including these new supercharged language models, struggle with such questions. But that's the point: Even a model trained on the entire corpus of the internet, running up millions of dollars in server costs—requiring acres of computers on some unknown server farm—*still* struggles to answer commonsense questions, those presumably answerable by even a middle-school human.

Of course, reasoning about things by simulating also comes with problems. Suppose I asked you the following question:

Tom W. is meek and keeps to himself. He likes soft music and wears glasses. Which profession is Tom W. more likely to be?

1) Librarian
2) Construction worker

If you are like most people, you answered *librarian*. But this is wrong. Humans tend to ignore base rates—did you consider the *base number* of construction workers compared to librarians? There are probably one hundred times more construction workers than librarians. And because of this, even if 95 percent of librarians are meek and only 5 percent of construction workers are meek, there still will be far more meek construction workers than meek librarians. Thus, if Tom is meek, he is still more likely to be a construction worker than a librarian.

The idea that the neocortex works by rendering an inner simulation and that this is how humans tend to reason about things explains why humans consistently get questions like this wrong. We *imagine* a meek person and compare that to an imagined librarian and an imagined construction worker. Who does the meek person seem more like? The librarian. Behavioral economists call this the representative heuristic. This is the origin of many forms of unconscious bias. If you heard a story of someone robbing your friend, you can't help but render an imagined scene of the robbery, and you can't help but fill in the robbers. What do the robbers look like to you? What are they wearing? What race are they? How old are they? This is a downside of reasoning by simulating—we fill in characters and scenes, often missing the true causal and statistical relationships between things.

It is with questions that require simulation where language in the human brain diverges from language in GPT-3. Math is a great example of this. The foundation of math begins with declarative labeling. You hold up two fingers or two stones or two sticks, engage in shared attention with a student, and label it *two*. You do the same thing with three of each and label it *three*. Just as with verbs (e.g., *running* and *sleeping*), in math we label operations (e.g., *add* and *subtract*). We can thereby construct sentences representing mathematical operations: *three add one*.

Humans don't learn math the way GPT-3 learns math. Indeed, humans don't learn *language* the way GPT-3 learns language. Children do not simply listen to endless sequences of words until they can predict what comes next. They are shown an object, engage in a hardwired nonverbal mechanism of shared attention, and then the object is given a name. The foundation of language learning is not sequence learning but the tethering of symbols to components of a child's already present inner simulation.

A human brain, but not GPT-3, can check the answers to mathematical operations using mental simulation. If you add one to three using your fingers, you notice that you always get the thing that was previously labeled *four*.

You don't even need to check such things on your actual fingers; you can imagine these operations. This ability to find the answers to things by simulating relies on the fact that our inner simulation is an accurate rendering of reality. When I mentally imagine adding one finger to three fingers, then count the fingers in my head, I count four. There is no reason why that must be the case in my imaginary world. But it is. Similarly, when I ask you what you see when you look toward the ceiling in your basement, you answer correctly because the three-dimensional house you constructed in your head obeys the laws of physics (you can't see through the ceiling), and hence it is obvious to you that the ceiling of the basement is necessarily between you and the sky. The neocortex evolved long before words, already wired to render a simulated world that captures an incredibly vast and accurate set of physical rules and attributes of the actual world.

To be fair, GPT-3 can, in fact, answer many math questions correctly.

GPT-3 will be able to answer 1 + 1 =___ because it has seen that sequence a billion times. When you answer the same question without thinking, you are answering it the way GPT-3 would. But when you think about *why* 1 + 1 =, when you prove it to yourself again by mentally imagining the operation of adding one thing to another thing and getting back two things, then you know that 1 + 1 = 2 in a way that GPT-3 does not.

The human brain contains both a language prediction system *and* an inner simulation. The best evidence for the idea that we have both these systems are experiments pitting one system against the other. Consider the cognitive reflection test, designed to evaluate someone's ability to inhibit her reflexive response (e.g., habitual word predictions) and instead actively think about the answer (e.g., invoke an inner simulation to reason about it):

> *Question 1: A bat and a ball cost $1.10 in total. The bat costs $1.00 more than the ball. How much does the ball cost?*

If you are like most people, your instinct, without thinking about it, is to answer ten cents. But if you thought about this question, you would realize this is wrong; the answer is five cents. Similarly:

> *Question 2: If it takes 5 machines 5 minutes to make 5 widgets, how long would it take 100 machines to make 100 widgets?*

Here again, if you are like most people, your instinct is to say "One hundred minutes," but if you think about it, you would realize the answer is still five minutes.

And indeed, as of December 2022, GPT-3 got both of these questions wrong in exactly the same way people do, GPT-3 answered ten cents to the first question, and one hundred minutes to the second question.

The point is that human brains have an automatic system for predicting words (one probably similar, at least in principle, to models like GPT-3) and an inner simulation. Much of what makes human language powerful is not the syntax of it, but its ability to give us the necessary information to render a simulation about it and, crucially, to use these sequences of words to render *the same inner simulation as other humans around us.*

The Paper-Clip Problem

In his 2014 book *Superintelligence: Paths, Dangers, Strategies*, the philosopher Nick Bostrom poses a thought experiment. Suppose a superintelligent and obedient AI, designed to manage production in a factory, is given a command: "Maximize the manufacture of paper clips." What might this AI reasonably do?

Well, it might start by optimizing the internal operations of the factory, doing things any factory manager might: simplifying processes, bulk-ordering raw materials, and automating various steps. But eventually this AI would reach the limit of how much production it could squeeze out of these tamer optimizations. It would then set its sights on more extreme improvements in production, perhaps converting nearby residential buildings into factory floors, perhaps disassembling cars and toasters for raw materials, perhaps forcing people to work longer and longer hours. If this AI were truly superintelligent, we humans would have no way to outsmart or stop this cascading escalation of paper-clip manufacture.

The result would be catastrophic. In Bostrom's words, this would end with the AI "converting first the earth and then increasingly large chunks of the observable universe into paper clips." This imagined demise of human civilization did not require any nefariousness on the part of this superintelligent AI; it was entirely obedient to the command given to it by humans. And yet clearly, this superintelligent AI failed to capture some notion of human intelligence.

This has been called the paper-clip problem. When humans use language with each other, there is an ungodly number of assumptions not to be found in the words themselves. We *infer* what people actually mean by what they say. Humans can easily infer that when someone asks us to maximize the production of paper clips, that person *does not mean* "convert Earth into paper clips." This seemingly obvious inference is, in fact, quite complex.

When a human makes a request like "Maximize the production of paper clips" or "Be nice to Rima" or "Eat breakfast," he or she is not actually providing a well-defined goal. Instead, both parties are guessing what is going on in the other's head. The requester simulated a desired end state, perhaps high profit margins or Rima being happy or a healthy well-fed child, and then the requester attempted to translate this desired simulation into the mind of another with language. The listener must then infer what the requester wants based on what was said. The listener can assume the requester doesn't want him to break the law or do anything that would lead to bad press or pledge his life in servitude to Rima or eat breakfast endlessly into oblivion. So, the path one picks, even when being fully obedient, contains constraints far more nuanced and complex than the command itself.

Or consider a different example of this, presented by the linguist Steven Pinker. Suppose you overheard the following dialogue:

> Bob: I'm leaving you.
> Alice: Who is she?

If you heard this and thought about it for just a second, it would be obvious what it means: Bob is breaking up with Alice for another woman. The response "Who is she?" seems like a complete non sequitur that has nothing to do with Bob's statement. And yet when you imagine why Bob might say, "I'm leaving you," and why Alice might respond, "Who is she?" the interaction and maybe even a backstory begins to form in your mind.

Humans do all of this with our primate trick of mentalizing; the same way we can render an inner three-dimensional world, we can render a simulation of another mind to explore how different actions will make someone feel. When I am told to maximize paper clips, I can explore possible outcomes and simulate how I believe this other mind will feel about it. When I do this, it is incredibly obvious that the person will be unhappy if I convert Earth into paper clips. When I do this, it is obvious why Alice asked, "Who is she?"

The intertwining of mentalizing and language is ubiquitous. Every conversation is built on the foundation of modeling the other minds you are conversing with—guessing what one means by what he said and guessing what should be said to maximize the chance the other knows what you mean.

The relationship between mentalizing and language can even be seen in the brain. Wernicke's area, presumably the place where words are learned and stored, is *right in the middle* of the primate mentalizing regions. Indeed, the specific subarea of the left primate sensory cortex (called temporoparietal junction), which is highly selective for modeling other people's intentions, knowledge, and beliefs, is entirely overlapping with Wernicke's area—which is, as we have learned, required for people to understand speech and produce meaningful speech.

Consistent with this, mentalizing skills and language skills in children are deeply interconnected. In preschoolers, there is a significant correlation between the development of language skills and performance in mentalizing tasks such as false-belief tests. Disorders that impair mentalizing cause similar impairments in language.

We are capable of puppeteering other minds because language is, it seems, built right on top of a direct window to our inner simulation. Hearing sentences directly and automatically triggers specific mental imagery. This is why if someone is saying something that upsets us, we cannot simply "not listen"; we must cover our ears, otherwise the words directly trigger simulations whether we like it or not.

But Wait . . . What About GPT-4?

In March 2023, OpenAI released GPT-4, its new and upgraded large language model. GPT-4 works in largely the same way as GPT-3—it predicts the next word based solely on the prior sequence of words. But unlike GPT-3, it is an even bigger model trained on even more data.

Amazingly, each question that I designed in this chapter to demonstrate a lack of common sense and physical intuition in GPT-3 was answered flawlessly by GPT-4. GPT-4 knew that if you looked upward in a basement, you would see the ceiling not the sky. GPT-4 knew that if someone threw a baseball one hundred feet above your head, you would not be able to catch it. GPT-4 could even answer theory-of-mind questions like the Sally-Ann Test.

How does GPT-4 do it? OpenAI did not add an inner world model or the ability to model the mind of others. Instead, in response to GPT-3's frequent mistakes on commonsense and reasoning questions, OpenAI began training GPT-4 specifically on questions of commonsense and reasoning. They used a technique called "reinforcement learning from human feedback" in which humans punished GPT-4 when it answered questions incorrectly, and rewarded it when it answered correctly. They even pushed the GPT-4 to answer certain questions in specific ways to improve its performance. For example, OpenAI trained GPT-4 to think about commonsense questions by writing out each step, a trick called chain-of-thought prompting. Indeed, contrast how GPT-3 and GPT-4 each answer the bat and ball question:

Me: A bat and a ball cost $1.10 in total. The bat costs $1.00 more than the ball. How much does the ball cost?

GPT-3:	GPT-4:
The ball costs $0.10 because if the bat costs $1.00 more than the ball, and the bat and the ball cost $1.10 in total, then the ball must cost $1.00 - $1.10 = $0.10 (incorrect)	Let's denote the cost of the ball as x. Then, the cost of the bat would be x + $1.00, since it costs $1.00 more than the ball. The total cost of both items is $1.10. So, we can set up an equation: x (ball) + x + $1.00 (bat) = $1.10 Combining the terms for x, we get 2x + $1.00 = $1.10 Now, we'll subtract $1.00 from both sides: 2x = $0.10 Finally, we'll divide both sides by 2: x = $0.05 So, the ball costs $0.05 or 5 cents. (correct)

By training GPT-4 to not just predict the answer, but to predict the next step in *reasoning* about the answer, the model begins to exhibit emergent properties of *thinking*, without, in fact, thinking—at least not in the way that a human thinks by rendering a simulation of the world.

Even though GPT-4 correctly answers the simpler questions I outlined in this chapter, you can still find plenty of examples of GPT-4 failing on commonsense and theory-of-mind questions. GPT-4's lack of a world model can be seen by probing deeper with more complicated questions. But it is becoming progressively more painstaking to find these examples. In some ways, this has become a game of Whac-A-Mole; everytime a skeptic publishes examples of commonsense questions that LLMs answer incorrectly, companies like OpenAI simply use these examples as training data for the next update of their LLMs, which thereby answer such questions correctly.

Indeed, the massive size of these models, along with the astronomical quantity of data on which they are trained, in some ways obscures the underlying differences between how LLMs think and how humans think. A calculator performs arithmetic better than any human, but still lacks the same understanding of math as a human. Even if LLMs correctly answer commonsense and theory-of-mind questions, it does not necessarily mean it reasons about these questions in the same way.

As Yann Lecun said, "the weak reasoning abilities of LLMs are partially compensated by their large associative memory capacity. They are a bit like students who have learned the material by rote but haven't really built deep mental models of the underlying reality." Indeed, these LLMs, like a supercomputer, have a gargantuan memory capacity, having read more books and articles than a single human brain could consume in a thousand lifetimes. And so what seems like commonsense reasoning is really more like pattern matching, done over an astronomically enormous corpus of text.

But still, these LLMs are an incredible step forward. What is most amazing about the success of LLMs is how much they seemingly understand about the world despite being trained on nothing but language. LLMs can correctly reason about the physical world without ever having experienced that world. Like a military cryptanalyst decoding the meaning behind encrypted secret messages, finding patterns and meanings in what was originally gibberish, these LLMs have been able to tease out aspects of a world they have never seen or heard, that they have never touched or experienced, by merely scanning the entire corpus of our uniquely human code for transferring thoughts.

It is possible, perhaps inevitable, that continuing to scale up these language models by providing them with more data will make them even better at answering commonsense and theory-of-mind questions.* But without incorporating an inner model of the external world or a model of other minds—without the breakthroughs of simulating and mentalizing—these LLMs will fail to capture something essential about human intelligence. And the more rapid the adoption of LLMs—the more decisions we offload to them—the more important these subtle differences will become.

In the human brain, language is the *window* to our inner simulation. Language is the interface to our mental world. And language is built on the foundation of our ability to model and reason about the minds of others—to infer what they mean and figure out exactly which words will produce the desired simulation in their mind. I think most would agree that the humanlike artificial intelligences we will one day create will not be LLMs; language models will be merely a window to something richer that lies beneath.

---

* And incorporating other modalities directly into these models. Indeed, already newer large language models like GPT-4 are now being designed to be "multimodal," whereby they are also trained in images in addition to text.

Summary of Breakthrough #5: Speaking

Early humans got caught in an unlikely perfect storm of effects. The dying forests of the African savannah pushed early humans into a tool-making meat-eating niche, one that required the accurate propagation of tool use across generations. Proto-languages emerged, enabling tool use and the manufacture of skills to successfully propagate across generations. The neurological change that enabled language was not a new neurological structure but an adjustment to more ancient structures, which created a learning program for language; the program of proto-conversations and joint attention that enables children to tether names to components of their inner simulation. Trained with this curriculum, older areas of the neocortex were repurposed for language.

From here, humans began experimenting with using this proto-language with unrelated individuals, and this kicked off a feedback loop of gossip, altruism, and punishment, which continuously selected for more sophisticated language skills. As social groups expanded and ideas began hopping from brain to brain, the human hive mind emerged, creating an ephemeral medium for ideas to propagate and accumulate across generations. This would have begged for bigger brains to store and share more accumulated knowledge. And perhaps due to this, or enabling it, cooking was invented, offering a huge caloric surplus that could be spent on tripling the size of brains.

And so, from this perfect storm emerged the fifth and final breakthrough in the evolutionary story of the human brain: language. And along with language came the many unique traits of humans, from altruism to cruelty. If there is anything that truly makes humans unique, it is that the mind is no longer singular but is tethered to others through a long history of accumulated ideas.

Conclusion

The Sixth Breakthrough

WITH THE EMERGENCE of the modern human brain in our ancestors around one hundred thousand years ago, we have reached the conclusion of our four-billion-year evolutionary story. Looking back, we can begin to make out a picture—a framework—for the process by which the human brain and intelligence emerged. We can consolidate this story into our model of five breakthroughs.

Breakthrough #1 was *steering*: the breakthrough of navigating by categorizing stimuli into good and bad, and turning *toward* good things and *away* from bad things. Six hundred million years ago, radially symmetric neuron-enabled coral-like animals reformed into animals with a bilateral body. These bilateral body plans simplified navigational decisions into binary turning choices; nerve nets consolidated into the first brain to enable opposing valence signals to be integrated into a single steering decision. Neuromodulators like dopamine and serotonin enabled persistent states to more efficiently relocate and locally search specific areas. Associative learning enabled these ancient worms to tweak the relative valence of various stimuli. In this very first brain came the early affective template of animals: pleasure, pain, satiation, and stress.

Breakthrough #2 was *reinforcing*: the breakthrough of learning to repeat behaviors that historically have led to positive valence and inhibit behaviors that have led to negative valence. In AI terms, this was the breakthrough of model-free reinforcement learning. Five hundred million years ago, one lineage of ancient bilaterians grew a backbone, eyes, gills, and a heart, becoming the first vertebrates, animals most similar to modern fish. And their brains formed into the basic template of all modern vertebrates: the cortex to recognize patterns and build spatial maps and the basal ganglia to learn by trial and error. And both were built on top of the more ancient vestiges of valence machinery housed in the hypothalamus. This model-free reinforcement learning came with a suite of familiar intellectual and affective features: omission learning, time perception, curiosity, fear, excitement, disappointment, and relief.

Breakthrough #3 was *simulating*: the breakthrough of mentally simulating stimuli and actions. Sometime around one hundred million years ago, in a four-inch-long ancestral mammal, subregions of the cortex of our ancestral vertebrate transformed into the modern neocortex. This neocortex enabled animals to internally render a simulation of reality. This enabled them to *vicariously* show the basal ganglia what to do before the animal actually did anything. This was learning by *imagining*. These animals developed the ability to plan. This enabled these small mammals to re-render past events (episodic memory) and consider alternative past choices (counterfactual learning). The later evolution of the motor cortex enabled animals to plan not only their overall navigational routes but also specific body movements, giving these mammals uniquely effective fine motor skills.

Breakthrough #4 was *mentalizing*: the breakthrough of modeling one's own mind. Sometime around ten to thirty million years ago, new regions of neocortex evolved in early primates that built a model of the older mammalian areas of neocortex. This, in effect, meant that these primates could simulate not only actions and stimuli (like early mammals), but also their own mental states with differing intent and knowledge. These primates could then apply this model to anticipating their own future needs, understanding the intents and knowledge of others (theory of mind), and learning skills through observation.

Breakthrough #5 was *speaking*: the breakthrough of naming and grammar, of tethering our inner simulations together to enable the accumulation of thoughts across generations.

Each breakthrough was possible only because of the building blocks that came prior. Steering was possible only because of the evolution of neurons earlier. Reinforcement learning was possible only because it bootstrapped on the valence neurons that had already evolved: without valence, there is no foundational learning signal for reinforcement learning to begin. Simulating was possible only because trial-and-error learning in the basal ganglia existed prior. Without the basal ganglia to enable trial-and-error learning, there would be no mechanism by which imagined simulations could affect behavior; by having *actual* trial-and-error learning evolve in vertebrates, *vicarious* trial and error could emerge later in mammals. Mentalizing was possible only because simulating came before; mentalizing is just simulating the older mammalian parts of the neocortex, the same computation turned inward. And speaking was possible only because mentalizing came before; without the ability the infer the intent and knowledge in the mind of another, you could not infer what to *communicate* to help transmit an idea or infer what people mean by what they say. And without the ability to infer the knowledge and intent of another, you could not engage in the crucial step of shared attention whereby teachers identify objects for students.

Thus far, humanity's story has been a saga of two acts. Act 1 is the evolutionary story: how biologically modern humans emerged from the raw lifeless stuff of our universe. Act 2 is the cultural story: how societally modern humans emerged from largely biologically identical but culturally primitive ancestors from around one hundred thousand years ago.

While act 1 unfolded over billions of years, most of what we have learned in history class unfolded during the comparatively much shorter time of act 2—all civilizations, technologies, wars, discoveries, dramas, mythologies, heroes, and villains unfolded in this time window that, compared to act 1, was a mere blink of an eye.

An individual *Homo sapiens* one hundred thousand years ago housed in her head one of the most awe-inspiring objects in the universe; the result of over a billion years of hard—even if unintentional—evolutionary work. She would have sat comfortably at the top of the food chain, spear in hand, warmed in manufactured clothing, having tamed both fire and countless gargantuan beasts, effortlessly invoking these many intellectual feats, utterly unaware of the past by which these still yet-to-be-understood abilities came to be and also, of course, unaware of the simultaneously magnanimous, tragic, and wonderful journey that would eventually unfold in her *Homo sapiens* descendants.

And so here you are, reading this book. An almost impossibly vast number of events led to this exact moment: the first bubbling cells in hydrothermal vents; the first predatory battles of single-celled organisms; the birth of multicellularity; the divergence of fungi and animals; the emergence of the first neurons and reflexes in ancestral corals; the emergence of the first brains with valence and affect and associative learning in ancient bilaterians; the rise of vertebrates and the taming of time, space, patterns, and prediction; the birth of simulation in minuscule mammals hiding from dinosaurs; the construction of politics and mentalizing of tree-living primates; the emergence of language in early humans; and, of course, the creation, modification, and destruction of countless ideas that have accumulated throughout the billions of language-enabled human brains over the past hundreds of thousands of years. These ideas have accumulated to the point that modern humans can now type on computers, write words, use cell phones, cure diseases, and, yes, even construct new artificial intelligences in our image.

Evolution is still unfolding in earnest; we are not at the end of the story of intelligence but at the very beginning. Life on Earth is only four billion years old. It will be another seven billion years before our sun dies. And thus life, at least on Earth, has another seven or so billion years to tinker with new biological forms of intelligence. If it took only four and a half billion years for raw molecules on Earth to transform into human brains, how far can intelligence get in another seven billion years of evolution? And assuming life does, somehow, make its way out of the solar system, or at least life independently shows up elsewhere in the universe, there will be astronomically more time for evolution to get to work: it will be over a trillion years before the universe has expanded so greatly that new stars cease to form, and a quadrillion before the last galaxy breaks apart. It can be hard to conceptualize just how young our fourteen-billion-year-old universe actually is. If you took the quadrillion-year timeline of our universe and squished it into a single calendar year, then we would find ourselves, today, at only the first seven minutes of the year, not even at the dawn of the very first day.

If our modern understanding of physics is correct, then about a quadrillion years from now, after the the last galaxy has finally broken apart, the universe will begin its slow process of fading meaninglessly into an inevitable heat death. This is the unfortunate result of the inexorable trend of entropy, that raw unstoppable force of the universe that the first self-replicated DNA molecules began their war against four billion years ago. By self-replicating, DNA finds respite from entropy, persisting not in matter but in information. All the evolutionary innovations that followed the first string of DNA have been in this spirit, the spirit of *persisting*, of fighting back against entropy, of refusing to fade into nothingness. And in this great battle, ideas that float from human brain to human brain through language are life's newest innovation but will surely not be its last. We are still at the base of the mountain, only on the fifth step on a long staircase to something.

Of course, we don't know what breakthrough #6 will be. But it seems increasingly likely that the sixth breakthrough will be the creation of artificial superintelligence; the emergence of our progeny in silicon, the transition of intelligence—made in our image—from a biological medium to a digital medium. From this new medium will come an astronomical expansion in the scale of a single intelligence's cognitive capacity. The cognitive capacity of the human brain is hugely limited by the processing speed of neurons, the caloric limitations of the human body, and the size constraints of how big a brain can be and still fit in a carbon-based lifeform. Breakthrough #6 will be when intelligence unshackles itself from these biological limitations. A silicon-based AI can infinitely scale up its processing capacity as it sees fit. Indeed, individuality will lose its well-defined boundaries as AIs can freely copy and reconfigure themselves; parenthood will take on new meaning as biological mechanisms of mating give way to new silicon-based mechanisms of training and creating new intelligent entities. Even evolution itself will be abandoned, at least in its familiar form; intelligence will no longer be entrapped by the slow process of genetic variation and natural selection, but instead by more fundamental evolutionary principles, the purest sense of variation and selection—as AIs reinvent themselves, those who select features that support better survival will, of course, be the ones that survive.

And whichever intellectual strategies end up evolving next, they will surely contain hints of the human intelligence from which they came. While the underlying medium of these artificial superintelligences may retain none of the biological baggage of brains, these entities will still irrevocably be built on the foundation of the five breakthroughs that came before. Not only because these five breakthroughs were the foundation of the intelligence of their human creators—creators cannot help but imbue their creations with hints of themselves—but also because they will be designed, at least at first, to interact with humans, and thereby will be seeded with a recapitulation, or at least a mirror, of human intelligence.

And so we stand on the precipice of the sixth breakthrough in the story of human intelligence, at the dawn of seizing control of the process by which life came to be and of birthing superintelligent artificial beings. At this precipice, we are confronted with a very *un*scientific question but one that is, in fact, far more important: What should be humanity's goals? This is not a matter of *veritas*—truth—but of *values*.

As we have seen, past choices propagate through time. And so how we answer this question will have consequences for eons to come. Will we spread out across galaxies? Explore the hidden features of the cosmos, construct new minds, unravel the secrets of the universe, find new features of consciousness, become more compassionate, engage in adventures of unthinkable scope? Or will we fail? Will our evolutionary baggage of pride, hatred, fear, and tribalism rip us apart? Will we go down as just another evolutionary iteration that came to a tragic end? Perhaps it will be some later species on Earth, millions of years after humans have gone extinct, that will make another stab at taking the next step up the mountain— perhaps the bonobos or octopuses or dolphins or Portia spiders. Perhaps they will uncover our fossils as we have uncovered those of dinosaurs and ponder what lives we must have lived and write books about our brains. Or even worse, perhaps we humans will end the grand four-billion-year experiment of life on Earth through ravaging the planet's climate or blowing our world into oblivion with nuclear warfare.

As we look forward into this new era, it behooves us to look backward at the long billion-year story by which our brains came to be. As we become endowed with godlike abilities of creation, we should learn from the god—the unthinking process of evolution—that came before us. The more we understand about our own minds, the better equipped we are to create artificial minds in our image. The more we understand about the process by which our minds came to be, the better equipped we are to choose which features of intelligence we want to discard, which we want to preserve, and which we want to improve upon.

We are the stalwarts of this grand transition, one that has been fourteen billion years in the making. Whether we like it or not, the universe has passed us the baton.

Acknowledgments

Writing this book was a case study in human generosity. It was only possible because of the remarkable kindness of many people who helped me bring it life. There are many people who deserve thanks.

First and foremost, my wife, Sydney, who edited many pages and helped me think through many conceptual snags. She woke up countless mornings to find me long gone because I had already snuck out to read and write; and she came home from work countless days to find me tucked away in my office. Thank you for supporting this endeavor despite how much mental space it consumed.

I want to thank my initial readers, who gave me feedback and encouragement: Jonathan Balcome, Jack Bennett, Kiki Freedman, Marcus Jecklin, Dana Najjar, Gideon Kowadlo, Fayez Mohamood, Shyamala Reddy, Billy Stein, Amber Tunnell, Michael Weiss, Max Wenneker, and, of course, my parents, Gary Bennett and Kathy Crost; and my stepmother, Alyssa Bennett.

In particular, I want to thank my father-in-law, Billy Stein, who has no intrinsic interest in AI or neuroscience, but nonetheless dutifully read and annotated every single page, questioned every concept and idea to make sure it made sense, and provided invaluable input and guidance on structure, understandability, and flow. Dana Najjar, Shyamala Reddy, and Amber Tunnell, who have far more writing experience than I, gave me essential input on early drafts. And Gideon Kowaldo, who gave me useful input on the AI history and concepts.

I am extremely grateful to the scientists who took time out of their busy lives to respond to my emails where I peppered them with innumerable questions. They helped me understand their research and think through many of the concepts in this book: Charles Abramson, Subutai Ahmed, Bernard Balleine, Kent Berridge, Culum Brown, Eric Brunet, Randy Bruno, Ken Cheng, Matthew Crosby, Francisco Clasca, Caroline DeLong, Karl Friston, Dileep George, Simona Ginsburg, Sten Grillner, Stephen Grossberg, Jeff Hawkins, Frank Hirth, Eva Jablonka, Kurt Kotrschal, Matthew Larkum, Malcolm MacIver, Ken-ichiro Nakajima, Thomas Parr, David Redish, Murray Sherman, James Smith, and Thomas Suddendorf. Without their willingness to respond to the questions of a complete stranger, it would have been impossible for someone like me to learn a new field.

I want to especially thank Karl Friston, Jeff Hawkins, and Subutai Ahmed, who read some of my early papers and generously took me under their wing and brought me into their labs to share my ideas and learn from them.

Joseph LeDoux, David Redish, and Eva Jablonka were astoundingly generous with their time. Not only did they read and annotate multiple drafts of the manuscript, but they provided essential feedback on concepts I had missed, areas of the literature I had failed to consider, and helped me expand on the framework and story. They became my de facto neuroscience editors and advisers. They deserve much of the credit for whatever aspects of this book are deemed valuable (and none of the blame for aspects deemed otherwise).

One of my favorite parts of this book is the art, and for this, Rebecca Gelernter and Mesa Schumacher deserve all the credit. They are the incredibly talented artists who produced the beautiful art herein.

As a first-time author, I am grateful to the people in the book industry who gave me guidance. Jane Friedman gave me tough and useful feedback. The writer Jonathan Balcome read one of the earliest drafts and gave feedback and encouragement. The writers Gerri Hirshey and Jamie Carr each helped me with my book proposal and gave me feedback on early chapters.

Lisa Sharkey at HarperCollins made this book real. I spoke to her before I decided to write it and asked her whether it was even worth attempting to write this book given I was a first-time author and not a formally trained neuroscientist. Despite the obvious fact that there was a good chance the book wouldn't see the light of day, she encouraged me to pursue it regardless. I am deeply grateful for that conversation, and her advice and support. It is wonderfully fitting that she was the one, over a year after that conversation, who ended up deciding to publish this book.

I want to thank my agent, Jim Levine, who was willing to read the book from nothing but a single introduction (thanks to Jeff Hawkins). Jim read the entire book in one day, and took a bet on it the next day. I want to thank my U.S. editor, Matt Harper, and my U.K. editor, Myles Archibald, who also took a bet on this book, and helped me work through countless drafts and navigate the many ups and downs of writing. I want to thank my copyeditor, Tracy Roe, who methodically fixed my many typos and grammatical mishaps.

There are also folks who helped me in less direct but equally important ways. My guitar teacher, Stephane Wrembel, who I turned to for advice on numerous occasions. My friend Ally Sprague (who tends to double as my coach), who helped me make the decision to take a year off to write this book. My friends Dougie Gliecher and Ben Eisenberg, who connected me to people they knew in the book industry. My brothers, Adam Bennett and Jack Bennett, who bring joy and play to my life, and are always a source of inspiration. And my parents, Gary Bennett and Kathy Crost, who fostered in me a love of learning, showed me how to follow my curiosity, and taught me to finish things I start.

This book was only possible because of many other prior works whose ideas, stories, and writing shaped this book in fundamental ways. *The Alignment Problem* by Brian Christian. *Behave* by Robert Sapolsky. *The Deep History of Ourselves* by Joseph LeDoux. *The Evolution of the Sensitive Soul* by Eva Jablonka and Simona Ginsburg. *How Monkeys See the World* by Dorothy Cheney and Robert Seyfarth. *The Mind within the Brain* by David Redish. *On Intelligence* and *A Thousand Brands* by Jeff Hawkins. *Why Only Us* by Robert Berwish and Noam Chomsky.

There were also numerous textbooks that became essential resources for me. *Brains Through Time* by Georg Striedter and R. Glenn Northcutt. *Brain Structure and Its Origins* by Gerald Schneider. *Deep Learning* by Ian Goodfellow, Yoshua Bengio, and Aaron Courville. *Evolutionary Neuroscience* by Jon H. Kaas. *The Evolution of Language* by W. Tecumseh Fitch. *Fish*

Cognition and Behavior by Culum Brown, Kevin Laland, and Jens Krause. *Neuroeconomics* by Paul Glimcher. *The Neurobiology of the Prefrontal Cortex* by Richard Passingham and Steven Wise. *The New Executive Brain* by Elkhonon Goldberg. *Reinforcement Learning* by Richard Sutton and Andrew Barto.

Lastly, I want to thank my dog, Charlie, whose begging for treats and playful nudges forced me to reenter the world of the living from numerous bleary-eyed sessions of reading papers and textbooks. As I write this paragraph, she is lying next to me fast asleep, twitching away in some dream, her neocortex surely rendering a simulation of something. Of what, of course, I will never know.

Glossary

acquisition (in relation to associative learning): the process by which a new association between a stimulus and a response is formed (i.e., "acquired") based on new experience

associative learning: the ability to associate a stimulus with a reflexive response, such that the next time that stimulus occurs that same reflexive response is more likely to occur

adaptation (in relation to the responses of neurons): the property of neurons whereby they change the relationship between a given stimulus strength and the resulting firing rate; for example, neurons will gradually decrease their firing rate in response to a constant stimulus over time

affect/affective state: a way to categorize the behavioral state of an animal along the dimensions of valence (either positive valence or negative valence) and arousal (either high arousal or low arousal)

agranular prefrontal cortex (aPFC): the region of frontal neocortex that evolved in early mammals. It is called "agranular" because it is a region of neocortex that is missing layer 4 (the layer that contains "granule cells")

auto-association: a property of certain networks of neurons whereby neurons automatically build associations with themselves, enabling the network to automatically complete patterns when given an incomplete pattern

backpropagation: an algorithm for training artificial neural networks; computes the impact of changing the weight of a given connection on the error (a measure of the difference between the actual output and the desired output) at the end of the network, and nudges each weight accordingly to reduce the error

bilaterian: a group of species with a common ancestor around 600 million years ago, in whom bilaterial symmetry emerged as well as the first brains

bilateral symmetry: animal bodies that contain a single plane of symmetry, which divides the animal into roughly mirror image right and left halves

blocking (in relation to associative learning): one of the solutions to the credit assignment problem that evolved in early bilaterians; once an animal has established an association between a predictive cue and a response, all further cues that overlap with the predictive cue are inhibited (i.e., "blocked") from making associations with that response

catastrophic forgetting: an outstanding challenge of sequentially training neural networks (as opposed to training them all at once); when you teach a neural network to recognize new patterns, it tends to lose the memory of previously learned old patterns

continual learning: the ability to automatically learn and remember new things as new data is provided

convolutional neural network: a type of neural network designed to recognize objects in images by looking for the same features in different locations

credit assignment problem: when an event or outcome occurs, what cue or action do you give "credit" for being predictive of that event or outcome?

extinction (in relation to associative learning): the process by which previously learned associations are inhibited (i.e., "extinguished") due to a conditional stimulus no longer occurring alongside a subsequent reflexive response (i.e., a buzzer sounding that used to occur before food, but no longer occurs before food)

firing rate (also spike rate): the number of spikes per second generated by a neuron

generative model: a type of probabilistic model that learns to generate its own data, and recognizes things by comparing generated data with actual data (a process some researchers call "perception by inference")

granular prefrontal cortex (gPFC): the region of frontal neocortex that evolved in early primates. It is called "granular" because it is a region of prefrontal neocortex that contains a layer 4 (the layer that contains "granule cells")

Helmholtz machine: an early proof of concept of Helmholtz's idea of perception by inference

mentalizing: the act of rendering a simulation of one's own inner simulation (i.e., thinking about your own thinking)

Model-based reinforcement learning: the type of reinforcement learning whereby possible future actions are "played out" (i.e., simulated) ahead of time before selecting an action

model-free reinforcement learning: the type of reinforcement learning whereby possible future actions are *not* "played out" (i.e., simulated) ahead of time; instead, actions are automatically selected based on the current situation

neuromodulator: a chemical released by some neurons ("neuromodulatory neurons") that has complex and often long-lasting effects on many downstream neurons. Famous neuromodulators include dopamine, serotonin, and adrenaline

overshadowing (in relation to associative learning): one of the solutions to the credit assignment problem that evolved in early bilaterians; when animals have multiple predictive cues to use, their brains tend to pick the cues that are the strongest (i.e., strong cues *overshadow* weak cues).

primate sensory cortex (PSC): the new regions of sensory neocortex that evolved in early primates, these include the superior temporal sulcus (STS) and temporoparietal junction (TPJ)

reacquisition (in relation to associative learning): one of the techniques to deal with changing contingencies in the world and enable continual learning in early bilaterians; old-extinguished associations are reacquired faster than entirely new associations

sensory neocortex: the back half of the neocortex, the area in which a simulation of the external world is rendered

spontaneous Recovery (in relation to associative learning): one of the techniques to deal with changing contingencies in the world and enable continual learning in early bilaterians; broken associations are rapidly suppressed but not, in fact, unlearned; given enough time, they reemerge

superior temporal sulcus (STS): a new region of sensory neocortex that evolved in early primates

synapse: the connection between neurons through which chemical signals are passed

temporal credit assignment problem: when an event or outcome occurs, what previous cue or action do you give "credit" for being predictive of that event or outcome? This is a subcase of the credit assignment problem when having to assign credit between things separated in time

temporal difference learning (TD learning): the model-free reinforcement learning process whereby AI systems (or animal brains) reinforce or punish behaviors based on changes (i.e., "temporal differences") in predicted future rewards (as opposed to actual rewards)

temporal difference signal (TD signal): the change in predicted future reward; this signal is used as the reinforcement/punishment signal in temporal difference learning systems

temporoparietal junction (TPJ): a new region of sensory neocortex that evolved in early primates

theory of mind: the ability to infer another animal's intent and knowledge

valence: the goodness or badness of a stimulus, behaviorally defined by whether an animal will approach or avoid the stimulus

Notes

Introduction

2 *"just like one of the family"*: "Rosey's Boyfriend." *The Jetsons*, created by William Hanna and Joseph Barbera. Season 1, episode 8, 1962.
2 *"on the verge of achieving human-level AI."*: Cuthbertson, 2022.
5 *over one* billion *connections:* width of a synapse of about 20 nanometers (Zuber et al., 2005). Within a cubic millimeter there are about 1 billion connections (Faisal et al., 2005).
6 *"to the human brain"*: quote from Hinton reported in "U of T computer scientist takes international prize for groundbreaking work in AI." U of T News. January 18, 2017, https://www.utoronto.ca/news/u-t-computer-scientist-takes-international-prize-groundbreaking-work-ai.
8 *"motor, and other functions"*: MacLean, 1990.
8 *and how it works:* Cesario et al., 2020, provides a good overview of the current view of MacLean's triune brain model. Although, to be fair to MacLean, it seems to me that most of the issues with his triune brain model are in its popular success. If one actually reads MacLean's work, he readily acknowledges many of the challenges with his framework.
10 *"so on to Artificial Human–level Intelligence (AHI)"*: Yann LeCun (@ylecun) tweeted this on December 9, 2019.
15 *faster than a human:* Healy et al., 2013.

Chapter 1: The World Before Brains

17 *an unremarkable hydrothermal vent:* For reviews of hydrothermal-vent theory and the timing of life, see Bell et al., 2015; Dodd et al., 2017; Martin et al., 2008; McKeegan et al., 2007.
17 *it duplicated itself:* For a review of RNA world and the evidence that RNA could originally duplicate itself without proteins, see Neveu et al., 2013.
18 *of a modern boat:* Bacterial flagellum is proton-driven and works via rotary motor that turns. See Lowe et al., 1987; Silverman and Simon, 1974.
19 *synthesis, lipids, and carbohydrates:* For evidence LUCA had DNA, see Hassenkam et al., 2017. For evidence LUCA was performing protein synthesis, see Noller, 2012.
20 *finance these many processes:* J. L. E. Wimmer et al., 2021.
20 *blue-green algae:* Note scientists don't like the term *blue-green algae* anymore since the word *algae* is reserved for a type of single-celled plant.

20 *Figure 1.1:* Figure from https://www.scienceimage.csiro.au/image/4203. CC BY 3.0 license Photograph by Willem van Aken on March 18, 1993.
20 *converted into cellular energy:* The ability to generate oxygen via photosynthesis likely first appeared in the ancestors of cyanobacteria; see K. L. French et al., 2015.
20 *and endlessly reproducing:* Cardona et al., 2015; Schirrmeister et al., 2013.
21 *produced a pollutive exhaust:* Note that earlier life may have used a less efficient and more primitive version of photosynthesis, one that produced less energy and did not produce oxygen as exhaust. See Raymond and Segrè, 2006.
21 *oxygen levels skyrocketed*: T. W. Lyons et al., 2014.
21 *the Oxygen Holocaust:* Margulis and Sagan, 1997.
21 *extinction events in Earth's history:* The Great Oxygenation Event occurred at 2.4B BCE; see Anbar et al., 2007. For evidence it killed off many species on Earth, see Hodgskiss et al., 2019.
21 *carbon dioxide as exhaust:* Technically, what evolved is aerobic respiration, the version of respiration that uses oxygen. For evidence that aerobic respiration evolved after oxygenetic photosynthesis in cyanobacteria, see Soo et al., 2017.
23 *oxygen-based approach (aerobic respiration):* O'Leary and Plaxton, 2016.
23 *much more internal complexity*: The complexity of eukaryotes can be seen under a microscope; the term *eukaryote* comes from the twentieth-century observation that the descendants of eukaryotes all had good (*eu*) kernels (*karyon*), while bacteria and bacteria-like life had no such inner structures and hence were called *prokaryotes*: before (*pro*) kernels (*karyon*).
23 *microbial killing machines yet:* For evidence eukaryotes were the first cells to engulf and internally digest life for food, see Cavalier-Smith, 2009.
24 *The tree of life:* Timing emergence of eukaryotes to around 2B years ago; see Knoll et al., 2006.
25 *mushroom-like fungi began growing:* Bengtson et al., 2017.
26 *Figure 1.5*: Illustration from Reichert, 1990. Used with permission.
27 *whom all neurons descend:* There may be a single exception to this: comb jellies might have independently evolved neurons.
28 *on Earth than animals:* Bar-On et al., 2018.
28 *as much as jellyfish embryos:* Technau, 2020.
29 *neuron-enabled animal ancestor:* Henceforth when I say *animals*, I'm referring to Eumetazoa, the "true" metazoans.
29 *gastrula-shaped creature with neurons:* Arendt et al., 2016.
29 *similar to today's corals:* Penny et al., 2014; Wan et al., 2016.
31 *false starts and wrong turns:* For a review of the historical discovery of nervous systems, see McDonald, 2004.
32 *not ether but electricity:* Piccolino, 1997; Schuetze, 1983.
32 *generate their own signals:* O'Brien, 2006.
32 *Adrian the Nobel Prize:* Garson, 2015; Pearce, 2018.
33 *spikes or action potentials:* The discovery of action potentials actually happened more gradually, some suggesting it happened as early as 1848. See du Bois-Reymond, E. 1848.
33 *the signals of neurons:* Garson, 2003.
34 *of the spike itself:* As with most everything in biology, there are exceptions: some areas of brains seem to use other coding strategies, such as temporal coding.
34 *from jellyfish to humans:* For rate coding in hydra, see Tzouanas et al., 2021. For rate coding in *C. elegans*, see Q. Liu et al., 2018; O'Hagan et al., 2005; Suzuki et al., 2003.
34 *in their firing rate:* J. T. Pearson and D. Kerschensteiner, 2015.
34 concentration *in their firing rate:* Parabucki et al., 2019.
34 *force of the muscles:* This has been shown even in *C. elegans*; see S. Gao and M. Zhen, 2011.
35 *a page in moonlight:* MacEvoy, B. 2015
34 *Figure 1.10*: Figure from B. MacEvoy, 2015. Used with permission (personal correspondence).

35 *five hundred spikes per second:* Wang et al., 2016.
37 *John Eccles, and others:* Eccles discovered inhibition; Dale discovered chemical neurotransmission (Todman, 2008); Sherrington discovered synapses (R. E. Brown et al., 2021).
38 *swallow reflexes to work:* Evidence of lateral inhibition through synaptic inhibition has been found in the hydra; see Kass-Simon, 1988. But some have argued that synaptic inhibition is absent in Cnidaria (Meech and Mackie, 2007).
38 *and another must relax:* See Bocharova and Kozevich, 2011, for details on mouth muscles of sea anemones.

Chapter 2: The Birth of Good and Bad

46 *a grain of rice:* For fossil evidence of early bilaterians, see Z. Chen et al., 2013, 2018; Evans et al., 2020.
47 *almost definitely very simple:* For a great review of early bilaterians, see Malakhov, 2010.
47 *a human's 85 billion:* To be more precise, *C. elegans* has 302 neurons in its entire body, while a human has 85 billion neurons in his or her brain; there are other neurons in the human nervous system outside of the brain.
47 *finds the food:* Henceforth whenever I use the term *nematode*, I am referring to the specific species of nematode *Caenorhabditis elegans*.
47 *in on the food:* For an example with flatworms, see Pearl, 1903. For nematodes, see Bargmann et al., 1993; Ward, 1973.
48 *bodies might get wounded:* For a good review of these types of behaviors in *C. elegans*, see Hobert, 2003. For thermal-gradient behavior in *C. elegans*, see Cassata et al., 2000; Hedgecock and Russell, 1975; L. Luo et al., 2014.
50 *"be gained from it":* Brooks, 1991.
50 *"grail within their grasp":* Ibid.
51 *over forty million units: History | iRobot.* (n.d.). Retrieved March 5, 2023, from https://about.irobot.com/History.
51 *or to recognize objects:* Note that future versions of the Roomba did add features that allowed it to learn maps of the house.
51 *Figure 2.8:* Photograph by Larry D. Moore in 2006. Picture published on Wikipedia at https://en.wikipedia.org/wiki/Roomba under a CC-BY license.
53 *to different downstream neurons:* Garrity et al., 2010.
55 *cold when they're too cold:* L. Luo et al., 2014. For additional details see the great review by Garrity et al., 2010. The "AFD" neuron only responds to "too hot" temperatures when temperature is above a threshold (see Goodman and Sengupta, 2018).
55 *to cross the barrier:* Hobert, 2003; Ishihara et al., 2002.
56 *across different sensory modalities:* Inoue et al., 2015.
57 *copper barrier:* The exact circuit for this mutual inhibition is more complex but similar in principle. In *C. elegans* there is a sensory neuron named AWC, which gets excited by positive valence smells. There are four downstream neurons in *C. elegans* that get input from sensory neurons; these downstream neurons are named AIZ, AIB, AIY, and AIA. AIZ and AIB promote turning, while AIY and AIA promote forward movement (discussed in Garrity et al., 2010). There is mutual inhibition among these downstream neurons: AIY inhibits AIB (Chalasani et al., 2007), AIY inhibits AIZ (Z. Li et al., 2014), and AIA inhibits AIB (Wakabayashi et al., 2009). Some of the mutual inhibition occurs further downstream; for example, by integrating inhibitory output from AIY with excitatory input from AIB on another neuron RIB, which promotes turns itself (Garrity et al., 2010; J. M. Gray et al., 2005). The circuit is messy, but the effect is the same; there is mutual inhibition between votes for moving forward and votes for turning.
57 *hungry a nematode is:* Davis et al., 2017; Lau et al., 2017.

Chapter 3: The Origin of Emotion

59 *separate words for each:* Jackson et al., 2019.
60 *two attributes of emotions:* Barrett and Russell, 1999; Russell, 2003.
60 *adrenaline, and blood pressure:* Heilman, 1997; Lang, Bradley, and Cuthbert, 1997.
61 *of specific brain regions:* Gerber et al., 2008.
61 *of valence and arousal:* Jackson et al., 2019; Wierzbicka, 1992.
61 *(e.g., crying and smiling):* Bridges, 1932; Graf, 2015; Huffman, 1997; Oster, 2012; Saarni et al., 2006.
63 *the food is gone:* Hills et al., 2004; D. Jones and Candido, 1999; Z. Liu et al., 2018.
65 *the brain with dopamine:* Chase et al., 2004.
65 *the state of exploitation:* If you destroy these dopamine neurons, exploitation behavior in response to food goes away (Sawin et al., 2000). Hills et al., 2004, shows this and also shows that if you leave those neurons intact but prevent dopamine signaling between neurons, exploitation behavior similarly goes away, and if you inject dopamine into the brain of *C. elegans*, it immediately shows exploitation (slowing down and increasing turning frequency as if it had detected food). And if you inject dopamine into *C. elegans* even after dopamine neurons have been destroyed, exploitation behavior returns.

Dopamine generates this persistent state by remaining in the extracellular fluid long after the dopamine neurons fired. Dopamine accomplishes this by directly modulating the responses of a whole suite of neurons. For example, dopamine modulates the responses of specific motor neurons (Chase et al., 2004) and the responses of the steering neurons (Hills et al., 2004), and it modulates valence neurons directly (Sanyal et al., 2004). The consequence of all this orchestrated modulation is that you get a new affective state of exploitation in which worms move more slowly and turn more frequently.

65 *food in their throats:* Rhoades et al., 2019.
65 *released, it triggers satiety:* For evidence serotonin is released by the detection of food in the stomach, see Gürel et al., 2012.

If you destroy the two food-related serotonin neurons, hungry worms no longer additionally slow down when they encounter food; see Sawin et al., 2000. If you prevent serotonin signaling, worms spend barely any additional time resting when full than when hungry (Churgin et al., 2017). Without serotonin signaling, worms spend far more time in escape/roaming behavior when hungry, as if it takes much longer for them to stop looking for food once they get full (Churgin et al., 2017; Flavell et al., 2013). One interpretation of this is that without serotonin, worms struggle to get satisfied. If you inject serotonin into the brain of *C. elegans*, it spends much less time moving around looking for food when hungry (Churgin et al., 2017). Serotonin also increases egg laying (Waggoner et al., 1998), mating behavior (Loer and Kenyon, 1993; Ségalat et al., 1995), and pharyngeal pumping, the equivalent of swallowing (Ségalat et al., 1995).

66 *pursuit of rewards (satiation):* The role of serotonin is similar across Bilateria (Gillette, 2006; Tierney, 2020). Serotonin is released when food is in the mouth and triggers swallowing in mollusks (Kabotyanski et al., 2000; Yeoman et al., 1994), nematodes (Hobson et al., 2006; Szø et al., 2000), and annelids (Groome et al., 1995). In vertebrates, the experience of a positive-valanced stimulus, even if expected, triggers serotonin release (Z. Liu et al., 2020; Zhong et al., 2017). The role of serotonin on aggression seems to be conserved as well, as serotonin *decreases* aggressiveness in rats (Nikulina et al., 1991), chickens (Dennis et al., 2008), and crustaceans (Kravitz, 2000). Serotonin consistently plays a role in satiation, although there are some differences. Serotonin induces satiety and reduces feeding in rats (Blundell and Leshem, 1975; Grinker et al., 1980), nonhuman

primates (Foltin and Moran, 1989), humans (McGuirk et al., 1991; Rogers and Blundell, 1979), flies (Long et al., 1986), cockroaches (Haselton et al., 2009), ants (Falibene et al., 2012), honeybees (A. S. French et al., 2014), and mosquitoes (Ling and Raikhel, 2018). However, in annelids and mollusks this seems to be different; serotonin seems to induce hunger and increase feeding in annelids (Lent et al., 1991) and mollusks (Hatcher et al., 2008; Yeoman et al., 1994) and lowers feeding thresholds across both (Palovcik et al., 1982).

66 *brain releases serotonin:* For a great review of serotonin, see Z. Liu et al., 2020.
66 *with whomever they see:* Musselman et al., 2012.
66 *willing to delay gratification:* For evidence that raising serotonin decreases eating, see Sharma and Sharma, 2012. For evidence that raising serotonin increases willingness to delay gratification, see Linnoila et al., 1983.
67 *dulling the responses of valence neurons:* For turning off dopamine responses, see Valencia-Torres et al., 2017. For dulling valence responses, see Lorrain et al., 1999.
67 *Figure 3.5:* Images from Kent Berridge (personal correspondence). Used with permission.
68 *food and starve to death:* Reviewed in Berridge and Robinson, 1998.
68 *of times an hour:* Heath, 1963.
68 *"reach the end point":* Admittedly, Heath believed that the evidence he found showed that septal stimulation was pleasurable, and he did claim that patients seemed to feel "good" (Heath, 1963). But other experimenters did the same experiments and found "there were no 'liking' effects during stimulation, in contrast to findings reported by Heath" (Schlaepfer et al., 2008).
69 *both liking and disliking reactions:* Treit and Berridge, 1990.
69 *15 million annual suicide attempts:* Morgan et al., 2018.
69 *and engage in life:* Depression. World Health Organization, 13 September 2021. Accessed on March 5, 2023, at https://www.who.int/news-room/fact-sheets/detail/depression.
70 *sleep, reproduction, and digestion:* Norepinephrine is highly arousing across many, if not all, vertebrates, including fish (Singh et al., 2015). Octopamine (a related compound) similarly increases arousal in diverse protostomes such as annelids (Crisp et al., 2010), arthropods (Crocker and Sehgal, 2008; Florey and Rathmayer, 1978), and nematodes (Churgin et al., 2017). Norepinephrine increases aggression across many vertebrates, including mice (Marino et al., 2005). Octopamine similarly increases aggression in flies (C. Zhou et al., 2008). Norepinephrine is released by starvation in vertebrates (P. J. Wellman, 2000). Octopamine is released by starvation and increases food consumption in arthropods (Long and Murdock, 1982), mollusks (Vehovszky and Elliott, 2002), and nematodes (Guo et al., 2015; Suo et al., 2006). Octopamine suppresses courtship conditioning in arthropods (C. Zhou et al., 2012), and suppresses egg laying in arthropods (Sombati and Hoyle, 1984) and nematodes (Alkema et al., 2005; Guo et al., 2015; Horvitz et al., 1982).

The specific valence of octopamine/norepinephrine and dopamine may have been flipped in arthropods (reviewed in Barron et al., 2010): Octopamine mediates appetitive signals in crickets (Mizunami et al., 2009; Mizunami and Matsumoto, 2017), honeybees (Farooqui et al., 2003; Hammer, 1993), flies (Schwaerzel et al., 2003), and crabs (Kaczer and Maldonado, 2009). Dopamine may instead mediate aversive signals in crickets (Mizunami et al., 2009; Mizunami and Matsumoto, 2017) and flies (Schwaerzel et al., 2003). However, the story is not so clear; rewards create dopamine-dependent positive-affective states in bumblebees (Perry et al., 2016). Further, lack of octopamine impairs *aversive* learning in arthropod flies (Mosca, 2017), and different subsets of octopamine neurons seem to trigger either approach or aversion in arthropod flies (Claßen and Scholz, 2018).

Admittedly, there are many arousing chemicals with slightly different effects (D.

Chen et al., 2016). But this is still an instructive "first pass" and it is remarkable how well individual neuromodulators map to specific affective states in nematodes. If you block norepinephrine, worms spend dramatically less time in their escape behavioral repertoire and far more time immobilized, even if exposed to noxious stimuli (Churgin et al., 2017). Worms lose the ability to get into the mode of "I have to get out of here and find food!" Norepinephrine accomplishes this the same way any other neuromodulator does—by persistently modulating various neurons that control movement and turning (Rengarajan et al., 2019). Like other neuromodulators, norepinephrine also modulates valence neurons. Without norepinephrine, worms fail to shift their behavior from CO_2 avoidance to attraction when starved (Rengarajan et al., 2019).

70 *rest and be content:* Churgin et al., 2017; Rex et al., 2004; Suo et al., 2006.
70 *energetic resources to muscles:* Specifically, the suite of adrenaline related compounds (norepinephrine, octopamine, and epinephrine).
70 *cannot go on indefinitely:* See Sapolsky et al., 2000, for a great overview of the stress response. This analogy was inspired by Sapolsky, 2004.
70 *in response to stressors:* Park et al., 2020.
71 *inhibited by acute stressors:* Staub et al., 2012.
71 *prolonged bouts of feeding:* Cheong et al., 2015.
71 *inhibited pain responses:* Mills et al., 2016; Nieto-Fernandez et al., 2009; Pryor et al., 2007.
71 *and inhibited reproductive behavior:* Ow and Hall, 2020; Seidel and Kimble, 2011.
71 *their normally hungry peers:* You et al., 2008.
73 *pause bodily functions:* Nath et al., 2016.
73 *they give up:* A. J. Hill et al., 2014.
73 *state of chronic stress:* Following exposure to a stressor like heat, the worms exhibit a period of quiescence that aids survival. See Fry et al., 2016; A. J. Hill et al., 2014; Konietzka et al., 2019; van Buskirk and Sternberg, 2007. The same is true of starvation; see Park et al., 2020; Skora et al., 2018.
73 *response, appetite, and reproduction:* Adamo and Baker, 2011.
73 *stress starts activating serotonin:* In genetically edited *C. elegans* worms with reduced insulin signaling (which triggers a depression-like state), this is what you see: if serotonin is blocked, these seemingly permanently depressed worms cease to be depressed (Dagenhardt et al., 2017). They return to moving around and responding to food cues. Fascinatingly, this serotonin depression cure seems to work all the way from *C. elegans* to humans. The primary medications to treat depression in humans are selective serotonin reuptake inhibitors (SSRIs), such as Prozac, which evidence suggests reduce the level of serotonin in the brain. There is complexity here. At first, SSRIs actually increase the level of serotonin in the brain by blocking the reuptake of serotonin from synapses. But over the course of weeks, this blocking changes the responses of serotonin neurons and makes them reduce their signaling, hence the net effect is to reduce serotonin levels. This is why SSRIs can make depression worse at first, but over the course of two to six weeks, people start to feel better. Of course, there is still controversy around whether this story is exactly right.

Giving opioids to depressed mice reduces their signs of depression (Berrocoso et al., 2013; Zomkowski et al., 2005).

74 *presence of inescapable stressors:* As further evidence for this, consider: What accelerates the speed at which a worm gives up? Energy reserves. If energy reserves are low, as signaled by low levels of insulin, worms give up much faster (Skora et al., 2018). This may seem too simplistic to be informative about depression or stress in humans, but you might be surprised by the connections. Insulin signaling has a well-known, albeit still mysterious, connection to depression and chronic stress in humans: People with diabetes (a disease where insulin signaling is disrupted) have a three times higher rate of depression than the general population (Anderson et al., 2001; Gavard et al., 1993); people with depression even without diabetes still show higher insulin resistance than

the general population (Hung et al., 2007; Kan et al., 2013), and people with diabetes even without classical depression still report apathy/anhedonia as a common symptom (Bruce et al., 2015; Carter and Swardfager, 2016). Further, diabetic mice display more signs of depression than nondiabetic mice, an effect that is completely reversed by insulin administration (Gupta et al., 2014). If you genetically edit *C. elegans* to reduce its insulin signaling, it becomes permanently depressed, showing dramatically reduced movement even in response to food signals (Dagenhardt et al., 2017). Anhedonia is an evolutionary beneficial feature of chronic stress—it is giving up to save energy in the face of hardship. And insulin seems to be a key signal for how much energy reserve an animal has, hence how likely an animal is to become chronically stressed in the face of hardship.

74 *stimuli have gone away*: A. J. Hill et al., 2014.
74 *slugs, and fruit flies*: For learned helplessness in cockroaches, see G. E. Brown, Anderson, et al., 1994; G. E. Brown and Stroup, 1988. For slugs, see G. E. Brown, Davenport, et al., 1994. For fruit flies, see G. E. Brown et al., 1996.
74 *don't even have serotonin neurons at all*: It is likely that the locomotive circuits within Cnidarians are also modulated by various neuropeptides, but in Cnidarians, such decisions don't seem to be driven by the same neuromodulators as in Bilateria. In Cnidaria, dopamine has been shown to inhibit the feeding response (Hanai and Kitajima, 1984) and tentacle contraction (Hudman and McFarlane, 1995) and even trigger sleep (Kanaya et al., 2020). Serotonin is not present across many Cnidarians (Carlberg and Anctil, 1993; Takeda and Svendsen, 1991), and in the few species where it is found, it seems to primarily induce spawning (Tremblay et al., 2004). Norepinephrine increases the likelihood of nematocyst release (Kass-Simon and Pierobon, 2007) and changes the speed of peristalsis waves in medusa (Pani et al., 1995). For the most part, these neuromodulators seem involved in modulating reflexes, and not, as in Bilateria, directly triggering behavioral repertoires for navigation.

Chapter 4: Associating, Predicting, and the Dawn of Learning

76 *"it we are nothing"*: Seeger, 2009.
77 *"source of error"*: Todes, 2014.
77 *"elicit in the dog thoughts about food"*: Ibid.
78 *kick in response to just the sound*: Irwin, 1943; Twitmyer, 1905.
78 *simple circuits in their spinal cords*: Illich et al., 1994.
79 *away from the salt*: Amano and Maruyama, 2011; Saeki et al., 2001; Tomioka et al., 2006; Wen et al., 1997.
80 *from that food smell*: Morrison et al., 1999; Y. Zhang et al., 2005.
80 *toward that temperature*: Hedgecock and Russell, 1975.
80 *will last for days*: This effect is not just a sensitization of the withdrawal response; if you shock a slug but do not pair the shock with the tap, it does not withdraw as much to the tap despite the same number of prior shocks. (Carew et al., 1981a; 1981b; Walters et. al. 1981). For a review of simple learning circuits see Hawkins & Kandel 1984.)
80 *capable of learning associations*: I am aware of only one report of associative learning in Cnidaria, where a sea anemone learned to contract its tentacles in response to a light that predicted shock, an experiment by Haralson et al., 1975. Other attempts have not replicated this result (Rushforth, 1973). Torley, 2009, performed a literature search and engaged in numerous personal inquiries with experts in Cnidarian behavior and did not find studies confirming classical conditioning in Cnidaria. Ginsburg and Jablonka, 2019, similarly conclude that Cnidaria do not exhibit associative learning.
82 *a process called* extinction: Pavlov, 1927.
84 *ancient Ediacaran Sea*: Ruben and Lukowiak 1982.
85 *credit assignment problem was blocking*: Kamin, 1969.

85 *in the animal kingdom:* Latent inhibition has been shown in honeybees (Abramson and Bitterman, 1986), mollusks (Loy et al., 2006), fish (Mitchell et al., 2011), goats (Lubow and Moore, 1959), and rats (Ackil et al., 1969; Boughner and Papini, 2006). Overshadowing and blocking have been observed in flatworms (Prados, Alvarez, Howarth, et al., 2013), honeybees (Couvillon and Bitterman, 1989), mollusks (Acebes et al., 2009; Sahley et al., 1981), rats (Prados, Alvarez, Acebes, et al., 2013), humans (Prados, Alvarez, Acebes, et al., 2013), rabbits (Merchant and Moore, 1973), and monkeys (Cook and Mineka, 1987).

86 *blocking, and overshadowing:* Illich et al., 1994.

87 *memories were these impressions:* Burnham, 1888.

87 *"has once been folded":* Levy, 2011.

87 *were persistent "vibrations":* Burnham, 1889.

87 *how learning might work:* For a good review of the history of different discoveries on synaptic learning, in particular timing-based learning rules, see Markram et al., 2011.

89 *involved in Hebbian learning:* Ramos-Vicente, D. et al., 2018; Stroebel, D. 2021.

Chapter 5: The Cambrian Explosion

96 *are incredibly similar:* Grillner and Robertson, 2016.

98 *many animal species:* P. Chance, 1999.

98 *Figure 5.5: One of Thorndike's puzzle boxes:* Picture is from Thorndike, 1898 (figure 1).

99 *"again in that situation":* P. Gray, 2011.

101 *specific button to get food:* Adron et al., 1973.

101 *getting caught in a net:* C. Brown, 2001; C. Brown and Warburton, 1999.

Chapter 6: The Evolution of Temporal Difference Learning

104 *not going to work:* Minsky, 1961.

106 *behaviors using* predicted *rewards:* To be fair, some of these ideas were already present in dynamic programming in the operations research world. Sutton's contribution was realization you could solve the policy and the value function simultaneously.

109 *backgammon using temporal difference learning:* Note that TD-Gammon was not an actor-critic model but a simpler version of temporal difference learning that learned what is called the *value function* directly. But the principle of bootstrapping on temporal differences was the same.

109 *"staggering level of performance":* Tesauro, 1994.

110 *for twenty-four hours straight:* Olds, 1956; Olds and Milner, 1954. These experiments actually stimulated the septal area, which triggered dopamine release. Later experiments confirmed that it was, in fact, dopamine that mediated the effect of septal stimulation; if you inject a rat with dopamine-blocking drugs, the rat will no longer push a lever for septal stimulation (reviewed in Wise, 2008).

110 *favor of dopamine stimulation:* "A hungry rat often ignored available food in favor of the pleasure of stimulating itself electrically" (Olds, 1956).

110 *(repeatedly removed from water):* Kily et al., 2008.

110 *such dopamine-enhancing chemicals:* Cachat et al., 2010; Gerlai et al., 2000, 2009.

112 *temporal difference learning signal:* Schultz et al., 1997.

113 *food in sixteen seconds:* Kobayashi and Schultz, 2008.

113 *"A Neural Substrate of Prediction and Reward":* Schultz et al., 1997.

113–14 *monkey, and human brains:* Grillner and Robertson, 2016; J. M. Li, 2012; Vindas et al., 2014.

115 *for a failed prediction of future reward:* For more information on this idea, see Joseph LeDoux's great review of avoidance learning in LeDoux et al., 2017.

NOTES

115 *expected rewards or punishments:* It can be hard to differentiate Pavlovian learning from learning from omission, and there are ongoing debates about the mechanisms of avoidance learning. For a great study showing that fish do truly learn from the omission of shock, see Zerbolio and Royalty, 1982.

116 *five seconds after light:* M. R. Drew et al., 2005; A. Lee et al., 2010. Reviewed in Cheng et al., 2011.

116 *cycle of the day:* Eelderink-Chen et al., 2021, shows circadian rhythm in prokaryotes; McClung, 2006, shows circadian rhythm in plants.

116 *intervals between events:* Abramson and Feinman, 1990; Craig et al., 2014. Reviewed in Abramson and Wells, 2018.

118 *that maximize dopamine release:* The connections between the cortex and basal ganglia contain different types of dopamine receptors. Rapid spikes of dopamine drive a strengthening between the cortical neurons and specific basal ganglia neurons that drive doing (or *ungating*) specific actions, and a *weakening* between cortical neurons and a different set of basal ganglia neurons that drive stopping (or *gating*) specific actions. Rapid declines of dopamine have the opposite effect. Through these parallel circuits, dopamine bursts reinforce recent behaviors and make them *more* likely to reoccur and dopamine declines punish behaviors and make them *less* likely to reoccur.

119 *basal ganglia of dopamine:* Cone et al., 2016, shows evidence that changing valence of sodium appetite derives from changes in input from lateral hypothalamus to midbrain dopamine neurons.

120 *dopamine neurons directly:* The purported actor circuit flows from the *matrix* of the striatum (the input structure of the basal ganglia); the purported critic circuit flows from the *striosomes* of the striatum.

Chapter 7: The Problems of Pattern Recognition

123 *specific types of molecules:* Niimura, 2009, provides evolutionary evidence that the olfactory receptors in modern vertebrates originated just before early vertebrates in the first chordates. Amphioxus (an animal often used as a model of the first chordates) has thirty-one functional vertebrate-like olfactory receptor genes and the lamprey (often used as a model for the early vertebrates) has thirty-two vertebrate-like olfactory receptor genes. Note that different lineages expanded the number of olfactory receptors; some modern fish have over one hundred, and rats have over one thousand.

As discussed in Niimura, 2012, while invertebrates also have olfactory receptors, their olfactory receptors seem to have independently evolved: "[Olfactory receptor] genes were also identified from other invertebrates including insects, nematodes, echinoderms, and mollusks . . . however, their evolutionary origins are distinct from that of vertebrate [olfactory receptor] genes. The neuroanatomical features of insect and vertebrate olfactory systems are common, but insect and vertebrate [olfactory receptor] genes are strikingly different to each other and share no sequence similarities."

126 *are similar but not the same:* D. A. Wilson, 2009, and Laurent, 1999, proposed a similar model of olfactory coding. Barnes et al., 2008, found evidence of pattern completion in olfactory cortex of rats. Yaksi et al., 2009, found evidence of similar types of pattern separation and completion in fish. For one of the original papers suggesting that the three-layered cortex performs this type of auto-association, see Marr, 1971.

127 *An artificial neural network:* Picture from https://en.wikipedia.org/wiki/Artificial_neural_network#/media/File:Colored_neural_network.svg.

129 *three layers of neurons:* Teleost fish may not have clear layers, but the lamprey has a layered cortex, as do reptiles, hence I proceed with the assumption that the cortex of early vertebrates was layered (Suryanarayana et al., 2022).

131 *networks to do math:* McCloskey and Cohen, 1989. For reviews of current challenges in continual learning see Parisi et al., 2019, and Chen and Liu, 2018.
133 *the year before:* Brown, 2001.
133 *learn this new pattern:* Grossberg, 2012.
135 *presented the cats with different visual stimuli:* Hubel and Wiesel, 1959, 1962, 1968.
136 *Figure 7.9:* Figure from Manassi et al., 2013. Used with permission.
137 *discovered by Hubel and Wiesel:* Fukushima, 1980.
139 *tap pictures to get food:* Wegman et al., 2022.
140 *rotated and translated objects:* Worden et al., 2021. The cortex and thalamus are densely interconnected. It was originally believed that the thalamus was merely a "relay" of input to the cortex. But new research is beginning to call this into question. Three observations give us hints that the interactions between the thalamus and the cortex might be important for the invariance problem. First, most sensory input to the cortex flows through the thalamus—input from eyes, ears, and skin all first go to the thalamus and then branch out to various regions of the cortex. However, there is one exception to this rule: *smell*. The single sense that skips the thalamus and connects directly to the cortex is smell. Perhaps this is because smell is the only sense that does not have the invariance problem; it does not need to interact with the thalamus to recognize objects at different scales, rotations, and translations. Second, the thalamus and the cortex evolved together. Even in the most distant vertebrates, like the lamprey fish, there is both a cortex and a thalamus with similar interactions between them as in other vertebrates. This suggests that their function might be emergent from interactions between them. And third, the circuitry of the thalamus seems to precisely gate and route connections between different areas of the cortex.

Chapter 8: Why Life Got Curious

142 *Montezuma's Revenge:* This story is retold in Christian, 2020.
143–44 *to human infants:* For curiosity in fish, see Budaev, 1997. For curiosity in mice, see Berlyne, 1955. For curiosity in monkeys, see Butler and Harlow, 1954. For curiosity in human infants, see Friedman, 1972.
144 *no "real" reward:* Matsumoto and Hikosaka, 2009.
144 *cephalopods, show curiosity:* Reviewed in Pisula, 2009. For cockroach curiosity, see Durier and Rivault, 2002. For curiosity in ants, see Godzinska, 2004. For cephalopod curiosity, see Kuba et al., 2006.
144 *wasn't present in early bilaterians:* In the review on curiosity by Pisula, 2009, on page 48, he concludes that "parallels in exploratory behaviors must therefore be a result of convergence, i.e. a similar response to similar challenges presented by the environment, rather than common ancestry."
144 *Fish exhibit this effect as well:* This partial reinforcement effect was found in goldfish in Wertheim and Singer, 1964. It was also shown in Gonzales et al., 1962, although with some differences from how the effect works in mammals.

Chapter 9: The First Model of the World

147 *directly to the container with the food:* Durán et al., 2008, 2010.
147 *one thing relative to another thing:* Larsch et al., 2015. For planarians see Pearl, 1903. For bees see Abramson et al., 2016.
148 *through the entire loop again:* Wehner et al., 2006.
148 *Figure 9.1:* Image by Carlyn Iverson / Science Source. Used with permission.
149 *facing a certain direction:* Bassett and Taube, 2001.
149 *ventral cortex, and medial cortex:* Note there is controversy around how to subdivide

parts of the cortex in fish and amphibians. Some argue there are four areas, an additional one being the dorsal cortex (see Striedter and Northcutt, 2019).

149 *cortex of early vertebrates:* Note that the location of these areas of cortex is shifted in some modern vertebrates, which leads to different naming of the same functional structure, a complexity I have intentionally omitted for readability. In teleost fish, for example, the cortex does not fold in the same ways as in the lamprey, reptiles, and mammals (in teleost fish, the cortex goes through *evagination*, folding outward, instead of *invagination*, folding inward). So the location of the same functional structure (the hippocampal part of the cortex) ends up in different places in the brain, and thus the names anatomists give these areas differ. When I refer to the medial cortex of early vertebrates, I am referring to the functional structure in early vertebrates that would later become the hippocampus in mammals. In the lamprey and reptiles, this part of the cortex is the *medial* cortex, whereas in teleost fish, the identical part of the cortex is the *lateral* cortex. For simplicity I refer to this area of cortex as hippocampus and graphically represent the cortex only in its invaginated state (hence the hippocampus structure shows up in the *medial* part), since we use the lamprey as the model organism for early vertebrates.

150 *are facing specific directions:* Fotowat et al., 2019; Vinepinsky et al., 2020.
150 *a spatial map:* Petrucco et al., 2022, shows evidence of head-direction cells in the hindbrain of fish. For a review of the overall network of head-direction cells and input to the hippocampus in rodents, see Yoder and Taube, 2014.
150 *ability to remember locations:* Broglio et al., 2010; Durán et al., 2010; López et al., 2000; Rodríguez et al., 2002.
150 *turn in a maze:* Rodríguez et al., 2002.
150 *space to get food:* Durán et al., 2010.
150 *given different starting locations:* Broglio et al., 2010.
150 *similarly impairs spatial navigation:* Naumann and Laurent, 2016; Peterson, 1980; Rodríguez et al., 2002.

Chapter 10: The Neural Dark Ages

157 *to previously inhospitable areas:* Algeo, 1998.
157–58 *were thirty meters tall:* Beck, 1962.
158 *The Late Devonian Extinction:* Algeo et al., 1995; McGhee, 1996.
164 *than you can underwater:* Mugan and MacIver, 2020.
164 *and plan are birds:* Boeckle and Clayton, 2017.
166 *Figure 10.3:* I am leaving out the dorsal cortex from early vertebrates because there is still debate about its presence (Striedter and Northcutt, 2019). For the alignment between medial cortex to hippocampus, lateral cortex to olfactory cortex, and ventral cortex to associative amygdala, see Luzzati, 2015; Striedter and Northcutt, 2019.

Chapter 11: Generative Models and the Neocortical Mystery

168 *stimuli elicited what responses:* Talbot et al., 1968.
170 *test of Mountcastle's hypothesis:* von Melchner et al., 2000.
173 *Figure 11.4:* "Editor" image from Jastrow (1899). Other images from Lehar S. (2003), obtained from Wikipedia.
174 *Figure 11.5:* Staircase from Schroeder (1858). "Necker cube" from Louis Necker (1832). Duck or rabbit from Jastrow (1899).
175 *Figure 11.6:* Image from Fahle, et al., 2002. Used with permission by The MIT Press.
175 *with Helmholtz's theory:* Later incarnations of Helmholtz's ideas include analysis-by-

177 *the Helmholtz machine:* Dayan, 1997; Hinton et al., 1995.
177 *learned on its own:* Dayan, 1997; Ibid.
178 *Figure 11.8:* Image from Hinton et al., 1995. Used with permission.
179 *thispersondoesnotexist.com:* Uses a StyleGAN2 (Karras et al., 2020).
180 *Figure 11.10:* Figure from He et al., 2019. Used with permission.
181 *with a generative model:* Reichert et al., 2013.
181 *"constrained hallucination":* Seth, 2017.
182 *presence of REM sleep:* The only nonmammalian animals shown to also have mammal-like sleep states of alternated NREM and REM sleep (suggestive of dreaming) are birds (Johnsson et al., 2022; Lesku and Rattenborg, 2014; Rattenborg et al., 2022).
182 *process of forced generation:* For a nice overview of theories of why we dream, the connection to generative models, and alternative explanations for dreaming, see Prince and Richards, 2021.
183 *processing actual visual data:* van der Wel and van Steenbergen, 2018.
183 *imagine the same thing:* O'Craven and Kanwisher, 2000.
183 *moving the body parts:* Doll et al., 2015.
183 *(recording their brains):* For a good review, see Pearsona and Kosslynb, 2015. There is now strong evidence that when one visualizes, there is activity in area V1 (Albers et al., 2013; Slotnick et al., 2005; Stokes et al., 2009). For evidence that one can decode imagery from activation in visual neocortex, see Kay et al., 2008; Naselaris et al., 2015; Thirion et al., 2006.
183 *(the left visual field):* Bisiach and Luzzatti, 1978; Farah et al., 1992.
185 *that unfolds over time:* Jeff Hawkins has some great writing about this. See his books J. Hawkins, 2021, and J. Hawkins, 2004.
186 *"from AI systems today":* tweeted by Yann LeCun (@lecun) on December 19, 2021.

Chapter 12: Mice in the Imaginarium

188 *rotated, rescaled, or perturbed:* Volotsky et al., 2022.
189 *before choosing a direction:* Tolman, 1939, 1948.
189 *try the next option:* Steiner and Redish, 2014. As reviewed in Redish, 2016.
189 *to consider alternative paths:* Schmidt et al., 2013.
190 *turning right and getting:* Ibid.
191 *the barrier at all:* Beyiuc, 1938 (page 409), describes this unfolding in a betta fish. Gómez-Laplaza and Gerlai, 2010, is the only counterexample I am aware of where fish were reported to do latent learning of a map and make the right decision ahead of time. It is hard to interpret a single counterexample without further replications. But if it turns out that some fish can, in fact, perform latent learning, then this suggests either that early vertebrates could solve latent-learning tasks without planning, planning evolved independently in some fish, or planning in some form existed in early vertebrates.
191 *navigate around a barrier:* Lucon-Xiccato et al., 2017.
192 *runs toward the salt:* Tindell et al., 2009.
193 *they called "restaurant row":* Steiner and Redish, 2014. As reviewed in Redish, 2016. Also see Bissonette et al., 2014.
194 *play rock, paper, scissors:* Abe and Lee, 2011.
195 *did not occur either:* D. Lewis, 1973.
196 *(of reasoning in birds):* For evidence of causal reasoning in rats, see Blaisdell et al., 2006; M. R. Waldmann et al., 2012. Fischer and Schloegl, 2017, conclude that causal reasoning evolved in early mammals and independently evolved also in birds.
197 *you're remembering the past:* Addis et al., 2007.
197 *(saw with imagining things):* O'Craven and Kanwisher, 2000; J. Pearson et al., 2015.

198 *mistaken eyewitness testimony:* Shermer et al., 2011.
198 *the event did occur:* Garry et al., 1996.
198 *to get more food:* Crystal, 2013; W. Zhou et al., 2012.
198 *structures for rendering simulations:* Many neuroscientists don't like using the term *episodic memory* to refer to the process by which these simpler mammals, such as rats, recall past life events. The term *episodic memory* has been loaded with a baggage of concepts, such as the *conscious experience* of mentally time traveling or some notion of an autobiographical self. Many neuroscientists instead use the safer term *episodic-like memory*. Indeed, it isn't clear how experientially rich the recollections of rats are. But regardless, the evidence from rats suggests that the precursor to episodic memory was present in early mammals.
199 *state and the best actions:* Actions include not just motor movements but the "next target location in space." In other words, the "spatial map" of ancestral vertebrates is not considered model-based RL because it doesn't seem to be employed for the purpose of simulating future actions. But it could still be employed for place recognition and the construction of homing vectors.
200 *games were model-free:* Mnih et al., 2013.
200 *are model-free:* Wang, 2018.

Chapter 13: Model-Based Reinforcement Learning

201 *complex games like chess:* Baxter et al., 2000, offers a TD-learning approach to playing chess (which still uses tree search and hence is not model-free) and provides a nice summary of the struggles of using model-free approaches (no search) in chess.
201 *Go champion Lee Sedol:* Silver et al., 2016.
201 *Go than in chess:* M. James, 2016.
202 *The game of Go:* Figure from https://en.wikipedia.org/wiki/Go_(game)#/media/File:FloorGoban.JPG.
202 *board position in Go:* AlphaGo documentary (Kohs, 2020).
203 *(possible subsequent next moves):* Ibid.
203 *(body and navigational paths):* Ibid.
204 *about anything at all:* Devinsky et al., 1995; Németh et al., 1988; B. A. Vogt, 2009.
205 *"no 'will' to reply":* Damasio and van Hoesen, 1983.
206 *to be a mammal:* According to J. H. Kaas, 2011, early mammals had two main areas of frontal cortex: anterior cingulate and orbitofrontal cortex. When I refer to agranular prefrontal cortex of early mammals, I am referring to both of these regions. The anterior cingulate cortex in humans is considered homologous to the prelimbic, infralimbic, dorsal anterior cingulate cortex of rodents (Laubach et al., 2018; van Heukelum et al., 2020), all of which can be assumed to have been inherited from the anterior cingulate cortex of the first mammals.
207 *vicarious trial and error:* Lose head toggling behavior: Schmidt et al., 2019; lose goal representations in hippocampus: Ito et al., 2015.
207 *episodic-memory recall:* Frankland et al., 2004.
207 *or counterfactual learning:* J. L. Jones et al., 2012.
207 *effort is worth it:* Friedman et al., 2015.
207 *to repeat past mistakes:* Frankland et al., 2004.
207 *repeat already completed actions:* Goard et al., 2016; Kamigaki and Dan, 2017; Kopec et al., 2015.
207 *patient to get food:* Inactivating prelimbic area of rat frontal cortex (part of their aPFC) increases premature responses (e.g., lever-release before go stimulus). (Hardung et al., 2017; Narayanan et al., 2006). Inactivation of the aPFC (prelimbic and infralimbic cortex) in rats increases impatient trials and decreases waiting time (Murakami et al., 2017).

For a good review on the role of prefrontal areas in behavioral inhibition, see Kamigaki, 2019. Also see M. G. White et al., 2018.

209 *toward an imagined goal:* Procyk et al., 2000; Procyk and Joseph, 2001. The latter observed that neurons in the monkey anterior cingulate cortex (a part of aPFC) are sensitive to the order of actions executed in a sequence (even if the actual movements performed are the same), suggesting that this area of the brain is modeling the overall sequence an animal is within, not just individual movements. There is also evidence for this in rats (Cowen et al., 2012; Cowen and McNaughton, 2007) and humans (Koechlin et al., 2002).

210 *in an ongoing task:* Dehaene et al., 1994; MacDonald et al., 2000; Ridderinkhof et al., 2004; Totah et al., 2009. For an interesting synthesis, also see Shenhav et al., 2016.

210 *is reported to be:* For some further reading on this, see Gal, 2016, and Lakshminarayanan et al., 2017.

210 *of the basal ganglia:* The frontal cortex sends a direct projection to a part of the basal ganglia called the subthalamic nucleus that has been shown to be able to completely halt behavior (Narayanan et al., 2020).

210 *with levels of uncertainty:* E. R. Stern et al., 2010.

212 *activity of the sensory neocortex:* For a good review of this circuitry, see Kamigaki, 2019. Note that different frontal regions may be associated with different sensory modalities—some subregions of the aPFC (like the anterior cingulate cortex) mostly send output to visual and not somatosensory and auditory areas, whereas others mostly send output to auditory and somatosensory areas (S. Zhang et al., 2016).

212 *cortex become uniquely synchronized:* Benchenane et al., 2010; Spellman et al., 2015, show synchronization between the aPFC and hippocampus during vicarious trial and error, as reviewed in Redish, 2016. Hyman et al., 2010; M. W. Jones and Wilson, 2005, show synchronization between the hippocampus and PFC during episodic memory tasks. Sauseng et al., 2004; Sederberg et al., 2003; Xie et al., 2021, show this same synchronization between prefrontal and sensory neocortex during working memory and episodic-memory tasks.

212 *an action is selected:* Bogacz and Gurney, 2007; Krajbich et al., 2010.

213 *those that had not:* Dickinson, 1985.

214 *the food was devalued:* Ibid.

216 *calls this "active inference":* Adams et al., 2013.

216 *missing the fourth layer:* His theory was about the motor cortex, which is also agranular (see next chapter), but the logic applies equally to aPFC.

217 *(looking at the sky):* There is some evidence that neocortical columns oscillate between different states when different layers are suppressed and they oscillate at different rhythms when you're imagining versus when you're attending to things. I review the evidence in Bennett, 2020.

218 *duck or a rabbit:* S. Zhang et al., 2014, shows frontal circuits for modulating representations in the sensory neocortex.

220 *lizards* hundreds *of trials:* Wagner, 1932.

220 *if you damage a rat's aPFC:* Dias and Aggleton, 2000. Note that rats can still learn non-matching-to-position tasks at approximately normal rates because they do not have to overcome their instinctual tendency to avoid recently exploited foraging sites (avoiding the place they just experienced is instinctual). Discussed in Passingham and Wise, 2015.

Chapter 14: The Secret to Dishwashing Robots

222 *cortex damage:* Darling et al., 2011.

222 *other areas of the neocortex:* Although note that the motor cortex is "agranular," meaning it has a thin or missing layer four (just like agranular prefrontal cortex).

223 *evolution of the motor cortex:* Karlen and Krubitzer, 2007.

223 *bats, elephants, and cats:* Karlen and Krubitzer, 2007. The placental-marsupial divergence is believed to have occurred about 160 million years ago in the Jurassic period (Z. X. Luo et al., 2011).

223 *suffer from such paralysis:* Kawai et al., 2015; Whishaw et al., 1991. Neurons in the motor cortex of primates bypass older circuits and make direct connections with spinal neurons (Lemon, 2019). Although some evidence suggests such direct projections may also occur in rats (Elger et al., 1977; Gu et al., 2017; Maeda et al., 2016), new evidence shows that these direct projections disappear in adulthood (Murabe et al., 2018), unlike in primates (Armand et al., 1997; Eyre, 2007). The paralysis of motor-cortex damage in primates does not seem to be representative of the motor cortex in early mammals.

224 *small unevenly placed platform:* Alaverdashvili and Whishaw, 2008; T. Drew et al., 2008; T. Drew and Marigold, 2015; Grillner and el Manira, 2020.

224 *learn the lever sequence:* Kawai et al., 2015.

225 *Figure 14.3:* Art by Rebecca Gelernter; this particular figure was inspired by imagery from Grillner and el Manira, 2020.

225 *movements that require planning:* Beloozerova et al., 2010; Farrell et al., 2014; Grillner and el Manira, 2020.

225 *known to be there:* Andujar et al., 2010; Beloozerova and Sirota, 2003; T. Drew and Marigold, 2015.

225 *movement it presumably planned:* Lajoie et al., 2010.

225 *motor cortex becomes activated:* Malouin et al., 2003.

226 *unrelated to movement:* Kosonogov, 2011.

226 *and even surgical maneuvers:* Arora et al., 2011.

226 *the whole affair is:* Kohlsdorf and Navas, 2007; Olberding et al., 2012; Parker and McBrayer, 2016; Tucker and McBrayer, 2012.

226 *to get around platforms:* Kohlsdorf and Biewener, 2006; Olberding et al., 2012; Self, 2012.

228 *can cure drug addiction:* N. Li et al., 2013.

228 *lower-level subgoals:* Lashley, 1951; Yokoi and Diedrichsen, 2019.

228 *levels in the hierarchy:* Thorn et al., 2010.

229 *(trials they go through):* Yin et al., 2004.

229 *own without their control:* Brainin et al., 2008.

229 *damage to the premotor cortex:* P. Gao et al., 2003.

229 *that nearby stimuli suggest:* Lhermitte, 1983.

230 *rate dropped to 42 percent:* N. Li et al., 2013.

Chapter 15: The Arms Race for Political Savvy

238 *of land-living vertebrates:* Sahney and Benton, 2008.

240 *Figure 15.2:* Figure from ibid.

240 *bigger its social group:* Dunbar, 1998.

241 *for most other animals:* Pérez-Barbería et al., 2007; Shultz and Dunbar, 2007. Reviewed in Dunbar and Shultz, 2017.

242 *down and look away:* Stringham, 2011.

242 *and flatten their ears:* S. Curtis, 1998.

242 *multi-male groups:* Shultz and Dunbar, 2007. Original data from Nowak, 1999. Note these classifications are inexact, not all subspecies fall into one or the other, and there may be other categorizations as well. But these are the common ones and are the broad first approximations of different types of social organizations. Dunbar used these four categorizations in some of his seminal work, and they are standard in primate literature (B. B. Smuts et al., 1987).

243 *four common social structures found in mammals:* Shultz and Dunbar, 2007. Original data from Nowak, 1999.

243 *his own children:* While most harem-organizing mammals have a single male and multiple females, there are indeed cases where the roles are reversed. Some marmosets, monkeys, and marsupials have single females that mate with multiple males (Goldizen, 1988). And of course, in many insects, like bees, female-run social groups are less the exception and more the rule.
243 *avoidance of large groups:* R. A. Hill and Dunbar, 1998.
243 *they created in response:* Bettridge et al., 2010.
244 *living in a one-acre forest:* Menzel, 1974. Story summarized in Kirkpatrick, 2007.
245 *"accidental" and "intentional" actions:* Call and Tomasello, 1998.
246 *those who seemed unwilling:* Call et al., 2004.
246 *goggles wouldn't see them:* Kano et al., 2019.
246 *like some birds, dolphins:* Tomonaga, 2010.
247 *location the trainer knows about:* Bräuer, 2014; Kaminski et al., 2009.
247 *grooming and group size:* Dunbar, 1991, 1998; Lehmann et al., 2007.
247 *by grooming themselves more:* R. M. Seyfarth, 1980.
247 *by appearance and voice:* Snowdon and Cleveland, 1980.
247 *what the mother does:* D. L. Cheney and Seyfarth, 1980, 1982.
248 *A will submit to C:* Andelman, 1985; R. M. Seyfarth, 1980.
248 *many years, even generations:* D. Cheney, 1983.
248 *play back these recordings:* Bergman et al., 2003.
249 *on and so forth:* Berman, 1982; Horrocks and Hunte, 1983; Walters and Seyfarth, 1987.
249 *rank of her mother:* Berman, 1983; Lee, 1983.
249 *of higher-ranking families:* Datta, 1983.
249 *to die from disease:* Silk, 1987; Silk et al., 2003, 2010.
249 *hierarchy has been established:* M. R. A. Chance et al., 1977; Gouzoules, 1980.
249 *nonfamily members that come:* Reviewed in chapter 2 of D. L. Cheney and Seyfarth, 2019.
249 *to recruit such allies:* Chapais, 1988; also discussed in chapter 2 of D. L. Cheney and Seyfarth, 2019.
250 *formed grooming partnerships with:* D. Cheney, 1983.
250 *makes a "help me" vocalization:* R. M. Seyfarth and Cheney, 1984.
250 *to their own defense:* F. B. M. de Waal, 1982; Packer, 1977; B. B. Smuts, 2017.
250 *food just for themselves:* Engelmann and Herrmann, 2016.
250 *to deal with Keith:* Datta, 1983; Dunbar, 2012.
250 *more access to food:* Packer, 1977; Silk, 1982.
250 *ranked higher than themselves:* Cheney and Seyfarth, 2019.
250 *members of the group:* Gouzoules, 1975; Scott, 1984.
250 *with high-ranking individuals:* D. L. Cheney and Seyfarth, 2019.
250 *the higher-ranking individual:* D. Cheney, 1983; D. L. Cheney, 1977; R. M. Seyfarth, 1977.
250 *the most popular playmates:* P. Lee, 1983.
251 *you under my wing:* Stammbach, 1988b, 1988a.
251 *with nonfamily members:* Cheney and Seyfarth, 1989.
251 *have recently quarreled with:* Cheney and Seyfarth, 2019.
251 *eating, resting, and mating:* Dunbar et al., 2009.
252 *than most other mammals:* Dunbar, 1991.
252 *spend more time socializing:* Borgeaud et al., 2021.
252 *with social savviness:* Byrne and Corp, 2004.

Chapter 16: How to Model Other Minds

253 *about three hundred fifty grams:* Ginneken et al., 2017; Tobias, 1971; van Essen et al., 2019.

NOTES

254 *the same fundamental ways:* To be fair, primates do have a higher density of neurons in their neocortices, but this denser packing doesn't suggest a change to the overall architecture of a neocortical column, merely that it has been scaled up and packed into a smaller area. For a good review of brain scaling, see Herculano-Houzel, 2012.

255 *addition to the frontal cortex:* Preuss, 2009.

255 *and sensory neocortical regions:* Goldman-Rakic, 1988; Gutierrez et al., 2000.

255 *functional significance at all:* Hebb, 1945; Hebb and Penfield, 1940; H. L. Teuber and Weinstein, 1954.

256 *intellect or perception whatsoever:* Hebb and Penfield, 1940.

256 *cortex was a "riddle":* H. Teuber, 1964.

256 *gPFC lit up:* Gusnard et al., 2001.

257 *about yourself in general:* Christoff et al., 2009; Herwig et al., 2010; Kelley et al., 2002; Moran et al., 2006; Northoff et al., 2006; Schmitz et al., 2004.

257 *of the surrounding elements:* Kurczek et al., 2015.

257 *themselves in a mirror:* Breen et al., 2001; Postal, 2005; Spangenberg et al., 1998.

258 *the amygdala and hippocampus:* Morecraft et al., 2007; Insausti and Muñoz, 2001.

258 *from the older aPFC:* Ray and Price, 1993. Further evidence for this is seen in the fact that stimulation of the agranular cortex in primates elicits autonomic effects (changes in respiratory rate, blood pressure, pulse, pupillary dilation, and piloerection), while in the granular cortex, it does not (Kaada, 1960; Kaada et al., 1949).

260 *lit up with activation:* Brunet et al., 2000; Völlm et al., 2006. The comic-strip task is similar to the earlier storytelling work by Baron-Cohen et al., 1986.

261 *Figure 16.4:* Images from Brunet et al., 2000; Völlm et al., 2006; and personal correspondence with Dr. Eric Brunet-Gouet. Used with permission of Dr. Brunet-Gouet (personal correspondence).

261 *the age of four:* H. M. Wellman et al., 2001; H. Wimmer and Perner, 1983.

262 *Figure 16.5:* Photo from Frith, 2003. Reused with permission.

262 *the degree of activation:* Gweon et al., 2012; Otsuka et al., 2009; Saxe and Kanwisher, 2003; Young et al., 2007.

262 *such false-belief tasks:* Carrington and Bailey, 2009; van Overwalle and Baetens, 2009, implicate specifically two areas of the granular prefrontal cortex (dorsomedial prefrontal cortex and anteromedial prefrontal cortex, which roughly make up Brodmann areas 8, 9, and 10) as well as the temporoparietal junction and the superior temporal sulcus as areas that are uniquely activated by tasks that require the theory of mind.

262 *the Sally-Ann test:* Siegal et al., 1996; V. E. Stone et al., 1998.

262 *emotions in other people:* Shaw et al., 2005.

262 *with other people's emotions:* Shamay-Tsoory et al., 2003.

262 *distinguish lies from jokes:* Winner et al., 1998.

262–63 *that would offend someone:* Shamay-Tsoory et al., 2005; V. E. Stone et al., 1998.

263 *someone else's visual perspective:* Stuss et al., 2001.

263 *struggle to deceive others:* Ibid.

263 *effects in nonhuman primates:* Dehaene et al., 2005; D. I. Perrett et al., 1992; Ramezanpour and Thier, 2020.

263 *it does in humans:* T. Hayashi et al., 2020.

263 *social-network size in primates:* Sallet et al., 2011.

263 *theory-of-mind tasks:* J. Powell et al., 2012; Stiller and Dunbar, 2007; P. A. Lewis et al., 2011; J. L. Powell et al., 2010.

263 *to model other minds:* See Amodio and Frith, 2006, for a detailed review of the specific areas in the prefrontal cortex implicated in self-reference and thinking about others.

263 *or "social projection theory":* Gallese and Goldman, 1998; Goldman, 1992; Gordon, 2011; Harris, 1992. It should be noted that not everyone agrees that these are implemented by the same process. For some good reviews of this debate, see Dimaggio et al., 2008; Gallup, 1998.

264 *the same process:* When evaluating your own personality traits or receiving evaluations of yourself by others, the same mentalizing network in the gPFC activates—specifically, the medial area of the prefrontal cortex (Ochsner et al., 2005).
264 *of theory of mind:* Further consistent with this idea that theory of mind of others is bootstrapped on a generative model of yourself is the fact that the concept of self emerges in child development before the theory of mind emerges. See Keenan et al., 2005; Ritblatt, 2000; Rochat, 1998.
264 *about two years old:* Amsterdam, 1972.
264 *want, wish, and pretend:* Frith and Frith, 2003.
264 *"know it's a crocodile":* Shatz et al., 1983.
264 *with respect to other people:* H. M. Wellman et al., 2001.
264 *better at the other:* Gopnik and Meltzoff, 2011; Lang and Perner, 2002.
264 *themselves in a mirror:* Gallup et al., 1971. There is some evidence suggesting that elephants and dolphins can recognize themselves in a mirror (Plotnik et al., 2006; Reiss and Marino, 2001). Recognition is seen in apes, such as chimps, orangutans, and gorillas (Suarez and Gallup, 1981; Posada and Colell, 2007). Monkeys may also recognize themselves in a mirror (Chang et al., 2017). A great summary of these mirror tests can be found in chapter 3 of Suddendorf, 2013.
264 *these states in others:* Kawada et al., 2004; Niedenthal et al., 2000.
264 *thirstier than they are:* van Boven and Loewenstein, 2003.
264 *personality traits onto others:* Bargh and Chartrand, 2000.
265 *the minds of others:* Michael Graziano has some fascinating ideas about this and its relationship to consciousness. He argues that our ancestors evolved theory of mind to navigate their unique social lives and a side effect of this was that when they applied this theory of mind inward, consciousness emerged (Graziano, 2019).
265 *("This is jumping"):* Surís et al., 2021.
265 *(faces classified by emotions):* Crawford, 2021.

Chapter 17: Monkey Hammers and Self-Driving Cars

267 *to clean their ears:* Flossing: Pal et al., 2018. Lists of different techniques: Sanz and Morgan, 2007.
267 *flies and scratch themselves:* Hart et al., 2001.
267 *to break open nuts:* Müller, 2010.
268 *to the inner food:* Bernardi, 2012.
268 *different tool-using behaviors:* Sanz and Morgan, 2007.
268 *than those in Gombe:* Musgrave et al., 2020.
269 *and others for tearing:* di Pellegrino et al., 1992.
269 *had just happened:* Story told in Taylor, 2016 and Roche and Commins, 2009.
269 *(sticking one's tongue out):* di Pellegrino et al., 1992; Ferrari et al., 2003; Gallese et al., 1996.
269 *(parietal lobe, motor cortex):* di Pellegrino et al., 1992; Dushanova and Donoghue, 2010; Fogassi et al., 2005; Tkach et al., 2007.
269 *numerous species of primates:* Brass et al., 2007; Buccino et al., 2001; Mukamel et al., 2010.
269 *interpretations of mirror neurons:* For reviews of the current debate around mirror neurons, see Heyes and Catmur, 2022; Hickok, 2014; Jeon and Lee, 2018; Rozzi, 2015.
270 *in theory of mind:* Rizzolatti et al., 2001.
270 *see someone else do:* Gallese and Goldman, 1998.
270 *open (without seeing anything):* Kohler et al., 2002.
270 *(box behind the wall):* Umiltà et al., 2001.
271 *(phrase, like combing hair):* Pazzaglia et al., 2008; Tarhan et al., 2015; Urgesi et al., 2014.

271 *bouncing on its own:* Pobric and Hamilton, 2006. Consistent with this, if people actively pick up a light box, they become biased toward thinking that boxes they see others pick up are also light. This bias is far greater when an individual *actually* lifts the box as opposed to passively holding a light box, which demonstrates it isn't about associating a box with lightness but about the active experience of picking up a box yourself (A. Hamilton et al., 2004).

272 *(eating a burger, blowing out a candle):* Michael et al., 2014.

272 *intended to hold:* Thompson et al., 2019, provides a nice review of some of these ideas. But see Negri et al., 2007 and Vannuscorps and Caramazza, 2016 for counterexamples that suggest that impairments to action production don't always impair action perception.

273 *becomes way more activated:* S. Vogt et al., 2007.

273 *following the red dots:* Catmur et al., 2009; Heiser et al., 2003.

273 *learns to do it:* Humle et al., 2009; Lonsdorf, 2005.

273 *skill later in life:* Biro et al., 2003; Matsuzawa et al., 2008.

274 *(food in the cage):* M. Hayashi et al., 2005; Marshall-Pescini and Whiten, 2008; Tomasello et al., 1987; Subiaul et al., 2004.

274 *in the right way:* Whiten et al., 2005.

274 *way to get food:* Dindo et al., 2009.

274 *drawer to get food:* Gunhold et al., 2014.

274 *opening an artificial fruit:* E. van de Waal et al., 2015.

274 *down through multiple generations:* Haslam et al., 2016; Mercader et al., 2007; Whiten, 2017.

274 *lever and get water:* Zentall and Levine, 1972.

274 *technique of their parents:* Müller and Cant, 2010.

274 *other dolphins or humans:* Hermann, 2002.

274 *dog perform the act:* Range et al., 2007.

275 *those same navigational paths:* For observational learning in fish, see Lindeyer and Reader, 2010. For reptiles, see Kis et al., 2015; Wilkinson, Kuenstner, et al., 2010; Wilkinson, Mandl, et al., 2010.

276 *actively teach one another:* For arguments on this, see Hoppitt et al., 2008; Kline, 2014; Premack, 2007.

276 *hands of their young:* Boesch, 1991.

276 *down to help teach:* Masataka et al., 2009.

276 *swap tools with them:* Musgrave et al., 2016.

276 *tool to a youngster:* Musgrave et al., 2020.

277 *skipped the irrelevant steps:* Call et al., 2005; Horner and Whiten, 2005; Nagell et al., 1993.

279 *a handful of laps:* Story told in Christian, *The Alignment Problem*, 232.

279 *called "inverse reinforcement learning":* Abbeel, Coates, and Ng, 2004.

279 *a remote-controlled helicopter:* Abbeel et al., 2010.

280 *imitation learning in robotics:* For a nice review of challenges in inverse reinforcement learning, see Hua et al., 2021.

Chapter 18: Why Rats Can't Go Grocery Shopping

282 *less than seventy-two hours:* Milton, 1981.

283 *a less competitive fruit:* Janmaat et al., 2014.

283 *to be depleted more quickly:* Noser and Byrne, 2007.

283 *evolve to survive winters:* Barry, 1976.

283 *140 species of primates:* DeCasien et al., 2017.

284 *the "Bischof-Kohler hypothesis":* Suddendorf and Corballis, 1997.

284 before *they were cold:* F. B. M. de Waal, 1982.
284 *of that task:* Mulcahy and Call, 2006.
284 *have no suitable stones:* Boesch and Boesch, 1984.
284 *use in another location:* Goodall, 1986.
284 *change their behavior accordingly:* Naqshbandi and Roberts, 2006.
287 *"hunger will be satisfied":* Suddendorf and Corballis, 1997.
288 *and anticipating future needs:* Note that Suddendorf is still skeptical of studies suggesting that other animals can anticipate future needs (personal correspondence). In fact, Suddendorf is skeptical about whether any animals other than humans are capable of considering the future at all (see Suddendorf, 2013; Suddendorf and Redshaw, 2022). His fascinating book *The Invention of Tomorrow* describes his argument.
288 *that were well fed:* Mela et al., 1996; Nisbett and Kanouse, 1969.

Summary of Breakthrough #4: Mentalizing

289 *at similar developmental times:* Children seem to begin to anticipate future needs around age four (Suddendorf and Busby, 2005), which is same age that they begin to pass theory of mind tasks (H. M. Wellman et al., 2001). For a review of different theories of mental time travel, see Suddendorf and Corballis, 2007.
290 *ago in eastern Africa:* T. D. White et al., 2009.

Chapter 19: The Search for Human Uniqueness

295 *"and not of kind":* Darwin, 1871.
296 *in all the same ways:* Herculano-Houzel, S. 2009
296 *sounds and gestures:* Graham and Hobaiter, 2023; Hobaiter and Byrne, 2014; Hobaiter and Byrne, 2011.
299 *signed,* Finger bracelet: "Mission Part 1: Research." Koko.org.
300 *running from his trainer:* L. Stern, 2020.
300 *70 percent of the time:* Savage-Rumbaugh et al., 1993.
300 *(I want to be tickled):* Yang, 2013, compared the diversity of phrases between young children and the chimpanzee Nim Chimpsky. Yang, 2013, concluded that the children showed the level of diversity consistent with the use of grammar to construct novel phrases but Nim Chimpsky did not, thus concluding that Chimpsky's phrase diversity was more consistent with directly memorizing phrases.
301 *few noises or gestures:* Some of my favorite writing on this is Daniel Dor's book *The Instruction of Imagination* (New York: Oxford University Press, 2015).
303 *"money paid out in fees":* Harari, 2015.
304 *on different cooperation strategies:* Dunbar, 1993, estimated human group size at 150 (the famous Dunbar's number) by looking at the human neocortex ratio and examining tribal societies. B. B. Smuts et al., 1987, report approximately 50 as an average group size in chimpanzees and approximately 18 as an average group size in capuchin monkeys.
304 *be about one hundred fifty people:* Dunbar, 1992, 1993.
305 *even millions, of generations:* For reviews of the theory of cumulative culture see Tennie et al., 2009; Tomasello et al., 1993.
305 *as one hundred thousand years ago:* Toups et al., 2011.
306 *children are over-imitators:* D. E. Lyons et al., 2007.
306 *copy all the steps:* For example, Gergely et al., 2002, and Schwier et al., 2006, showed that twelve- and fourteen-month-old infants are more likely to copy an unusual component of a sequence when it was not clear why it was done and less likely to copy it if the teacher was "forced" to do the unusual action due to some physical limitation. For

example, Schwier et al., 2006, had teachers demonstrate putting a toy dog into a toy house that had two openings, one through a front door and another through a chimney. In cases where the front door was blocked and the teacher demonstrated putting the dog into the toy house through the chimney, infants were less likely to do the same when it was their turn (if the door was open for them); they just put it through the door (achieving the same goal through different means). In contrast, when teachers put the dog into the house through the chimney when the door was open (hence the teacher clearly *chose* the chimney for some reason), infants did the same and put it through the chimney.

306 *pull the toy apart:* Carpenter et al., 1998; Meltzoff, 1995.
306 *improves the accuracy and speed:* Chopra et al., 2019; Dean et al., 2012.
307 *discontinuity that changed everything:* My favorite writings about the idea of cumulative culture are in Tennie et al., 2009.
309 *can persist across generations:* Henrich, 2004.

Chapter 20: Language in the Brain

311 *"like you want before":* Aphasia. National Institute on Deafness and Other Communication Disorders. March 6, 2017. Accessed on March 5, 2023 at https://www.nidcd.nih.gov/health/aphasia.
311 *to* understand *speech:* For a review of Wernicke's area, see DeWitt and Rauschecker, 2013.
311 *selective for language in general:* Campbell et al., 2008.
311 *are in* writing *words:* Chapey, 2008.
311 *Broca's area is damaged:* Emmorey, 2001; Hickok et al., 1998; Marshall et al., 2004.
311 *spoken language and* written *language:* DeWitt and Rauschecker, 2013; Geschwind, 1970.
312 *watches someone sign:* Neville et al., 1997.
312 *are otherwise intellectually typical:* Lenneberg, 1967, showed that language capacity is radically dissociated from other cognitive capacities. For a more recent review, see Curtiss, 2013.
312 *over fifteen languages:* Smith and Tsimpli, 1995.
313 *a scaled-up chimpanzee brain:* Herculano-Houzel, 2012; Herculano-Houzel, 2009, show that the human brain is largely just a scaled-up primate brain. Semendeferi and Damasio, 2000, show that the prefrontal cortex of humans is *not* uniquely enlarged relative to other primates (it was just scaled up proportionally with the rest of the brain).

The few differences that have been found in human brains relative to other primate brains include the following. First, humans have a unique projection from the motor cortex to the area of the larynx that controls the vocal cords. So, yes, humans have unique control over their voice boxes, and this is clearly related to speech. But as we review later in the chapter, this is not what unlocked language, for there are many nonvocal languages that are as sophisticated without ever using this projection (such as the sign languages of people born deaf). Second, although there are no uniquely human areas of neocortex (all the same areas are found in other primates), there is indeed some evidence that the *relative space* devoted to different areas of prefrontal cortex might be somewhat different in humans (Teffer and Semendeferi, 2012). And third, minicolumns in the neocortex of humans might have greater width than those in other primates (Buxhoeveden and Casanova, 2002; Semendeferi et al., 2011), although this doesn't prove that there is anything fundamentally different about the neocortical microcircuit itself in humans. But if it were at some point discovered that the neocortical microcircuit of humans was, in fact, fundamentally wired differently (it would have to be a subtle enough change that it has eluded our detection thus far), this would surely require us to rethink this entire evolutionary story, as it would open the door for the possibility that

the human neocortex does enable something that is different "in kind."
313 *impact on monkey communication:* Aitken, 1981; Jürgens, 1988; Jürgens et al., 1982.
314 *controlled by the neocortex:* Burling, 1993.
314 *Figure 20.2:* Image from Trepel et al., 1996, used with permission.
315 *"impossible task [for chimpanzees]":* Goodall, 1986.
316 *system in the neocortex:* Note that this does *not* mean that emotional states themselves emerge only from the amygdala and brainstem. It merely means that the automatic emotional expressions—the smiles, frowns, and cries—are hardwired into these circuits. But emotional experiences and states in humans (and other mammals) are more complicated and do probably involve the cortex.
316 *normal gesture-call behavior:* Hammerschmidt and Fischer, 2013.
316 *of the same gestures:* Graham et al., 2018.
316 *to similar emotional states:* The extent to which emotional expressions and corresponding emotional states are universal versus culturally learned is controversial. Attempts to define explicit emotion categories using facial expressions has recently been challenged, as it turns out that much of what people define as one category in one culture does not always translate to another. I do not mean to suggest that emotions like *anger* and *happy* are universal. But even if many aspects of emotion categories are learned, this does not mean that there is no initial template with predefined emotional expressions that humans are born with. Indeed, I am unaware of any reports of a child who screams and cries in response to happy things and smiles and laughs in response to painful things. Babies born both deaf and blind still smile, laugh, frown, and cry normally (Eibl-Eibesfeldt, 1973). As mammals develop, the neocortex learns, and it can modulate and modify genetically prewired systems in the midbrain and hindbrain and thus complexify, modify, and expand on the emotional-expression template that we are born with. This is a general way mammal brains develop. For example, the midbrain and hindbrain of a baby have pre-defined wiring for basic motor behaviors (grasping). As the neocortex learns, it begins to take over and modulate these midbrain and hindbrain circuits to override them and take control over the hands. But this does not mean that there was not already a hardwired circuit for grasping.

For the evidence showing some universality in human emotional expressions across cultures, see Ekman, 1992; Ekman et al., 1969; Ekman and Friesen, 1971; Scherer, 1985. For evidence that emotion categories are not as universal as previously thought, see Lisa Feldman Barrett's wonderful book *How Emotions Are Made* and Barrett et al., 2019. Barrett's theory of constructed emotion, whereby the neocortex constructs emotion categories, is consistent with the idea presented in this book whereby the aPFC and gPFC construct a generative model of *the self* and construct explanations of an animal's own behaviors and mental states. An emergent property of this might be constructing the notion of an emotional state—using the notion of anger to explain the behavioral repertoire the animal performing. (Similar, or perhaps the same, to how we have suggested the aPFC constructs intent.)
316 *it later in life:* Andrei, 2021.
316 *but will never speak:* Lenneberg, 1967.
317 *given the previous words:* He introduced the simple recurrent neural network, also called the Elman network (Elman, 1990).
318 *as the model itself:* For a well-written discussion on curriculum, see Christian, 2020.
318 *facial expressions, and gestures:* Beebe et al., 1988, 2016.
318 *joint attention to objects:* Tomasello, 1995.
319 *back at her mother:* Carpenter and Call, 2013.
319 *and point again:* Liszkowski et al., 2004.
319 *same object they do:* Warneken et al., 2006.
320 *vocabulary is twelve months later:* Morales et al., 2000; Mundy et al., 2007.
320 *simplest questions about others:* Hauser et al., 2002.

Chapter 21: The Perfect Storm

320 *another's inner mental world:* Some claim that some nonhuman apes in these studies did ask some types of questions. It is still controversial.
320 *when asking yes/no questions:* D. L. Everett, 2005; Jordania, 2006.
320 *of our language curriculum:* MacNeilage, 1998; Vaneechoutte, 2014.

Chapter 21: The Perfect Storm

323 *Figure 21.1:* DeSilva et al., 2021.
324 *"runaway growth of the brain":* D. Everett, 2017, 128.
324 *on which the forest depended:* Davies et al., 2020.
325 *would eventually become human:* Coppens, 1994. Although it is undisputed that the climate changed in Africa around ten million years ago, how dramatic this change was and how relevant it was to our ancestors becoming bipedal is not settled.
325 *legs instead of four:* Set of cranial volumes from different fossils: Du et al., 2018.
325 *of a modern chimpanzee's:* Bipedalism appeared before tool use as well (see Niemitz, 2010).
326 *was scavenging carcasses:* Bickerton and Szathmáry, 2011. Before two million years ago, cut marks lie above bite marks, indicating that hominins accessed these bones only after other animals had. After two million years ago, bite marks more frequently lie above cut marks, indicating that hominins had first access to bones (Blumenschine, 1987; Blumenschine et al., 1994; Domínguez-Rodrigo et al., 2005; Monahan, 1996).
326 *came from meat:* Ben-Dor et al., 2021.
327 *was an apex predator:* Ibid.
327 *almost absurd 85 percent meat:* Ibid.
327 *began to go extinct:* From 1.8 to 1.5 million years ago had the highest per capita carnivoran extinction event (Bobe et al., 2007; M. E. Lewis and Werdelin, 2007; Ruddiman, 2008).
327 *three times as fast:* Perkins, 2013.
328 *must have invented cooking:* Wrangham, 2017.
328 *time and energy digesting:* Carmody and Wrangham, 2009.
328 *50 percent become temporarily infertile:* Koebnick et al., 1999.
328 *ash in ancient caves:* For evidence of fire use between 1.5 to 2 million years ago, see Gowlett and Wrangham, 2013; Hlubik et al., 2017; James et al., 1989.
329 *after they are born:* Garwicz et al., 2009.
329 *in today's human societies:* This is inferred based on the shift in *Homo erectus* toward less sexual dimorphism (difference in body size between males and females); see Plavcan, 2012.
329 *hunter-gatherer societies:* Original grandmother hypothesis proposed by Hawkes et al., 1998. For another review with more detail on hunter-gatherer societies, see Hawkes et al., 2018.
330 *full brain size achieved:* Malkova et al., 2006.
330 *origin of human languages:* Christiansen and Kirby, 2010.
330 *"own and my child":* Story from Terrace, 2019.
331 *"problem in all of science":* Christiansen and Kirby, 2010.
331 *language-ready vocal cords:* D'Anastasio et al., 2013.
333 *the survival of the species:* D. S. Wilson and Wilson, 2007, provide a modern interpretation of group selection. The key idea is that evolution operates under *multilevel* selection, so group effects and individual effects are always interacting. In other words, the simple survival of the species (group-only effect) is not how evolution works. Traits that hurt an individual's fitness but benefit the group overall are not necessarily selected for. Only in the right circumstances, where the individual cost is outweighed by the group benefit and competition with other groups.

Note that even these multilevel selection accounts of human evolution still acknowledge that defectors and cheaters were a problem that evolution would have had to account for. Thus, these multilevel accounts are still consistent with the evolutionary accounts in which mechanisms for the detection and punishment of violators would have been essential for stabilizing altruism and cooperation.

Once human groups had language, altruism, and the punishment of violators, the balance might have shifted toward strong group-selection effects, since altruism makes an individual's fitness differences *within a group* muted (as members support and help each other), thereby strengthening the effect of *across group* competition.

333 *where it is safest:* W. D. Hamilton, 1971.
335 *are around family members:* R. Seyfarth and Cheney, 1990; Sherman, 1977, 1985.
335 *called reciprocal altruism:* Trivers, 1971.
335 *them in the past:* Mitani, 2006; Mitani and Watts, 1999, 2001; Povinelli and Povinelli, 2001.
335 *that did not help them:* Olendorf et al., 2004.
336 *and language-enabled teaching:* Mesoudi and Whiten, 2008.
337 *among non-kin became possible*: This story and ordering described here is largely that proposed by Fitch, 2010.
337 *for defectors and liars:* For arguments that lying, defectors, and cheating was an important obstacle to overcome in the evolution of language see Dunbar, 2004; Fitch, 2010; Knight, 2008; Tomasello, 2016. Dor, 2017, provides a more nuanced take, suggesting that the challenge of cheating (i.e., lying) that emerged with language may have driven more than one type of evolutionary feedback loop, not only for the punishment of violators after their discovery but also the ability to detect when someone was lying. As one individual gets better at lying, instead of making language unstable (selecting for worse language skills), it may have created selection for better theory of mind to better identify when people were lying and engaging with malintent. And this, in turn, created more pressure for liars to further hide their intent through better emotional regulation, which then created pressure for better theory of mind to see through the tricks, thereby creating a feedback loop.
337 *human conversation is gossip:* Dunbar et al., 1997.
337 *The origins of language:* Dunbar, 1998; Dunbar, 2004.
338 *large group of individuals:* For work on the importance of punishment in supporting altruism and its importance in the evolution of humans, see Boyd et al., 2003.
340 *(hence no altruism required):* Bickerton and Szathmáry, 2011.
340 *language to evolve:* Tomasello, 2016, 2018; as summarized in Dor, 2017.
340 *trick for inner thinking:* Berwick and Chomsky, 2017.
342 *the last one million years:* Morwood et al., 1999.
342 *chimpanzee's, perhaps even smaller:* Falk et al., 2007.
342 *tool use as Homo erectus:* Sutikna et al., 2016.

Chapter 22: ChatGPT and the Window into the Mind

344 "*fear of being turned off*": N. Tiku, 2022.
345 "*nothing to fear from AI*": GPT-3, 2020. Note that the authors also did some editing of the article.
346 "*they would fall over*": K. Lacker, (2020). Giving GPT-3 a Turing Test. Kevin Lacker's blog. https://lacker.io/ai/2020/07/06/giving-gpt-3-a-turing-test.html.
348 *four questions to GPT-3:* This was in the GPT-3 sandbox, on text-davinci-002, test performed on Tuesday, June 28, 2022.
349 *this the representative heuristic:* Kahneman, 2013.
354 *and produce meaningful speech:* For TPJ in theory of mind, see Samson et al., 2004; Gallagher et al., 2000.

354 *as false-belief tests:* Cutting and Dunn, 1999; Hughes and Dunn, 1997; Jenkins and Astington, 1996. Note that the causation between mentalizing and language is controversial and unsettled. Astington and Jenkins, 1999, performed a longitudinal study that suggested that language skills predicted later performance on mentalizing tasks but not the reverse. But even if language dramatically improves mentalizing abilities, the basic ability to understand that other people have thoughts and agency seems to be a necessary foundation for joint attention to begin the process of naming objects in the first place (this is discussed in de Villiers, 2007).

354 *similar impairments in language:* Baron-Cohen, 1995.

356 *"models of the underlying reality":* Posted on LinkedIn by Yann LeCun in January 2023.

Bibliography

To save paper, the full bibliography can be found at briefhistoryofintelligence.com.

Over the course of my years of research for this book, there were hundreds of books, papers, and journals that I read—the vast majority of which are cited in the Notes section on page 373. The works below (in alphabetical order by title) were particularly important in formulating the framework in this book.

The Alignment Problem: Machine Learning and Human Values by Brian Christian
Behave: The Biology of Humans at Our Best and Worst by Robert Sapolsky
Brain Structure and Its Origins: In Development and in Evolution of Behavior and the Mind by Gerald E. Schneider
Brains Through Time: A Natural History of Vertebrates by Georg F. Striedter and R. Glenn Northcutt
Cerebral Cortex by Edmund Rolls
The Deep History of Ourselves: The Four-Billion-Year Story of How We Got Conscious Brains by Joseph LeDoux
Deep Learning by Ian Goodfellow, Yoshua Bengio, and Aaron Courville
Evolution of Behavioural Control from Chordates to Primates by Paul Cisek
The Evolution of Language by W. Tecumseh Fitch
The Evolution of Memory Systems by Elisabeth A. Murray, Steven P. Wise, and Kim S. Graham
The Evolution of the Sensitive Soul: Learning and the Origins of Consciousness by Simona Ginsburg and Eva Jablonka
Evolutionary Neuroscience by Jon H. Kaas
Fish Cognition and Behavior by Culum Brown, Kevin Laland, and Jens Krause
From Neuron to Cognition via Computation Neuroscience edited by Michael A. Arbib and James J. Bonaiuto
The Gap: The Science of What Separates Us from Other Animals by Thomas Suddendorf
How Emotions Are Made: The Secret Life of the Brain by Lisa Feldman Barrett
How Monkeys See the World: Inside the Mind of Another Species by Dorothy L. Cheney and Robert M. Seyfarth
How the Brain Might Work: A Hierarchical and Temporal Model for Learning and Recognition by Dileep George
The Invention of Tomorrow: A Natural History of Forethought by Thomas Suddendorf

Language Evolution edited by Morten H. Christiansen and Simon Kirby
The Mind Within the Brain: How We Make Decisions and How Those Decisions Go Wrong by A. David Redish
The Neurobiology of the Prefrontal Cortex: Anatomy, Evolution, and the Origin of Insight by Richard E. Passingham and Steven P. Wise
Neuroeconomics: Decision Making and the Brain by Paul Glimcher and Ernst Fehr
The New Executive Brain: Frontal Lobes in a Complex World by Elkhonon Goldberg
On Intelligence by Jeff Hawkins
Reinforcement Learning: An Introduction by Richard S. Sutton and Andrew G. Barto
Resynthesizing Behavior Through Phylogenetic Refinement by Paul Cisek
Superintelligence: Paths, Dangers, Strategies by Nick Bostrom
A Thousand Brains: A New Theory of Intelligence by Jeff Hawkins
Why Chimpanzees Can't Learn Language and Only Humans Can by Herbert S. Terrace
Why Only Us: Language and Evolution by Robert C. Berwick and Noam Chomsky

Art, Photo, and Figure Credits

Basics of Human Brain Anatomy: Original art by Mesa Schumacher
Our Evolutionary Lineage: Original art by Rebecca Gelernter
Figure 1: Figure by Max Bennett (inspired by similar figures found in MacLean's work)
Figure 2: Original art by Rebecca Gelernter
Figure 1.1: Photograph by Willem van Aken on March 18, 1993. Figure from www.scienceimage.csiro.au/image/4203 CC BY 3.0 license.
Figure 1.2: Original art by Rebecca Gelernter
Figure 1.3: Original art by Rebecca Gelernter
Figure 1.4: Original art by Rebecca Gelernter
Figure 1.5: Figure from Reichert, 1990. Used with permission.
Figure 1.6: Original art by Rebecca Gelernter
Figure 1.7: Original art by Rebecca Gelernter
Figure 1.8: Original art by Rebecca Gelernter
Figure 1.9: Original art by Rebecca Gelernter
Figure 1.10: Figure made by B. MacEvoy, 2015. Used with permission (personal correspondence).
Figure 1.11: Original art by Rebecca Gelernter
Figure 1.12: Original art by Rebecca Gelernter
Figure 1.13: Original art by Rebecca Gelernter
Breakthrough #1 Section Cover Image: Original art by Rebecca Gelernter
Figure 2.1: Original art by Rebecca Gelernter
Figure 2.2: Original art by Rebecca Gelernter
Figure 2.3: Original art by Rebecca Gelernter
Figure 2.4: Original art by Rebecca Gelernter
Figure 2.5: Original art by Rebecca Gelernter
Figure 2.6: Original art by Rebecca Gelernter
Figure 2.7: Original art by Rebecca Gelernter
Figure 2.8: Photograph by Larry D. Moore in 2006. Picture published on Wikipedia at https://en.wikipedia.org/wiki/Roomba.
Figure 2.9: Original art by Rebecca Gelernter
Figure 2.10: Original art by Rebecca Gelernter
Figure 3.1: Original art by Rebecca Gelernter
Figure 3.2: Original art by Rebecca Gelernter

ART, PHOTO, AND FIGURE CREDITS

Figure 3.3: Original art by Rebecca Gelernter
Figure 3.4: Original art by Rebecca Gelernter
Figure 3.5: Images from Kent Berridge (personal correspondence). Used with permission.
Figure 3.6: Original art by Rebecca Gelernter
Figure 3.7: Original art by Rebecca Gelernter
Figure 4.1: Figure designed by Max Bennett (with some icons from Rebecca Gelernter)
Figure 4.2: Original art by Rebecca Gelernter
Figure 4.3: Figure by Max Bennett
Breakthrough #2 Section Cover Image: Original art by Rebecca Gelernter
Figure 5.1: Original art by Rebecca Gelernter
Figure 5.2: Original art by Rebecca Gelernter
Figure 5.3: Original art by Mesa Schumacher
Figure 5.4: Original art by Mesa Schumacher
Figure 5.5: Image from Thorndike, 1898
Figure 5.6: Images from Thorndike, 1898
Figure 6.1: Original art by Rebecca Gelernter
Figure 6.2: Original art by Rebecca Gelernter
Figure 6.3: Figure by Max Bennett
Figure 6.4: Original art by Rebecca Gelernter
Figure 7.1: Original art by Rebecca Gelernter
Figure 7.2: Figure by Max Bennett
Figure 7.3: Figure by Max Bennett
Figure 7.4: Figure by Max Bennett
Figure 7.5: Original art by Rebecca Gelernter
Figure 7.6: Original art by Rebecca Gelernter
Figure 7.7: Free 3D objects found on SketchFab.com
Figure 7.8: Free 3D objects found on SketchFab.com
Figure 7.9: Figure from Manassi et al., 2013. Used with permission.
Figure 7.10: Figure designed by Max Bennett. The dog photo is from Oscar Sutton (purchased on Unsplash).
Figure 9.1: Image by Carlyn Iverson / Science Source. Used with permission.
Figure 9.2: Original art by Mesa Schumacher
Figure 9.3: Original art by Rebecca Gelernter
Breakthrough #3 Section Cover Image: Original art by Rebecca Gelernter
Figure 10.1: Original art by Rebecca Gelernter
Figure 10.2: Original art by Rebecca Gelernter
Figure 10.3: Original art by Rebecca Gelernter
Figure 11.1: Original art by Rebecca Gelernter
Figure 11.2: Original art by Rebecca Gelernter
Figure 11.3: Original art by Mesa Schumacher
Figure 11.4: "Editor" from Jastrow, 1899. Others from Lehar, 2003.
Figure 11.5: Staircase from Schroeder, 1858. "Necker cube" from Necker, 1832. Duck or rabbit from from Jastrow, 1899.
Figure 11.6: Image from Fahle et al., 2002. Used with permission by The MIT Press.
Figure 11.7: Original art by Rebecca Gelernter
Figure 11.8: Image from Hinton et al., 1995. Used with permission.
Figure 11.9: Pictures from thispersondoesnotexist.com
Figure 11.10: Figure from He et al., 2019. Used with permission.

Figure 12.1: Original figure by Max Bennett, with oversight and permission from David Redish

Figure 13.1: Picture from https://en.wikipedia.org/wiki/Go_(game)#/media/File:FloorGoban.JPG

Figure 13.2: Original art by Mesa Schumacher

Figure 13.3: Original art by Mesa Schumacher

Figure 13.4: Original art by Rebecca Gelernter

Figure 14.1: Original art by Rebecca Gelernter

Figure 14.2: Original art by Mesa Schumacher

Figure 14.3: Original art by Rebecca Gelernter

Figure 14.4: Original art by Rebecca Gelernter

Breakthrough #4 Section Cover Image: Original art by Rebecca Gelernter

Figure 15.1: Original art by Rebecca Gelernter

Figure 15.2: Original art by Rebecca Gelernter

Figure 16.1: Original art by Rebecca Gelernter

Figure 16.2: Original art by Mesa Schumacher

Figure 16.3: Original art by Rebecca Gelernter

Figure 16.4: Images from Brunet et al., 2000; Völlm et al., 2006; and personal correspondence with Dr. Eric Brunet-Gouet. Used with permission of Dr. Brunet-Gouet (personal correspondence).

Figure 16.5: Photo from Frith, 2003. Reused with permission.

Figure 18.1: Original art by Rebecca Gelernter

Figure 18.2: Original art by Rebecca Gelernter

Breakthrough #5 Section Cover Image: Original art by Rebecca Gelernter

Figure 19.1: Original art by Rebecca Gelernter

Figure 19.2: Original art by Rebecca Gelernter

Figure 19.3: Original art by Rebecca Gelernter

Figure 20.1: Original art by Rebecca Gelernter

Figure 20.2: Images from Trepel et al., 1996. Used with permission.

Figure 20.3: Original art by Rebecca Gelernter

Figure 21.1: Original art by Rebecca Gelernter

Figure 21.2: Original art by Rebecca Gelernter

Figure 21.3: Original art by Rebecca Gelernter

Figure 21.4: Original art by Rebecca Gelernter

Figure 21.5: Original art by Rebecca Gelernter

Figure 21.6: Original art by Rebecca Gelernter

Index

NOTE: *Italic page references* indicate figures.

Abbeel, Pieter, 279–80
abiogenesis, 19–22
accumulation, 305–6, *307*, 307–8, *308*
acquisition, 82, *83*, 84, 370
action potentials, *33*, 33–34, 37, 38
active inference, 216–17, 223
active learning and AI, 278–80
actor-critic reinforcement learning, 106–7, *107*, 118, 120, 121
acute stress response, 69–72
adaptation, 35–37, *36*, 54, 74, 370
adrenaline, 70–72, *71*
Adrian, Edgar, 32–37, *33*
aerobic respiration, 23, *24*, 374*n*
affect (affective states), *60*, 60–63, 90, 370
  of nematodes, 61–64, *62*, *63*
  role of neuromodulators in first bilaterians, 65–67, *66*
  stress and worms, 69–72
African origins of modern humans, 238–39, 290, 324–28, 341, 343
agranular prefrontal cortex (aPFC), 206, *207*, 208–9, 211–13, 216–20, 222–23, *223*, 224, 226–30, 232, 255–60, *259*, 370
AI. *See* artificial intelligence
akinetic mutism, 204–5, 206–7
"alien limb syndrome," 229
all-or-nothing law, 33, 38
AlphaZero, 201–4, 211, 318
altruism, 333–36, 337–40, *339*, 358
ALVINN (Autonomous Land Vehicle in a Neural Network), 278–79

American Sign Language, 299
amino acids, 18
amniotes, 159–60, 165*n*, 241
amphibians, 133, 159, 237
amygdala, 149, *150*, 165, *166*, 208, 219–20, 258, 286, 314–15, 321
anaerobic respiration, 23, *24*
anemones, *29*, *30*, 38, 74, 80, 93
anhedonia, 73–74
Anomalocaris, 122–23
antelopes, 241–42, 302, 303, 328
anticipating future needs, 285–88, 289–90, *290*, 295, 296, 360
  Suddendorf and Bischof-Kohler hypothesis, 284–88, 392*n*
  theory of mind and, 286–87, *287*
antidepressants, 65
antioxidants, 21
antipsychotics, 65
ants, 94, 147–48
anxiety, 59, 65, 69–70
aphasia, 314–15
Aquinas, Thomas, 86
archosaurs, 161, 163
Aristotle, 13–14, *14*, 295
arousal, 59, *60*, 60–61, 73–75
arthropods, 93–94, 114*n*, 157–58, 377*n*
artificial intelligence (AI), 2–5, 11–12, 363–64
  active teaching and, 278–80
  brain and, 9–10, 11
  challenge of pattern recognition, 127–28
  continual learning problem, 81–82

artificial intelligence (AI) (*cont.*)
   first robot, 49–52
   Minsky's SNARC, 103–5
   Mountcastle's theory, 171
   origins of term, 103
   paper-clip problem, 352–53
   theory of mind and, 265–66
artificial neural networks, *127,* 127–28
artificial superintelligence (ASI), 265, 352, 363–64
associative learning, 78–81, 87–88, 90, 370
   acquisition, 82, *83,* 84, 370
   blocking, *85,* 85–86, 90, 104, 195, 370
   continual learning problem, 81–84, *83*
   credit assignment problem, 84–86, *85*
   extinction, 82, *83,* 371
   overshadowing, *85,* 85–86, 90, 104, 195, 371
   reacquisition, 82–84, *83,* 86, 90, 371
   spontaneous recovery, 82–84, *83,* 86, 90, 371
   time course of, 82–84, *83*
attention, 218–20, 318–20, 321, 336, 350
audition, 171, 172, 174
auditory cortex, 167–68, 170
Australopithecus, 323, *341*
auto-association, 130–31, 135, 139, 151, 152, 176, 370
automation, *228,* 229–30
avoidance, 52–53, 63, 79–80, 115–17, 219
axons, 32, *37,* 130

baboons, 43, *243,* 248, 283
backpropagation, 128, 137n, 139, 370
bacteria, 18–20, 48–49
Bagnell, Drew, 279
Barto, Andrew, 105–6
basal ganglia, 95–96, *96,* 117–21, 152, 165, 208, 212–11, 215–16, 219–20, 229–30, 253–54
bees, 43, 94, 116, 116n, 147, 296
behavioral AI, 49–51
behavioral economics, 215
behavioral inhibition, 219–20
behavioral states, 62–63
Bentham, Jeremy, 43
Berridge, Kent, 67–68, 72
bilateral symmetry, 43, *44,* 45–46, 370

bilaterians, *xiv,* 43, *44, 45,* 45–56, 90, 370
   affect, 74–75
   associative learning, 80–81, 84–88, 132, 152, *302*
   chronic stress response, 72–75
   credit assignment in, 195–96, *196*
   dopamine and, 114
   early brain, 58, 80, 96, *153,* 184
   how they recognized things, 124–25, *125*
   prediction in, *184,* 185
   role of neuromodulators in affective states in, 65–67, *66*
   Roomba, 51–52
   steering, 46–49, *49,* 52–53, 81
   synapses, 87–88
   valence and, 52–55, *54,* 57–58, 119
bipedalism, 325–26, 329, 395n
bird feathers, 340
birds, *xiv,* 13, 160n, 163, 164, 182, 196, 198, 238, 268, 317, 322, 335, 340
Bischof, Doris, 284
Bischof, Norbert, 284
Bischof-Kohler hypothesis, 284–88
blindness, 167, 170–71, 181, 183, 204
blocking, *85,* 85–86, 90, 104, 195, 370
Boesch, Christophe, 276
bonobos, 244, 284, 297, 299–300, 313, 316, 364
bootstrapping, 107–9, 152, 259, 265, 361
Bostrom, Nick, 352
brain
   AI and, 9–10, 11
   anatomy, *xiii,* 5–6, *7,* 95–96, *96,* 253–54. *See also specific regions*
   evolution. *See* brain evolution
   first model of the world, 146–51
   five breakthroughs, 10–11. *See also specific breakthroughs*
   language in, 310–17, 338–40, *339*
   MacLean's triune brain hypothesis, 8–9, *9*
   similarities across animal kingdom, 6–8
   size of. *See* brain size
brain evolution, 6–8, 13–14, 93, *323,* 323–24, 359–61
   first mammals, 164–66, *166*
   five breakthroughs, 10–11, 39

social-brain hypothesis, 239–41
valence and nematodes, 52–55, *54*
vertebrate template, 94–96, *95, 97*
brain scaling, 253–55, 296
brain size, 239, 253–55, *254, 323,* 323–24, *330*
neocortex ratio, *240,* 240–41
brainstem, 117–18, 312, 315
breakthroughs, 10–11, 359–65
#1: steering. *See* steering
#2: reinforcing. *See* reinforcement learning
#3: simulating. *See* simulation
#4: mentalizing. *See* mentalizing
#5: speaking. *See* language
evolution of progressively more complex sources of learning, *302,* 302–3
Broca, Paul, 310–11
Broca's area, 310–12, *311,* 313–14, 316, 320–21
Brooks, Rodney, 49–51
Brunet-Gouet, Eric, 260
Buddhism, 192
buffalo, 241

Caenorhabditis elegans (C. elegans), 47, *47,* 58, 375*n*
caloric surplus, 328–29, 358
Cambrian explosion, 93–94, 95, 140
Cambrian period, 93, 122–23
can't-unsee property of perception, 174–75, *175*
carbon dioxide, 20–22, *22,* 57, 158
Carboniferous period, 159, *162*
Carnegie Mellon University, 278, 279
catastrophic forgetting, 131–33, 135, 140, 199, 371
cats, 43, 186
learning, 97–101, 115
motor cortex, 223, *224,* 224–25, 226
visual cortex, 135–36
causation vs. correlation, 195–96
cellular respiration, 21–23, *22*
cephalopods, 157
Charles Bonnet syndrome, 181
ChatGPT, 2–3, 132, 344
chauvinism, 13
cheating, 333–34, 337–38, 396*n*
chess, 2, 105, 109, 200, 201

chimpanzees, *xiv, 239*
brain and brain size, 6, *240, 254,* 290, *330,* 342, 393*n*
communication, 296, 297, 299–300, 313, *315,* 315–16, 319
diet and nesting locations, 282–84
grooming, 247, 249–50, 335
mating styles, 329
mental maps of, 244–46
motor cortex, 222
observational learning, 306
reciprocal altruism, 335
skill transmission, 273–77, 279
social structures, *243,* 244–47, 250
theory of mind, 264
tool use, 267–68, 273
Chomsky, Noam, 340
chronic stress response, 72–75
classical conditioning, 76–79, 80, 82, 85–86
climate change, 158
Cnidarians, 379*n*
Coates, Adam, 279–80
"cocktail-party effect," 174
Cohen, Neal, 131–32, 135
coincidence detection, 88*n*
communication, 296–99. *See also* language
altruism problem, 340
attempts to teach apes language, 299–301
emotional expressions, *314,* 314–17
transferring thoughts, 301–7
concepts, 61, 301–2
conditional reflexes, 77–78
connectionism, 97–100
consciousness, 309, 390*n*
constrained hallucinations, 181–82
content-addressable memory, 130–31
continual learning problem, 81–84, *83,* 371
catastrophic forgetting, 131–33
convolutional neural networks (CNNs), 137–40, 137*n, 138,* 139*n,* 371
cooking, 328–29, 358
cooperation strategies, 303–5, *304*
copper barrier, 55, 56*n,* 57, 374*n*
corals (coral polyps), 29–31, *30, 31,* 38, 47, 81, 90
correlation vs. causation, 195–96
cortex, 95–96, *96,* 117, *129,* 129–31, 133, 152

cortical columns, 168–72, *169,* 211, 216–17, 386*n*
   microcircuitry of, 171–72, *172*
counterfactual learning, 192–96, *193,* 232
cravings, 68, 219–20, 227–30
credit assignment problem, 84–86, 90, 104, 371
   evolution of, 195–96, *196*
   original four tricks for tackling, 84–86, *85*
   temporal, 105–7, 113, 120, 152, 200, 371
Cretaceous period, *162*
crows, 186, 267–68
cruelty, 12, 336, 340, 358
cultural bias and emotions, 59–60
curiosity, 142–45, 152, 382*n*
cyanobacteria, 19–21, *20,* 24, 158, 238
cynodonts, 161, *162*

Dale, Henry, 37
Damasio, Antonio, 204–5, 206, 217
dard, 59
Darwin, Charles, 7, 295, 330
Dawkins, Richard, 305
Dayan, Peter, 110, 112, 113, 175–77
DeCasien, Alex, 283–84
deception, 245, 252
declarative labels, 297–98, 300
decorrelation, 130
Deep Blue, 108–9
DeepMind, 142, 201
   AlphaZero, 201–4, 211
deliberative choices, 208–13, *210*
   step #1: triggering simulation, *210,* 210–11
   step #2: simulating options, *210,* 211–12
   step #3: choosing an option, *210,* 212–13
DeLong, Caroline, 139
Democritus, 86
dendrites, 32, 129
depression, 59, 65, 69, 73–74
Descartes, René, 86, 87
detour tasks, 190–92
Devonian period, 157–58, *162*
de Waal, Frans, 239–40
diabetes, 378–79*n*
Dickinson, Tony, 213–14
diet, 238–39, 251–52, 282–84, 326, 327, 328–29

digestion, 28–29, 76–77
dinosaurs, 159–60, 160*n,* 161, *162,* 163, 164, 233, 237–38, 241
disappointment, 115–17
discounting, 113
discrimination problem, 125–26, *126,* 129–30, *130*
dishwashing robots, 2, 4, 230
diurnal, 238
DNA, 18, 20, 304–5, 363
Dobzhansky, Theodosius, 7
dogs, *xiv,* 77–78, 82, 97, 186, *239,* 242, 246–47, 274
dolphins, *xiv,* 238, *239, 239,* 246, 274, 365
dominance, 242–43, 244, 247–48
dopamine, 64–69, *66,* 88, 118, 119, 152, 165, 359, 376*n,* 381*n*
dorsal cortex, 165*n,* 383*n*
dreams (dreaming), 182, 183
drug addiction, 110, 144, 227–30
dualism, 86–87
Dunbar, Robin, 239–41, 282, 290, 337–38

East Side apes, *325,* 325–26
Eccles, John, 37–38
ecological-brain hypothesis, 282–84, 290
Ediacaran period, *46,* 46–48, 84, 93–94, *94*
Edison, Thomas, 305
electricity, 4, 32, 305
electrophysiology, 32–33
elephants, *xiv,* 223, 238, *239,* 267–68, 326
eligibility traces, 84–86, *85,* 88, 90
Elman, Jeffrey, 317–18
"embodiment," 224
emotion, categories of, 59–60
emotion, origin of, 59–75
   the blahs and blues, 72–75
   dopamine and serotonin, 64–69, *66*
   steering in the dark, 61–64
   stress and worms, 69–72
emotional expression system, *315,* 315–16, 394*n*
empathizing, 262
endurance running, 328
entropy, 17–18, 20, 363
Epicurus, 86
episodic memory, 13, 196–99, 232–33, 303, 385*n*
ether, 32

eukaryotes, 23–24, *24*, 25, 28, 374*n*
euphoria, 68, 74
evagination, 383*n*
evolution, 359–62
   arms race for political savvy, 237–39, 251–52
   of the brain. *See* brain evolution
   Cambrian explosion, 93–96
   fungi, 27–31, *31*
   Homo Erectus and emergency of human hive mind, 336–41
   Homo Erectus and rise of humans, 326–30
   human lineage and proliferation, 13–15, *14*, *323*, 323–24, *341*, 341–43
   of language, *302*, 302–3, 330–33, *332*, 358–59
   of nervous system, 26–27
   neural dark ages, 157–66
   origins of life, 17–22
   Pavlov and origin of learning, 76–79
   of prediction, *184*, 185
   of progressively more complex sources of learning, *302*, 302–3
   shared developmental stages for all animals, 28–29, *29*
   of temporal difference learning, 103–21
   tension between the collective and the individual, 241–44
   tree of life. *See* tree of life
exaptation, 340
excitatory neurons, 38, 65
executive control, 218
expansion recoding, 129–30, *130*
exploitation, 66, 68, 376*n*
exploitation-exploration dilemma, 142–43, 152
extinction, 82, *83*, 371
extinction events, 158–59
   Late Devonian Extinction, 158–59, *162*, 238
   Permian-Triassic extinction event, 160–61, 237–38, 251
eye, 117, 135–37, 332–33
eyewitness testimonies, 197–98

Facebook, 144
facial expressions, *314*, 314–15, 394*n*
   dopamine and reward, *67*, 67–68

Fadiga, Luciano, 268–69
false-belief tests, 261–62, 354, 389*n*, 397*n*
   Sally-Ann test, 260–62, *261*, *262*, 264
fear, 61, 63, 117, 123, 125–26
female hierarchies, 248–49
ferrets, 170
Feynman, Richard, 10
field dependence, 229
fight-or-flight response, 70
filling-in property of perception, 173, *173*
fire, use of, 328–29
firing rate, *33*, 33–36, 371
first model of the world, 146–51
   inner compass, 148–49
   maps of fish, 146–48
   medial cortex, 149–51
first move, 163–64
fish, 100–102, 193, 233, 334
   avoidance tasks, 115, 116, 116*n*, 117
   brain, 132–33, 139–40, 164–65, 165*n*
   catastrophic forgetting, 132–33
   communication, 296
   evolution and tree of life, *xiv*, 157, 158–59, *162*, 164–65, 194, 237, 241
   invariable problem, 139–40
   maps of, 146–48, 190–91, 384*n*
   observational learning, 274–75, *275*
   reinforcement learning, 100–102, 110, 115, 144
   smell and nose, 123–24, *124*, 125–26
   vestibular sense, *148*, 148–49
flatworms, *49*, 85, 116, 125
Fogassi, Leonardo, 268–69
forebrain, 95–96, *96*, 119
Franklin, Benjamin, 4
freeloaders, 333, 335, 337
free time, 251–52
friendships, 250, 252
Friston, Karl, 216–17, 223–24
frugivores, 251–52, 282–84, 288
Fukushima, Kunihiko, 136–38
full signals, 58
fungi, *24*, 27–31, *31*, 31*n*

Gallese, Bittorio, 268–69
gambling, 144–45
gap junctions, 37, *37*
gastrulation, 28–29, *29*
generalization problem, 126, *126*

generative mode (generative models), 177–81, 371
  Helmholtz machine, 177–79, *178*
  neocortex as, 181–83, 188, 222, 258–60
  predicting everything, 183–87, *185*
  StyleGAN2, *179*, 179–81
genes, 18, 20, 304–5, 363
genome, 317
gestures, 296–97, 301, 310, 313–14, 315–16
Go (game), 2, 201–3, *202*
goal-driven behavior, 213–17
goal hierarchy, 226–31, *228*
Goodall, Jane, 267–68, 315–16
Google, 344
  DeepMind, 142, 201
gorillas, *239, 243,* 299–300, 313
gossip, 337–38, *339,* 358
GPT-3 (Generative Pre-trained Transformer 3), 3–4, 344–51, 354–55, *355*
GPT-4 (Generative Pre-trained Transformer 4), 354–56, *355*
grammar, 297–98, 300, 336
"grandmothering," 329
granular prefrontal cortex (gPFC), 206, 226, 255–60, *259,* 262, 263, 289, 290, 371
granule cells, 206
Great Ape Dictionary, 296
Great Oxygenation Event, 21, 238, 374*n*
Great Rift Valley, 324–25
grief, 59–60
grocery shopping, 284–88
grooming, 247, 249–50, 335
group living, 241–44
group selection, 333–36, 337, 395*n*

habitual behavior, 213–15
Haldane, J. B. S., 334
hallucinations, 181–83
Harari, Yuval, 303
harems, 242–44, *243,* 388*n*
Harvard University, 97
head-direction neurons, 149
Heath, Robert, 68
Hebb, Donald, 88
Hebbian learning, 88–89, 130
Helmholtz, Hermann von, 175–76, 180–82, 185

Helmholtz machine, 177–79, *178,* 180, 182, 371
heroin addiction, 230
hindbrain, 95–96, *96,* 149, 165
Hinton, Geoffrey, 6, 127–28, 175–77, 182
hippocampus, 149–51, 165, 190, 196, 198–99
Hippocrates, 31–32
Hobbes, Thomas, 86, 330
Homo erectus, 323, 326–30, 331–32, *341*
  emergence of the human hive mind, 336–41
Homo floresiensis, 341, 341–42
Homo neanderthalensis, 323, 331, *341,* 342–43
Homo sapiens, 297, 301, *323,* 331–32, *341,* 342–43, 361
horses, *xiv,* 223, 238, *239*
Hubel, David, 135–36, 137
human proliferation, *341,* 341–43
human uniqueness, 295–309
  attempts to teach apes language, 299–301
  communication, 296–99
  the singularity, 307–9
  transferring thoughts, 301–7
Humphrey, Nicholas, 239–40
hunger, 58, 62, *79,* 79–80, 119, 286, 287
hypothalamus, 95–96, *96,* 119–21

IBM Research, 108–9
ideas, 301–2, 305–6, 307–8
illusions, 172
imagination, 182–83, 186–87, 303
imitation (imitation learning), 98–99, 274–75, 277–81, 289–90, *290,* 306–7
  AI and, 278–81
imperative labels, 297, 300
inductive bias, 138, 140
inference, 175–77, 180–82, 185
inhibitory neurons, 38, 65
inner compass, 148–9
inner ear, 124, 135, 140, 148–49
Instagram, 144
intention, 205, 208–9, 245–47, 257, 260
internal models, 146, *147,* 151. *See also* models
intuitions, 60–61, 146
invariance problem, 133–40, *134,* 151

inverse reinforcement learning, 277–81
invertebrates, 94–95, 95, 114n, 116, 144, 151, 157, 237
involuntary associations, 78
iPhone, 127
iRobot, 51

jellyfish, *xiv,* 27, 28, *29,* 34, 38, 39, 43, 74, 80
Jennings, Ken, 109
Jetsons, The (TV show), 1–2, 132
Johns Hopkins University, 131, 135
Johnson, Adam, 190
joint attention, 318–20, 321, 336, 337, 358
Jurassic period, *162,* 233

Kahneman, Daniel, 215
Kanada, 86
Kandel, Eric, 76
kangaroos, *xiv, 223*
Kanzi (bonobo), 299–300, 320
Kasparov, Garry, 108–9
kin selection, 334–36, 337
knowledge, 132, 246–47, 257
koalas, *xiv, 223*

lamprey fish, 95, 118–19, 123, 129
language, 185–86, 297–99, 309, 318–19
  attempts to teach apes, 299–301
  in the brain, 310–17, 338–40, *339*
  breakthrough #5 summary, 358, 360–61
  emergence of the human hive mind, 336–41
  evolution of, *302,* 302–3, 330–33, *332,* 358–59, 360
  relationship between mentalizing and, 353–54
  transferring thoughts, 301–7
language curriculum, 317–21
large language models (LLMs), 2–3, 344–50, 356–57
  GPT-3, 3–4, 344–51, 354–55, *355*
  GPT-4, 354–56, *355*
last universal common ancestor (LUCA), 19–20, *24*
Late Devonian extinction, 158–59, *162,* 238
latent inhibition, *85,* 85–86, 90, 104, 195, 380n
Late Permian extinction event, 160–61, 237–38, 251

lateral cortex, 149–51, *150,* 165, *166*
law of effect, 99–100, 103, 144, 189, 213
layer four, *172,* 206, 206n, 216, 217
Leakey, Louis, 267
Leborgne, Louis Victor, 310
LeCun, Yann, 10, 137n, 186, 200, 356
Lemoine, Blake, 344
limbic system, 8–9, *9*
lizards, 159–60, 161
logic, 50, 185–86
luminance, 34–35, 35n
lying (liars), 334, 337, 396n

macaque monkeys, 222, *240, 243,* 244, *256,* 268, 313, 329, *330*
McCloskey, Michael, 131–32, 135
Machiavellian apes, 244–47
machine learning, 12, 84
MacLean, Paul, 8–9, 371n
mammals. *See also specific mammals*
  brain, *95,* 113–14, 135–36, 149–50, 163–66, *166,* 186–87, 203–4, *205,* 205–7, 232–33, 253–55
  control and, 218–20
  credit assignment in, 195–96, *196*
  Era of Mammals, 238–39, *239*
  evolution and tree of life, *xiv, 162,* 163, 238–39, *239*
  evolutionary tension between the collective and the individual, 241–44
  goals and habits, 213–15
  inner duality of, 213–15
  making choices, 209–13
  motor cortex, *223,* 223–26
  motor hierarchy, 226–28, *227, 228*
  neocortex, 206–8, *207,* 209, *209,* 232–33, *256*
  neocortex ratio, *240,* 240–41
  prediction in, *184,* 185
  primate politics, 247–52
  simulating actions, 163–64
  visual cortex, 135–38
materialism, 86–87
medial cortex. *See* hippocampus
memes, 305
memory, 76, 116
  attention and self-control, 218–20
  catastrophic forgetting, 131–33

memory (*cont.*)
  episodic, 196–99, 232–33
  working, 187, 218, 219–20
mentalizing, 289–91, *290,* 361, 371
  breakthrough #4 summary, 289–91, 360
  evolution of progressively more complex sources of learning, *302,* 302–3, 360
  relationship between language and, 353–54
Menzel, Emil, 244–45
Mestral, George de, 4
metacognition, 258
mice, 163–64, 226, 283, 296
midbrain, 95–96, *96,* 110, 117, 165
mind. *See* models; theory of mind
Minsky, Marvin, 2, 103–5, 120, 200
mirror neurons, 268–73
mirror self-recognition tests, 257, 264
mirror-sign syndrome, 257–58
models (modeling)
  first. *See* first model of the world
  frontal vs. sensory neocortex in first mammals, 209, *209*
  mind to model other minds, 263–65
  other minds, 260–63, *261*
  own mind, 258–60, *259*
model-based reinforcement learning, *199,* 199–200, 201–20, 371
  AlphaZero, 201–4, 211, 318
  attention, working memory, and self-control, 218–20
  evolution of first goal, 215–17
  goals and habits, 213–15
  mammals making choices, 209–13
  predicting oneself, 208–9
  prefrontal cortex and controlling the inner simulation, 204–8, *205, 207*
model-free reinforcement learning, *199,* 199–200, 201, 211, 212, 215–16, 318, 359–60, 371
Molaison, Henry, 196–97, 198
mongooses, 267–68, 274, 275
monkeys, *xiv,* 194, 247–48, 269–71, 284–85, 287–88, 316
Montague, Read, 110, 112, 113
Morse code, 33
motivation, 73–74

motor cortex, 206, 221–26, *222,* 232, 241, 360
  language and, 312
  leading theory on evolution of, 222–23, *223*
  mirror neurons, 268–73
  missing layer four, 206, 206*n*
  predictions, 223–26
motor hierarchy, 226–31, *227, 228*
motor planning, 224–26, 270, 271
Mountcastle, Vernon, 168–70, 289
multicellular organisms, *24,* 24–26, *25,* 28
multi-male groups, 242–44, *243,* 387*n*
myths, 303–4, *304*

Naqshbandi, Miriam, 284–85, 285*n,* 287–88
natural selection, 330, 340, 363
nature and intelligence, 4–6
'nduh, 59
negative-valence neurons, 53–55, *54,* 56–57, 61, 100
nematodes, *xiv,* 46–48, *47,* 94, 101, 147
  affective states of, 61–64, *62, 63*
  dopamine and serotonin, 64–69, *66,* 114
  problem of trade-offs, 55–57, *56*
  steering, 46–49, *48, 49,* 53–54, *54*
  stress, 69–71, 73–74
  temporal difference learning, 115–16, 116*n*
  tweaking goodness and badness of things, *79,* 79–80
  valence and, 52–55, *54*
neocortex, 8–9, *9*
  anatomy, 167–72, *168,* 205. *See also* agranular prefrontal cortex; cortical columns; granular prefrontal cortex; motor cortex
  counterfactual learning, 192–96, *193*
  episodic memory, 196–99
  evolution, 163–64, 165–66, *166,* 188, 289–90
  functions, 218–20, 289–90
  as a generative model, 181–83, 188, 222, 258–60
  language and, 312–17, *315*
  layers, *169,* 171–72, *172*
  MacLean's triune brain hypothesis, 8–9, *9*

new neocortical regions of early primates, 255–56, *256*
new regions in primates, 255–56, *256*, 263–64
perception, 172–75
prediction, 183–87, *185*
ratio, *240*, 240–41
sensory. *See* sensory neocortex
use of term, 167*n*
vicarious trial and error, 189–92
neocortical columns. *See* cortical columns
nepotism, 252
nerves, 32
nervous system, 26–27, 32
nervus, 32
neural circuits, 38–39, *39*, 56, 86, 90
Neurogammon, 109
neuromodulators, 64–69, *66*, 70–72, *71*, 88, 165, 359, 371. *See also specific neuromodulators*
role in affective states of first bilaterians, 65–67, *66*
neurons, 5, 7, 19, *26*, 26–27, 28–29, 31–32
Adrian's discoveries, 32–37, *33*
cortical column, 168–72, *169*
history of neuroscience, 31–39
negative-valence, 53–55, *54*, 56–57, 61
positive-valence, 53, *54*, 56–57
response of dopamine to predictive cues, rewards, and omissions, 110–14, *112*
neurotransmitters, 37–38, 87
Newton, Isaac, 32
New York University (NYU), 283–84
Ng, Andrew, 279–80
NMDA receptors, 88*n*
Nobel, Alfred, 76
Nobel Prize, 32, 37, 76
nocturnal, 238
nonassociative learning, 80*n*
norepinephrine, 70, 123, 377–78*n*

observational learning, 272–77, *275*, 280–81, 306, 360
"obstetric dilemma," 329
octopamine, 70, 377*n*
octopuses, *xiv*, 14, 15, 157, 267–68, *275*, 364
Oldowan tools, 326–27, *327*
olfactory neurons, 123–30, *124*, *129*, 135
expansion and sparsity, 129–30, *130*

olfactory receptors, 123–24, *124*, 381*n*
one-at-a-time property of perception, 173–74, *174*
On the Origin of Species (Darwin), 7, 330
OpenAI, 132, 354, 355, 356
opioids, 70–72, *71*, 74
opposable thumbs, 238
origin of emotion. *See* emotion, origin of
origins of life, 17–22
orthogonalization, 130
overshadowing, *85*, 85–86, 90, 104, 195, 371
oxygen, 21, 27
Oxygen Holocaust, 21

pair-bonding mammals, 242–44, *243*, 329
paper-clip problem, 352–53
parasitic strategy, 28*n*
Parkinson's disease, 118
pattern recognition, 122–41, 165
catastrophic forgetting, 131–33
computers and, 127–28
cortex, *129*, 129–31
discrimination problem, 125–26, *126*
generalization problem, 126, *126*
invariance problem, 133–40, *134*
problem of recognizing a smell, 123–26
pattern separation, 130, 133
Pavlov, Ivan, 76–79, 80, 82, 85–86, 98
Pellegrino, Giuseppe di, 268–69
perception, 172–75, 218
can't-unsee property of, 174–75, *175*
filling-in property of, 173, *173*
one-at-a-time property of, 173–74, *174*
Permian, 159, 160, 161, *162*, 169
Permian-Triassic extinction event, 160–61, 237–38, 251
persistence hunting, 328
phagotrophy, 23–24, 28
photosynthesis, 20–22, *22*, 23, *24*, 27
physics, 17–18, 195–96, 350, 363
Pinker, Steven, 353
placoderms, 157
Plato, 86, 87, 330
political power, 247–52
Pomerleau, Dean, 278–79
positive-valence neurons, 53, *54*, 56–57, 100, 119
predation, 93, 122–23, 243

predictions, 208–13, *210,* 223–26
  evolution of, *184,* 184–85
  motor commands and, 223–26, 271
  neocortex and, 183–87, *209*
  reward-prediction, 111, 113, 114*n,* 115, 213–14
  step #1: triggering simulation, *210,* 210–11
  step #2: simulating options, *210,* 211–12
  step #3: choosing an option, *210,* 212–13
predictive cues, 84–86, 111, 112, 121
prefrontal cortex, 209. *See also* agranular prefrontal cortex; granular prefrontal cortex
  controlling the inner simulation, 204–8, *205, 207*
premotor cortex, 226, 229, 230
  mirror neurons, 268–73
primates. *See also specific primates*
  acquiring novel skills through observation, 275–77
  anticipating future needs, 285–88
  counterfactual learning, 194–95
  ecological-brain hypothesis, 282–84, 290
  evolution and tree of life, *xiv,* 238–39, *239,* 243–44, 289–91
  evolution of progressively more complex sources of learning, *302,* 302–3
  modeling mind to model other minds, 263–65
  modeling other minds, 260–63, *261*
  modeling own mind, 256–60, *259*
  motor cortex, 206, 221, *222,* 222–23, *223,* 268–73
  neocortex, *240,* 240–41, 313–14, 360
  new neocortical regions of, 255–56, *256,* 263–64
  skill transmission, 273–77, *275*
  social-brain hypothesis, 239–41, 282
  social politics, 247–52, 281
  social structures, 242–44, *243*
  theory of mind. *See* theory of mind
  tool use, 267–68, 273–75
  visual cortex, 253–55, *254*
primate sensory cortex (PSC), 255, 258–59, 354, 371
procedural memory, 197
proteins, 18–19
protein synthesis, 18–19

proto-conversations, 318–20, 336–37
protolanguages, 331–32, 336, 358
psychedelics, 65
psychic stimulation, 77–78
punishment, 337–38, 358, 396*n*
puzzle boxes, *98,* 98–99, *99,* 101, 103, 115, 277, 306

radial symmetry (radiatans), 43, *44,* 45, 53, 54, 80
Ramón y Cajal, Santiago, 37
rate coding, 34–37, *36,* 38
rats, *xiv*
  anticipating future needs, 284–85, 285*n,* 287
  brain, 8, 78, 149, 150, 169, 189–90, 198–99, 206, 207, 213–14, 223, 224, 229
  detour tasks, 191–92
  dopamine and pleasure, 66
  dopamine and stimulation, 65, 66–69, 110
  episodic memory, 198–99
  observational learning, 274, 276–77
  regret in, *193,* 193–94
  role of habits, 213–14
  role of play, 241
  spinal cord, 78, 86
  variable-ratio reinforcement, 144
  vicarious trial and error, 189–90, 191–92, 209–10, 212, 220
reacquisition, 82–84, *83,* 86, 90, 371
reciprocal altruism, 335–36
reciprocity, 250, 252
recognition. *See also* pattern recognition
  mirror self-recognition tests, 257, 264
  neocortex and, 182–83, 188
recognition modes, 177–79, *178*
Redish, David, 190, *193,* 193–94
register-addressable memory, 130–31
regrets, 192, *193,* 193–94
reinforcement learning, 101–6, 164–65, 192–93, 359–61
  based on actual rewards, 107–8, *108*
  based on temporal differences in expected rewards, 107–8, *108*
  breakthrough #2 summary, 152–53, 359–60
  evolution of progressively more complex sources of learning, *302,* 302–3

importance of curiosity in, 142–45
model-based. *See* model-based reinforcement learning
model-free. *See* model-free reinforcement learning
Thorndike and, 96–101
relief, 115–17
REM sleep, 182, 384n
reptile brain, 8–9, *9*
reptiles, *xiv*, 159–61, *162*, 165, 165n, 296
respiration, 21–23, *22*, 27, 374n
ribosomes, 18
Rizzolatti, Giacomo, 268–69
Roberts, William, 284–85, 285n, 287–88
robotics
   first robot, 49–52
   imitation learning, 278–81
Rochester Institute of Technology, 139
rock, paper, scissors (game), 194–95
Roomba, *51*, 51–52, 53, 58, 64
Rosey the Robot, 1–2, 5, 51, 132
Ross, Stephane, 279
Rousseau, Jean-Jacques, 330
Rumelhart, David, 127–28
rumination, 192–93

salamanders, 159
Salk Institute, 110
Sally-Ann test, 260–62, *261*, *262*, 264
salt, *79*, 79–80, 81
Sapiens (Harari), 303
satiation, 62, *62*, *63*, 66, 69, 287
Savage-Rumbaugh, Sue, 300
"scale of nature," 14
Schultz, Wolfram, 111–13, *112*, 115
search problem, 200, 202, 203, 209, 211, 232
Searle, John, 303
second law of thermodynamics, 17–18
sehnsucht, 59–60
seizures, 196–97, 198
selective serotonin reuptake inhibitors (SSRIs), 378n
self-concept (sense of self), 217, 264, 390n
self-control, 219–20
self-driving cars, 278–79
Selfish Gene, The (Dawkins), 305
self-reference, 257
self-replication, 18, 19

semicircular canals, *148*, 148–49
sensitization, 80n
sensory neocortex, 197, 198, *205*, 205–6, 211–13, 216–17, 232, 258–59, 371
   in first mammals, 209, *209*
serotonin, 64–69, *66*, 71–72, 73, 88, 359, 376n, 378n
Sherrington, Charles, 37
sign language, 299, 311–12
simulation, 163–64, 361
   breakthrough #3 summary, 232–33, 360
   evolution of progressively more complex sources of learning, *302*, 302–3, 360
   GPT-3 and LLMS, 349–51
   hierarchy of goals, *228*, 229–30
   making choices and, *210*, 210–13
   survival by, 163–64
simulation theory, 263–64
skill transmission, 273–77, *275*
Skinner, B. F., 100, 144
sleep, 181, 182
smell, 34, 38, 47, 53–54, 123–26, 135. *See also* olfactory receptors
Smith, Neil, 312
snakes, 159–60, *162*
social-brain hypothesis, 239–41, 282, 290
social groups, 241–44
social hierarchy, 242–44, 247–52, 265–66
social media, 144–45
social projection theory, 263–64, 389n
solitary mammals, 242–44, *243*
"source of error," 77
spandrels, 340
spatial maps, 146–48
   vestibular sense, *148*, 148–49
speaking. *See* language
spiders, 93, 158, 364
spike (firing) rate, *33*, 33–36, 371
spontaneous recovery, 82–84, *83*, 86, 90, 371
squirrel monkeys, 284–85, 287–88
squirrels, 163, 226, 271
"squishing problem," 35–37
steering, 46–49, *49*, 57–58, 61–64, 64
   bilaterians, 46–49, *49*, 52–53
   breakthrough #1 summary, 90, 359
   Roomba, 51–52, 53, 64

"steer in the dark," 64
Steiner, Adam, 193–94
stimulants, 65
stimulus strengths, 33–34, *34*, *36*
Stochastic Neural-Analog Reinforcement Calculator (SNARC), 103–5
stress, 69–72, *71*, 90
  acute stress response, 69–72
  ancient stress cycle, 71–72, *72*
  chronic stress response, 72–75
stroke victims, 171, 204–5, 221, 222
StyleGAN2, *179*, 179–81
submission, 242–43, 247–48
Suddendorf, Thomas, 284, 285, 286–88, 392*n*
sugar, 20, 21–22, 27–28
Superintelligence (Bostrom), 352
superior temporal sulcus (STS), 255*n*, 371
supervised learning, 128, 176, 180
Sutton, Richard, 105–9, 113, 118, 120, 121, 142–43, 203
sweat (sweating), 328
symbolic AI, 49–51
symbols, 297–98, 300
synapses, *37*, 37–38, 87–89, *88*, 118, 371
system 1 thinking, 215
system 2 thinking, 215

TD-Gammon, 109, 110, 142, 201, 318, 380*n*
temperature navigation, 54–55
temporal credit assignment problem, 105–7, 113, 120, 152, 200, 371
temporal difference learning (TD learning), 103–21, *106*, 142–43, 152, 198–99, 203, 371
  basal ganglia, 117–21
  emergence of relief, disappointment, and timing, 115–17
  exploitation-exploration dilemma, 142–43, 152
  grand repurposing of dopamine, 110–14
  magical bootstrapping, 105–9
temporal difference signals (TD signals), 107, 111–14, 152, 372
temporoparietal junction (TPJ), 255*n*, *256*, 354, 372
terraforming of Earth, 19–22

Tesauro, Gerald, 108–9, 110
tetrapods, 159, *162*
thalamus, 95–96, *96*, 117, 133, 134, 139–40, 172, *172*, 382*n*
theory of mind, 246–47, 260–66, 268, 289–90, *290*, 372
  acquiring novel skills through observation, 275–77
  anticipating future needs and, 286–87, *287*
  childhood development and, 264, 390*n*
  modeling mind to model other minds, 263–66
  politicking and, 281
  Sally-Ann test for, 260–62, *261*, *262*, 264
therapsids, 160–61, *162*
Thinking, Fast and Slow (Kahneman), 215
thispersondoesnotexist.com, *179*, 179–80
Thorndike, Edward, 96–100, *98*, 101, 110, 111, 115, 189
Thorpe, Chuck, 278–79
thought transfer, 301–7
time perception, 173–74, *174*
timing, 116–17, 152
Tolman, Edward, 189–90, 244
tool use, 267–68, 273–75, 284, 327–28, 358
  Oldowan tools, 326–27, *327*
trade-offs, 55–57, *56*
translation, 139
transmissibility, 273–77, *275*
tree of life, *xiv*, 23–25, *24*, 43, 45, *162*
  Cambrian ancestors, 94–95, *95*
  humans, *341*, 341–43
  mammals, 238–39, *239*
  neuron-enabled animals, 29–30, *30*
  radial vs. bilateral symmetry, *44*
trial-and-error learning, *99*, 99–100, *101*, 103–4, 110–11, 142–43, 152
  vicarious, 189–92, 211, 212–13, 232–33, 360, *361*
Triassic period, *162*
  Permian-Triassic extinction event, 160–61, 237–38, 251
tribalism, 252, 364
triune brain hypothesis, 8–9, *9*, 373*n*

Tsimpli, Ianthi-Maria, 312
Turing, Alan, 103
turtles, 159–60, *162*, 319
Tyrannosaurus, 161

uncertainty, 210–11, 214
unconditional reflexes, 78
unconscious inference, 175–77, 180–82, 185
ungating, 117–18, *120*, 381*n*
University College London, 216
University of California, Berkeley, 189
University of California, San Diego, 317
University of Massachusetts Amherst, 105–6
University of Michigan, 67
University of Minnesota, 190
University of Parma, 268–69
University of Western Ontario, 284
unsupervised learning, 176
utilization behavior, 229

V1 (visual area 1), 135–37, *136*
V2 (visual area 2), 136, *136*
V4 (visual area 4), 136, *136*
valence, 4, 52–59, *54*, 90, 119, 372
variable-ratio reinforcement, 144
Velcro, 4
ventral cortex, 149–51, *150*, 165, *166*
Venus flytraps, 30*n*
vertebral column, 94
vertebrates. *See also specific vertebrates*
    brain, 94–96, *97*, 110–11, 118–19, *120*, 120–21, 122, 129, *129*, 132–33, 139–40, 140, 149, *153*, 164–66, 259
    cortex, *129*, 129–31, 149–51, *151*, 164–66, *166*
    credit assignment in, 195–96, *196*
    evolution and tree of life, *xiv*, 94–96, *95*, *96*, 100, *162*, 233, 360
    how they recognized things, 124–25, *125*
    prediction in, *184*, 185
    smell and nose of, 123–26, *124*
    temporal difference learning, 110, 114, 115–16, 118–19, 143–44, 192–93, 194
vervet monkeys, 247–48, 297, 301–2, 335
vestibular sense, *148*, 148–49
vicarious trial and error, 189–92, 211, 212–13, 232–33, 360, 361
vision, 34–35, 124, 172–75
    invariance problem, 133–40, *134*
    visual cortex, 134–37, 139, 167, 170, 253–55, *254*
    cortical column, 168–72, *169*
volition, *216*

"wake-sleep algorithm," 182
Wallace, Alfred, 330–32
wanting, 68–69, 111, 114, 114*n*
warm-bloodedness, 160–61, 160*n*, *162*, 163, 164–65
Washburn, Sherwood, 329
Wernicke, Carl, 311
Wernicke's area, *311*, 311–12, 313–14, 316, 320–21, 354
whales, *xiv*, 238, *239*, 322
Wiesel, Torsten, 135–36, 137
wildebeests, 334
Williams, Ronald, 127–28
willpower, 219–20
working memory, 187, 218, 219–20
world models, 186, 200, *209*, 232
Wrangham, Richard, 328
wrasses fish, 268